Physics of the Space Environment

This book provides a comprehensive introduction to the physical phenomena that result from the interaction of the Sun and the planets – often termed space weather. *Physics of the Space Environment* explores the basic processes in the Sun, in the interplanetary medium, in the near-Earth space, and down into the atmosphere. The first part of the book summarizes fundamental elements of transport theory relevant for the atmosphere, ionosphere, and the magnetosphere. In the rest of the book this theory is applied to physical phenomena in the space environment. The fundamental physical processes are emphasized throughout, and basic concepts and methods are derived from first principles. This book is unique in its balanced treatment of space plasma and aeronomical phenomena.

Students and researchers with a basic math and physics background will find this book invaluable in the study of phenomena in the space environment.

Cambridge Atmospheric and Space Science Series
Editors: J. T. Houghton, M. J. Rycroft, and A. J. Dessler

This series of upper-level texts and research monographs covers the physics and chemistry of different regions of the Earth's atmosphere, from the troposphere and stratosphere, up through the ionosphere and magnetosphere, and out to the interplanetary medium.

A native of Hungary, Professor Gombosi carried our postdoctoral research at the Space Research Institute in Moscow before moving to the United States to participate in theoretical work related to NASA's Venus exploration. He is currently Professor of Space Science and Professor of Aerospace Engineering at the University of Michigan. He has served as Senior Editor of the *Journal of Geophysical Research – Space Physics* and is the author of *Gaskinetic Theory*, published in 1994.

Cambridge Atmospheric and Space Science Series

EDITORS

Alexander J. Dessler

John T. Houghton

Michael J. Rycroft

TITLES IN PRINT IN THIS SERIES

M. H. Rees
Physics and chemistry of the
upper atmosphere

Roger Daley
Atmosphere data analysis

Ya. L. Al'pert
Space plasma, Volumes 1 and 2

J. R. Garratt
The atmospheric boundary layer

J. K. Hargreaves
The solar–terrestrial environment

Sergei Sazhin
Whistler-mode waves in a hot plasma

S. Peter Gary
Theory of space plasma
microinstabilities

Martin Walt
Introduction to geomagnetically
trapped radiation

Tamas I. Gombosi
Gaskinetic theory

Boris A. Kagan
Ocean–atmosphere interaction and
climate modelling

Ian N. James
Introduction to circulating
atmospheres

J. C. King and J. Turner
Antarctic meteorology and climatology

J. F. Lemaire and K. I. Gringauz
The Earth's plasmasphere

Daniel Hastings and Henry Garrett
Spacecraft–environment
interactions

Thomas E. Cravens
Physics of solar system plasmas

John Green
Atmospheric dynamics

Physics of the Space Environment

Tamas I. Gombosi
University of Michigan

CAMBRIDGE
UNIVERSITY PRESS

PUBLISHED BY THE PRESS SYNDICATE OF THE UNIVERSITY OF CAMBRIDGE
The Pitt Building, Trumpington Street, Cambridge, United Kingdom

CAMBRIDGE UNIVERSITY PRESS
The Edinburgh Building, Cambridge CB2 2RU, UK   http://www.cup.cam.ac.uk
40 West 20th Street, New York, NY 10011-4211, USA   http://www.cup.org
10 Stamford Road, Oakleigh, Melbourne 3166, Australia

First published 1998

Printed in the United States of America

Typeset in Times $10\frac{1}{4}/13\frac{1}{2}$ pt. and Joanna in LaTeX 2_ε [TB]

A catalog record for this book is available from the British Library

Library of Congress Cataloging in Publication Data
Gombosi, Tamas I.
Physics of the space environment / Tamas I. Gombosi.
      p.   cm.
ISBN 0-521-59264-X
1. Atmospheric physics.   2. Space environment.   3. Atmosphere,
Upper.   4. Solar wind.   5. Magnetosphere.   I. Title.
QC861.2.G64   1998
551.51′4 – dc21   97-51318
                                                                    CIP
ISBN 0 521 59264-X   hardback

To the memory of my mentor
Konstantin Iosifovich Gringauz,
a dedicated pioneer of space exploration

Contents

Appendices

Preface

This book provides a comprehensive introduction to the physics of the space environment for graduate students and interested researchers. The text is based on graduate level courses I taught in the Department of Aerospace Engineering and in the Department of Atmospheric, Oceanic, and Space Sciences of the University of Michigan College of Engineering. These courses were intended to provide a broad introduction to the physics of solar–planetary relations (or space weather, as we have started to call this discipline more recently).

The courses on the upper atmosphere and on the solar wind and magnetosphere have been taught for a long period of time by many of my friends and colleagues here at Michigan before I was fortunate enough to teach them. I greatly benefited from discussions with Drs. Thomas M. Donahue, Lennard A. Fisk, and Andrew F. Nagy here at the University of Michigan and Drs. Thomas E. Cravens (University of Kansas), Jack T. Gosling (Los Alamos National Laboratory), and József Kóta (University of Arizona). I am grateful for their advise, criticism, and physical insight.

I would also like to acknowledge the constructive criticism of Konstantin Kabin, my graduate student here at the University of Michigan. His mathematical rigor and helpful suggestions greatly helped me in producing the final version of the manuscript.

This book was intended to provide a comprehensive introduction to students with very different backgrounds and interests. Over the years my students came from physics, aerospace and electrical engineering, computational fluid dynamics, meterology, aeronomy, planetary science, astronomy, and astrophysics.

In the presentation of the material my approach was to emphasize the fundamental physical processes and not the morphology of the various phenomena. This means that, in some respect, the book is a survey of transport theory applied to the space environment. This approach is also reflected in the organization of the book: Part I summarizes fundamental elements of transport theory relevant for the space environment, whereas Part II and III apply the basic concepts to physical phenomena in the Sun, the helisphere, the upper atmosphere, the ionosphere, and the magnetosphere.

The reader is expected to have some familiarity with integral and differential calculus, vector and tensor algebra, complex variables, statistics, classical mechanics, and electricity and magnetism. It was my intent to produce a more or less self-contained text and introduce all important concepts and definitions used in the book.

In the preparation of this text I have consulted other textbooks, monographs, review articles, and research papers. When I used research papers in the preparation of this text, I usually referenced the source in a footnote. Textbooks and monographs, however, are listed in the bibliography at the end of the book.

During my scientific career I greatly benefited from many exceptional individuals. However, I would like to dedicate this book to the memory of my first scientific mentor, Konstantin Iosifovich Gringauz. Gringauz was a true pioneer of space science: He designed devices and instruments for all generations of space vehicles from *Sputnik-1* to the first missions of the Moon, to the first planetary probes, and finally, to the spacecraft that intercepted Halley's comet. He not only introduced me to world-class space science, but he also stood by me during difficult periods of my life. I greatly benefited from his scientific insight, from his no-nonsense approach to science, and from his personal friendship.

Ann Arbor, Michigan   **Tamas I. Gombosi**

Part I

Theoretical Description of Gases and Plasmas

Chapter 1

Particle Orbit Theory

In this chapter we investigate how single particles (charged or neutral) behave in gravitational, electric, and magnetic fields. It is assumed that these fields are *externally prescribed* and that they are not affected by the particles themselves. This approach is usually referred to as the "test particle" method.

1.1   Electromagnetic Fields

1.1.1   Maxwell's Equations in a Vacuum

In vacuum, electromagnetic fields are generated by electric charges and currents. These quantities are characterized by the net electric charge density, ρ, and by the electric current density vector, $\mathbf{j}$. The relation between the source quantities, ρ and $\mathbf{j}$, and the resulting electric and magnetic field vectors ($\mathbf{E}$ and $\mathbf{B}$) are described by *Maxwell's equations*. Maxwell's equations can be written in differential or in integral form. For the sake of completeness, we present here both forms.

Maxwell's equations for vacuum contain three universal constants: the speed of light in vacuum ($c = 2.9979 \times 10^8$ m/s), the magnetic permeability of vacuum ($\mu_0 = 4\pi \times 10^{-7}$ henry/m), and the electric permittivity of vacuum ($\varepsilon_0 = 8.8542 \times 10^{-12}$ farad/m). These constants are not independent of each other: They are related by the $\mu_0 \varepsilon_0 = 1/c^2$ relation.

Differential Form of Maxwell's Equations   The four equations in differential form are:

Poisson's equation

$$\nabla \cdot \mathbf{E} = \frac{\rho}{\varepsilon_0}, \tag{1.1}$$

Absence of magnetic monopoles

$$\nabla \cdot \mathbf{B} = 0, \tag{1.2}$$

Faraday's law

$$\nabla \times \mathbf{E} = -\frac{\partial \mathbf{B}}{\partial t}, \tag{1.3}$$

and *Ampère's law*

$$\nabla \times \mathbf{B} = \frac{1}{c^2} \frac{\partial \mathbf{E}}{\partial t} + \mu_0 \mathbf{j}. \tag{1.4}$$

Integral Form of Maxwell's Equations   The integral form of Maxwell's equations can be obtained by integrating Eqs. (1.1) and (1.2) over a finite volume, $\mathcal{V}$, bounded by a closed surface, $\mathcal{A}$. One can now use Stokes's theorem to obtain the following forms of the two source equations:

Gauss's law

$$\oint_{\mathcal{A}} \mathbf{E} \cdot \mathbf{n} \, d\mathcal{A} = \frac{1}{\varepsilon_0} \int_{\mathcal{V}} \rho \, d\mathcal{V}, \tag{1.5}$$

Absence of magnetic monopoles

$$\oint_{\mathcal{A}} \mathbf{B} \cdot \mathbf{n} \, d\mathcal{A} = 0. \tag{1.6}$$

In these integrals $d\mathcal{A}$ is an infinitesimal element of area on $\mathcal{A}$, and $\mathbf{n}$ is the unit normal vector to the surface element $d\mathcal{A}$ pointing outward from the enclosed volume.

The integral form of Faraday's and Ampère's laws can be obtained by integrating Eqs. (1.3) and (1.4) over an open surface area, $\mathcal{F}$, bounded by the closed curve, $\mathcal{C}$. After invoking Gauss's theorem, one obtains the following relations:

Faraday's law

$$\oint_{\mathcal{C}} \mathbf{E} \cdot d\mathbf{s} = -\int_{\mathcal{F}} \frac{\partial \mathbf{B}}{\partial t} \cdot \mathbf{n}' \, d\mathcal{F}, \tag{1.7}$$

Ampère's law

$$\oint_{\mathcal{C}} \mathbf{B} \cdot d\mathbf{s} = \frac{1}{c^2} \int_{\mathcal{F}} \frac{\partial \mathbf{E}}{\partial t} \cdot \mathbf{n}' \, d\mathcal{F} + \mu_0 \int_{\mathcal{F}} \mathbf{j} \cdot \mathbf{n}' \, d\mathcal{F}. \tag{1.8}$$

In these integrals $\mathbf{n}'$ is the unit normal vector to the surface element $d\mathcal{F}$ in the direction given by the right-hand rule from the sense of integration around $\mathcal{C}$.

1.1.2 Lorentz Transformation

Consider two inertial reference frames K and K' moving with respect to each other with a constant relative velocity, $\mathbf{u}$. The time and space coordinates are (t, x, y, z) and (t', x', y', z') in the frames of reference K and K', respectively. For the sake of simplicity, we choose the x axis of our coordinate systems (in both frames) to be parallel to the relative velocity. In this case $\mathbf{u} = (u, 0, 0)$.

The time and space coordinates in K' are related to those in K by the

Lorentz transformation:

$$
\begin{aligned}
ct' &= \gamma \left(ct - \frac{u}{c}x \right), \\
x' &= \gamma \left(x - \frac{u}{c}ct \right), \\
y' &= y, \\
z' &= z,
\end{aligned}
\tag{1.9}
$$

where the Lorentz factor, γ, is defined as

$$
\gamma = \frac{1}{\sqrt{1 - \frac{u^2}{c^2}}}.
\tag{1.10}
$$

The inverse Lorentz transformation is

$$
\begin{aligned}
ct &= \gamma \left(ct + \frac{u}{c}x' \right), \\
x &= \gamma \left(x' + \frac{u}{c}ct' \right), \\
y &= y', \\
z &= z'.
\end{aligned}
\tag{1.11}
$$

The relativistic momentum and energy of a particle moving with velocity $\mathbf{v}$ are the following:

$$
\mathbf{p} = \gamma m \mathbf{v}
$$

and

$$
\varepsilon = \gamma m c^2,
\tag{1.12}
$$

where m is the particle's rest mass and ε is its total energy. It can be readily seen that

$$
\frac{\varepsilon^2}{c^2} - \mathbf{p} \cdot \mathbf{p} = (mc)^2.
\tag{1.13}
$$

The transformation of the electric and magnetic field vectors between K and K' are given by

$$\mathbf{E}' = \gamma(\mathbf{E} + \mathbf{u} \times \mathbf{B}) - \frac{\gamma^2}{\gamma + 1}\left(\frac{\mathbf{u}}{c} \cdot \mathbf{E}\right)\frac{\mathbf{u}}{c},$$

$$\mathbf{B}' = \gamma\left(\mathbf{B} - \frac{\mathbf{u}}{c^2} \times \mathbf{E}\right) - \frac{\gamma^2}{\gamma + 1}\left(\frac{\mathbf{u}}{c} \cdot \mathbf{B}\right)\frac{\mathbf{u}}{c}. \tag{1.14}$$

The inverse Lorentz transformation can be found by interchanging the primed and unprimed quantities and replacing $\mathbf{u}$ by $-\mathbf{u}$.

In this book we will restrict ourselves to nonrelativistic situations, when $u/c \ll 1$. In this case the above expressions can be expanded into a Taylor series in $\mathbf{u}/c$ and all terms containing any power of $\mathbf{u}/c$ can be neglected. This yields the following nonrelativistic transformations (also called a *Gallilean transformation*) for t, $\mathbf{r}$, $\mathbf{E}$, and $\mathbf{B}$:

$$\begin{aligned}
t' &= t, \\
\mathbf{r}' &= \mathbf{r} - \mathbf{u}\,t, \\
\mathbf{E}' &= \mathbf{E} + \mathbf{u} \times \mathbf{B}, \\
\mathbf{B}' &= \mathbf{B}.
\end{aligned} \tag{1.15}$$

1.1.3   Lorentz Force

Electromagnetic fields exert the following force (called the *Lorentz force*) on a charged particle:

$$\mathbf{F} = q(\mathbf{E} + \mathbf{v} \times \mathbf{B}), \tag{1.16}$$

where q is the particle charge and $\mathbf{v}$ is the particle velocity.

In this book we assume that particles are nonrelativistic and that they move under the influence of electromagnetic and gravitational forces. In this case the equation of motion of the individual particles (neutral or electrically charged) is given by

$$\frac{d\mathbf{p}}{dt} = \frac{d}{dt}(m\,\mathbf{v}) = m\,\mathbf{g} + q\,(\mathbf{E} + \mathbf{v} \times \mathbf{B}), \tag{1.17}$$

where $\mathbf{g}$ is the gravitational acceleration of the particle.

1.2   Particles in Constant External Fields

1.2.1   Neutral Particles

The simplest case is when a neutral test particle ($q = 0$) moves in a gravitational field of a large body (such as the Sun or a planet). If we consider nonrelativistic neutral particles and assume constant gravitational acceleration, Eq. (1.17) simplifies to

$$\frac{d\mathbf{v}}{dt} = \mathbf{g}. \tag{1.18}$$

The solution of this equation is well-known:

$$\mathbf{v} = \mathbf{g}\,t + \mathbf{v}_0, \tag{1.19}$$

$$\mathbf{r} = \frac{1}{2}\mathbf{g}\,t^2 + \mathbf{v}_0\,t + \mathbf{r}_0, \tag{1.20}$$

where the initial conditions, $\mathbf{r}_0$ and $\mathbf{v}_0$, represent the location and velocity of the particle at $t = 0$.

1.2.2   Uniform Electric Field

Consider the case of a nonrelativistic charged particle moving in a uniform electric field, $\mathbf{E}_0$, with no gravitational or magnetic fields present. In this case the equation of motion becomes the following:

$$m\frac{d\mathbf{v}}{dt} = q\mathbf{E}_0. \tag{1.21}$$

Mathematically, this equation is equivalent to Eq. (1.18). The solution is again

$$\mathbf{v} = \frac{q}{m}\mathbf{E}_0\,t + \mathbf{v}_0, \tag{1.22}$$

$$\mathbf{r} = \frac{1}{2}\frac{q}{m}\mathbf{E}_0\,t^2 + \mathbf{v}_0\,t + \mathbf{r}_0. \tag{1.23}$$

1.2.3   Uniform Magnetic Field

Gyrofrequency   Next, we consider the case of a single nonrelativistic charged particle moving in a uniform magnetic field, $\mathbf{B}_0$, in the absence of electric or gravitational fields. In this case the equation of motion can be written as

$$\frac{d\mathbf{v}}{dt} = \frac{q}{m}\,(\mathbf{v} \times \mathbf{B}). \tag{1.24}$$

The same equation can be written in the following explicit vector form:

$$\begin{pmatrix} \dot{v}_x \\ \dot{v}_y \\ \dot{v}_z \end{pmatrix} = \frac{q}{m} \begin{vmatrix} \mathbf{e}_x & \mathbf{e}_y & \mathbf{e}_z \\ v_x & v_y & v_z \\ 0 & 0 & B_0 \end{vmatrix} = \frac{q\,B_0}{m} \begin{pmatrix} v_y \\ -v_x \\ 0 \end{pmatrix}, \tag{1.25}$$

where the unit vectors, $\mathbf{e}_x$, $\mathbf{e}_y$, and $\mathbf{e}_z$, are perpendicular to each other, forming a right-handed coordinate system, and $\mathbf{e}_z$ is in the direction of the uniform magnetic field vector, $\mathbf{B}_0$.

One can take the time derivative of Eq. (1.25) and obtain the following second-order differential equation:

$$\begin{pmatrix} \ddot{v}_x \\ \ddot{v}_y \\ \ddot{v}_z \end{pmatrix} = -\left(\frac{q\,B_0}{m}\right)^2 \begin{pmatrix} v_x \\ v_y \\ 0 \end{pmatrix}. \tag{1.26}$$

Equation (1.26) describes a simple harmonic oscillator at the *gyrofrequency* (or cyclotron frequency)

$$\Omega_c = \frac{|q|B_0}{m}. \tag{1.27}$$

Note that by convention the gyrofrequency, Ω_c, is always nonnegative. However, the sign of q is different for positively and negatively charged particles. As can be seen from Eq. (1.25) the sign of q determines the direction of oscillation of the velocity components: Positively charged particles gyrate in the right-hand sense (the angular velocity vector is antiparallel to the magnetic field vector, $\mathbf{B}_0$), whereas the direction of gyration of particles with negative charge is to the left (the angular velocity vector is parallel to the magnetic field vector, $\mathbf{B}_0$).

Guiding Center and Gyroradius Equation (1.25) represents three second-order differential equations describing the evolution of the particle location vector, $\mathbf{r}$, and the solution contains six integration constants, $\mathbf{r}_0$ and $\mathbf{v}_0$. For the sake of simplicity, we choose our coordinate system in a way so that $\mathbf{v}_0 = (v_\perp, 0, v_\parallel)$, where $v_\parallel$ and $v_\perp$ represent the initial velocity components parallel and perpendicular to the magnetic field vector, $\mathbf{B}_0$. With this choice of the integration constants the solution of Eq. (1.25) is the following:

$$\begin{pmatrix} x - x_0 \\ y - y_0 \\ z - z_0 \end{pmatrix} = \begin{pmatrix} 0 \\ \mp v_\perp/\Omega_c \\ v_\parallel t \end{pmatrix} + \frac{v_\perp}{\Omega_c} \begin{pmatrix} \sin \Omega_c t \\ \pm \cos \Omega_c t \\ 0 \end{pmatrix}, \tag{1.28}$$

where $\pm$ refers to the sign of charge q.

Equation (1.28) shows that the motion of the particle can be decomposed into two simple motions: a circular motion along the perimeter of a circle with angular velocity Ω_c and a motion with constant velocity $v_\parallel$ along the magnetic field line. In other words, Eq. (1.28) can be written as

$$\mathbf{r} - \mathbf{r}_0 = \mathbf{r}_g + r_c \begin{pmatrix} \sin \Omega_c t \\ \pm \cos \Omega_c t \\ 0 \end{pmatrix}, \tag{1.29}$$

where we introduced the radius vector of the *guiding center* (i.e., the center of gyration), $\mathbf{r}_g$, and the *gyroradius*, r_c (sometimes also called the *Larmor radius*):

$$\mathbf{r}_g = \begin{pmatrix} 0 \\ \mp r_c \\ v_\parallel t \end{pmatrix} \quad \text{and} \quad r_c = \frac{v_\perp}{\Omega_c}. \tag{1.30}$$

It is important to note that the direction of gyration is always such that the magnetic field generated by the charged particle motion is opposite to the externally imposed field (gyration *reduces* the magnetic field and therefore gases composed of charged particles

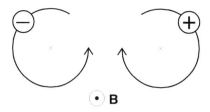

Figure 1.1 Circular trajectories of charged particles in a uniform and constant magnetic field (directed out of the page).

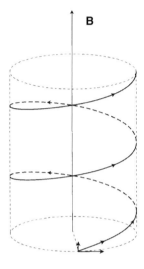

Figure 1.2 Helical trajectory of a negatively charged particle in a uniform and constant magnetic field (directed upward in the z direction).

are *diamagnetic*). The direction of gyration is opposite for positively and negatively charged particles (positive ions and electrons). Positive ions gyrate counterclockwise, whereas electrons gyrate clockwise around the magnetic field. Figure 1.1 shows this gyration. Note that the magnetic field points outward from the page; therefore clockwise and counterclockwise directions with respect to the $\mathbf{B}_0$ vector are just the opposite as they look on the page. In addition to gyration, there is also an arbitrary motion along the magnetic field (with velocity $v_\parallel$), which is not affected by $\mathbf{B}_0$. The trajectory of a charged particle, therefore, is a helix (Figure 1.2).

1.2.4   Guiding Center Drifts

Let us consider the situation when none of the external fields are zero but $\mathbf{g}$, $\mathbf{E}$, and $\mathbf{B}$ are all uniform: $\mathbf{g} = \mathbf{g}_0$, $\mathbf{E} = \mathbf{E}_0$, and $\mathbf{B} = \mathbf{B}_0$. Choosing a coordinate system with the z axis pointing along $\mathbf{B}_0$, the equation of motion for a nonrelativistic particle can be written in the following form:

$$\frac{d}{dt}\begin{pmatrix} v_x \\ v_y \\ v_z \end{pmatrix} = \begin{pmatrix} a_x \\ a_y \\ a_z \end{pmatrix} \pm \Omega_c \begin{pmatrix} v_y \\ -v_x \\ 0 \end{pmatrix}, \tag{1.31}$$

where $\mathbf{a} = \mathbf{g}_0 + (q/m)\mathbf{E}_0$, while the $\pm$ symbol again refers to the sign of the charge of the particle. Equation (1.31) can be readily solved with initial conditions of $\mathbf{r_0} = (x_0, y_0, z_0)$, $\mathbf{v_0} = (v_\perp, 0, v_\parallel)$:

$$\mathbf{r} - \mathbf{r_0} = \begin{pmatrix} a_x/\Omega_c^2 \\ a_y/\Omega_c^2 \mp v_\perp/\Omega_c \\ v_\parallel t + \frac{1}{2}a_z t^2 \end{pmatrix} + r_c' \begin{pmatrix} \sin(\Omega_c t - \delta) \\ \pm\cos(\Omega_c t - \delta) \\ 0 \end{pmatrix} \pm \begin{pmatrix} a_y t/\Omega_c \\ -a_x t/\Omega_c \\ 0 \end{pmatrix},$$

(1.32)

where the modified Larmor radius, r_c', and the phase angle, δ, are given by

$$r_c' = \frac{\sqrt{(v_\perp \Omega_c \mp a_y)^2 + a_x^2}}{\Omega_c^2}$$

(1.33)

and

$$\tan \delta = \frac{a_x}{(v_\perp \Omega_c \mp a_y)}.$$

(1.34)

The three vectors on the right-hand side of Eq. (1.32) represent three components of the particle motion: an accelerating motion of the guiding center along the magnetic field line, gyration around the guiding center with a modified gyroradius, r_c', and finally a slow drift of the guiding center described by the last vector.

The new elements of the solution are the drift and the modification of the gyroradius. These two phenomena are closely interrelated, and we will explain them together. First of all, one can express the drift in a more general form. It is easy to see that the drift is the displacement of the guiding center with a constant velocity, $\mathbf{v}_{\mathrm{drift}}$:

$$\mathbf{v}_{\mathrm{drift}} = \pm \frac{1}{\Omega_c} \begin{pmatrix} a_y \\ -a_x \\ 0 \end{pmatrix} = \frac{m}{q} \frac{\mathbf{a} \times \mathbf{B_0}}{B_0^2}.$$

(1.35)

The physical reason for the drift and the change of the gyroradius is that during one half-cycle of the particle orbit the particle is accelerated by the external force field (when $\mathbf{v} \cdot \mathbf{a} > 0$), whereas during the other half-cycle of the orbit the particle is decelerated. This difference in r_c on the two halves of the orbit causes the drift and the modification of the gyroradius.

Next we examine two different drifts, one caused by an constant electric field, $\mathbf{E_0}$, and the other related to gravity, $\mathbf{g_0}$.

Drift Due to $\mathbf{E_0}$ In this case $\mathbf{a} = (q/m)\mathbf{E_0}$ and consequently

$$\mathbf{v}_E = \frac{\mathbf{E_0} \times \mathbf{B_0}}{B_0^2}.$$

(1.36)

It is important to note that in this case the drift velocity is independent of all characteristics of the individual particles, q, m, and $v_\perp$. Both ions and electrons drift in the same direction and with the same velocity; consequently, this drift cannot create net electric current.

Drift Due to gravity In this case the drift velocity becomes

$$\mathbf{v}_g = \frac{m}{q} \frac{\mathbf{g}_0 \times \mathbf{B}_0}{B_0^2}. \tag{1.37}$$

This drift is different from the electric drift in a very important way: $\mathbf{v}_g$ is proportional to the mass over charge ratio. This means that electrons and ions not only drift with different speeds, but they drift in opposite directions. This is a very important feature of the gravitational drift, a feature that may result in the generation of net electric currents. This drift is usually negligible, but it serves as an important analogy for other, nonnegligible drifts.

1.3 Nonuniform Magnetic Fields

In the previous section we introduced the concept of guiding center drift and derived a general expression for the drift velocity due to a constant external acceleration (see Eq. 1.35). Next, we discuss some important drifts occurring in inhomogeneous fields.

It is customary to expand the inhomogeneous field in the smallness parameter, r_c/ℓ, where ℓ is the characteristic length of the inhomogeneity. We consider problems when r_c/ℓ is small (i.e., the field changes very little during one gyration). In this approximation the inhomogeneous field can be expanded into a Taylor series around the location of the guiding center, $\mathbf{r}_g$, and the series can be truncated after a low-order term (typically after the first-order term). This approximation is called *orbit theory*, and some simple examples will be examined here.

1.3.1 Magnetic Mirror Force

Let us consider a slowly varying magnetic field pointing primarily in the z direction. By slow variation we mean that the gyroradius, r_c, is small compared to the characteristic scale length of the magnetic field variation, ℓ:

$$\frac{1}{\ell} = \max_{i=1,3} \left\{ \left| \frac{1}{B_i} \nabla B_i \right| \right\}, \tag{1.38}$$

where B_i refers to the components of the magnetic field. For the sake of mathematical simplicity, we consider an axially symmetric magnetic field configuration around the z axis (see Figure 1.3) with $B_z = B_0(z)$, $B_r = B_r(z, r)$, and $B_\Phi = 0$, where B_z, B_r, and B_Φ are the components of the magnetic field vector in cylindrical coordinates, and $|B_0| \gg |B_r|$. Furthermore, it is assumed that all derivatives in the azimuthal direction vanish, $\partial/\partial\Phi = 0$. In this configuration the divergenceless nature of the magnetic field results in the following relation coupling the z and r components of the magnetic field:

$$\frac{1}{r} \frac{\partial}{\partial r}(r B_r) + \frac{\partial B_z}{\partial z} = 0. \tag{1.39}$$

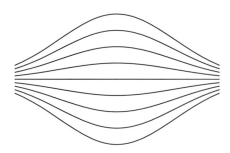

Figure 1.3 Simple
magnetic mirror
configuration.

Assuming that $|\partial B_0/\partial z| \gg |\partial B_0/\partial r|$, one can readily integrate Eq. (1.39) to obtain

$$B_r = -\frac{r}{2}\left(\frac{\partial B_0}{\partial z}\right). \tag{1.40}$$

The Lorentz force acting on a particle can now be explicitly written in the following way:

$$\begin{pmatrix} \dot{v}_z \\ \dot{v}_r \\ \dot{v}_\Phi \end{pmatrix} = \frac{q}{m}\begin{pmatrix} \mathbf{e}_z & \mathbf{e}_r & \mathbf{e}_\Phi \\ v_z & v_r & v_\Phi \\ B_z & B_r & 0 \end{pmatrix} = \frac{qB_0}{m}\begin{pmatrix} 0 \\ v_\Phi \\ -v_r \end{pmatrix} + \frac{1}{2}\frac{q}{m}r\left(\frac{\partial B_0}{\partial z}\right)\begin{pmatrix} v_\Phi \\ 0 \\ -v_z \end{pmatrix}. \tag{1.41}$$

The first term on the right-hand side of Eq. (1.41) describes gyration around the large-scale magnetic field, $B_0\,\mathbf{e}_z$, whereas the second term is due to the slow spatial variation of the field.

Particle drift is a slow process compared to gyration. In order to derive the average force resulting in this slow drift one has to average the acceleration given by Eq. (1.41) over a particle gyration. For the sake of simplicity, let us consider a particle whose guiding center lies on the axis of symmetry. Since the dominant component of the magnetic field is $B_0\,\mathbf{e}_z$, for this particle $v_r = 0$ and both v_Φ and v_z are approximately constant during a gyration. The field-aligned velocity is $v_z = v_\parallel$, while $v_\Phi = \mp v_\perp$ depending on the sign of the charge q. Because the guiding center is located at the axis, the particle is gyrating at a radial distance of $r = r_{\rm c}$. Now the force acting on the particle can be written as

$$\mathbf{F} = \mp\frac{1}{2}qr_{\rm c}v_\perp\frac{\partial B_0}{\partial z}\mathbf{e}_z \mp qB_0v_\perp\mathbf{e}_r - \frac{1}{2}qr_{\rm c}v_\parallel\frac{\partial B_0}{\partial z}\mathbf{e}_\Phi. \tag{1.42}$$

Next, we take the average of this force for a gyration (i.e., we average expression (1.42) over Φ). As the particle gyrates the $\mathbf{e}_z$ unit vector remains the same, but $\mathbf{e}_r$ and $\mathbf{e}_\Phi$ rotate $360°$; therefore $\langle\mathbf{e}_r\rangle_\Phi = 0$ and $\langle\mathbf{e}_\Phi\rangle_\Phi = 0$. This means that the average force acting on the particle is

$$\langle\mathbf{F}\rangle_\Phi = \mp\frac{1}{2}q\,r_{\rm c}\,v_\perp\frac{\partial B_0}{\partial z}\mathbf{e}_z = -\frac{1}{2}\frac{m\,v_\perp^2}{B_0}\frac{\partial B_0}{\partial z}\mathbf{e}_z. \tag{1.43}$$

In general, this force acts along the slowly varying magnetic field vector and it can be written as

$$F_\parallel = m \frac{d v_\parallel}{dt} = -\mu_m \frac{dB}{ds},$$
(1.44)

where we introduced the following notations:

$$\mu_m = \frac{1}{2} \frac{m v_\perp^2}{B},$$
(1.45)

$$\frac{d}{ds} = \frac{(\mathbf{B} \cdot \nabla)}{B}.$$
(1.46)

Here s is distance along the magnetic field line and ds is the line element along $\mathbf{B}$. In general, $\mathbf{F}_\parallel$ is called the magnetic mirror force (the reason will become obvious in the following sections).

1.3.2   Magnetic Moment

Equation (1.44) has several interesting consequences. First of all, it can be easily shown that the newly introduced quantity, μ_m, is the magnitude of the magnetic moment represented by the gyrating particle. The definition of the magnetic moment of a current loop with area A and current I is AI. In the case of our gyrating charged particle, $I = q/T_c$, where $T_c = 2\pi/\Omega_c$ is the gyration period of the particle. The area of the loop is $A = \pi r_c^2 = \pi v_\perp^2/\Omega_c^2$. Thus the magnitude of the magnetic moment becomes

$$A |I| = \frac{\pi v_\perp^2}{\Omega_c^2} \frac{|q| \Omega_c}{2\pi} = \frac{1}{2} \frac{m v_\perp^2}{B} = \mu_m.$$
(1.47)

An important consequence of Eq. (1.44) is the conservation of the magnetic moment, as the particle moves along the field line into regions of stronger or weaker magnetic field. To prove this statement, let us first multiply both sides of Eq. (1.44) by $v_\parallel$ and recognize that $v_\parallel = ds/dt$:

$$m v_\parallel \frac{d v_\parallel}{dt} = \frac{d}{dt}\left(\frac{1}{2} m v_\parallel^2\right) = -\mu_m \frac{dB}{ds}\frac{ds}{dt} = -\mu_m \frac{dB}{dt}.$$
(1.48)

Here the d/dt operator represents the convective (full) derivative; for instance dB/dt is the variation of the magnetic field magnitude as seen by the moving particle.

In this simple example the particle is moving in a magnetic field only, and its equation of motion is given by

$$\frac{d\mathbf{v}}{dt} = \frac{q}{m} (\mathbf{v} \times \mathbf{B}).$$
(1.49)

Next, we take the scalar product of this vector equation with the particle momentum vector, $m\mathbf{v}$, to obtain

$$m\mathbf{v} \cdot \frac{d\mathbf{v}}{dt} = \frac{d}{dt}\left(\frac{1}{2}mv^2\right) = q\mathbf{v} \cdot (\mathbf{v} \times \mathbf{B}) = 0. \tag{1.50}$$

Equation (1.50) tells us that the total energy of the particle is conserved as it moves in a pure magnetic field. In our case the total energy can be written as the sum of the parallel and perpendicular energies, and therefore one obtains

$$\frac{d}{dt}\left(\frac{1}{2}mv_\parallel^2 + \frac{1}{2}mv_\perp^2\right) = \frac{d}{dt}\left(\frac{1}{2}mv_\parallel^2 + \mu_\mathrm{m}B\right) = 0. \tag{1.51}$$

Here we used the definition of μ_m (Eq. 1.45) to replace the perpendicular energy with $\mu_\mathrm{m}B$. Finally, we substitute Eq. (1.48) into Eq. (1.51) to obtain

$$-\mu_\mathrm{m}\frac{dB}{dt} + \frac{d}{dt}(\mu_\mathrm{m}B) = B\frac{d\mu_\mathrm{m}}{dt} = 0. \tag{1.52}$$

The magnitude of the magnetic field is obviously nonzero. Therefore, we conclude that the magnetic moment of the particle is conserved as it moves along slowly converging or diverging magnetic field lines:

$$\frac{d\mu_\mathrm{m}}{dt} = 0. \tag{1.53}$$

This result plays a very fundamental role in space plasma physics, and we will refer to it quite frequently.

1.3.3 Magnetic Mirroring

Equation (1.53) has very important consequences for both laboratory and space plasmas. As a gyrating particle moves toward increasing magnetic field magnitudes (i.e., into a region of converging magnetic field) its perpendicular velocity, $v_\perp$, must increase in order to keep μ_m constant. However, the total kinetic energy of the particle must be conserved (see Eq. 1.50), and therefore the magnitude of the parallel velocity, $|v_\parallel|$, decreases. At some value of the magnetic field magnitude, B_m, the entire kinetic energy of the particle is converted to the perpendicular direction in order to keep the magnetic moment constant. The particle cannot move toward regions of stronger magnetic field any longer, because that would violate the conservation of energy. At this point the particle has no field-aligned velocity, $v_\parallel = 0$, but the field-aligned force acting on the particle, $F_\parallel = -\mu_\mathrm{m}dB/ds$, points back toward the region of weaker field (diverging field lines). As a consequence, the particle will turn around at this point and move back toward the region of weakening field.

The conservation of the magnetic moment of the particle means that the variation of the magnitude of the magnetic field controls the *pitch angle* of the particle, Θ (the angle between the particle velocity vector, $\mathbf{v}$, and the magnetic field vector, $\mathbf{B}$). It

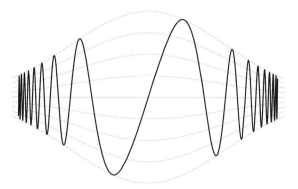

Figure 1.4 Mirroring particle trajectory.

follows from the conservation of μ_m and of the magnitude of the particle velocity, v, that

$$\mu_m = \frac{1}{2}\frac{mv_\perp^2}{B} = \frac{1}{2}mv^2\frac{\sin^2\Theta}{B} = const. \Rightarrow \frac{\sin^2\Theta}{B} = const. \qquad (1.54)$$

Equation (1.53) is the basis of one of the primary ways to confine plasma both in the laboratory and space. Let us assume that at a reference point the magnetic field magnitude is B_0 and the pitch angle of a particle is Θ_0. If the particle is moving toward a region of converging field lines (increasing magnetic field magnitude), its pitch angle, Θ, will vary as

$$\sin^2\Theta = \frac{B}{B_0}\sin^2\Theta_0. \qquad (1.55)$$

At the mirror point the particle pitch angle is $\Theta_m = \pi/2$; therefore, one can express the magnetic field magnitude necessary to reflect this given particle as

$$B_m = \frac{B_0}{\sin^2\Theta_0}. \qquad (1.56)$$

A magnetic mirror geometry and a typical particle orbit is shown in Figure 1.4.

It is obvious that for smaller pitch angles larger magnetic field strength is required at the mirror point to reflect the particle. This also means that all magnetic mirrors are "leaky": Particles with very small initial pitch angle will never be reflected and can escape. For a given magnetic mirror configuration with maximum magnetic field strength of B_{max}, one can define the minimum initial pitch angle of reflected particles:

$$\sin^2\Theta_{min} = \frac{B_0}{B_{max}}. \qquad (1.57)$$

The angle, Θ_{min}, is called the loss-cone angle, since particles with $\Theta_0 < \Theta_{min}$ are lost from the magnetic mirror.

1.3.4   Gradient and Curvature Drifts

Next, we consider a somewhat more complicated configuration, where the electric field is zero everywhere and the slowly varying magnetic field vector points in the z direction everywhere, $\mathbf{B} = [0, 0, B]$. Furthermore, it is assumed that the magnetic field magnitude depends only on the x coordinate, $B = B(x)$, and that the scale length of the magnetic field variation, ℓ, is very large compared to the particle gyroradius, $\ell \gg r_c$.

Let us consider the motion of a charged particle in the vicinity of a reference point $\mathbf{r}_0 = (x_0, y_0, z_0)$, where $\mathbf{B}_0 = (0, 0, B_0)$. In the vicinity of point $\mathbf{r}_0$, the magnetic field vector can be expanded into a Taylor series, and one can stop after the first-order terms to get

$$\mathbf{B} = \begin{pmatrix} 0 \\ 0 \\ B_0 + (x - x_0)\frac{\partial B_0}{\partial x} \end{pmatrix}, \tag{1.58}$$

where

$$\frac{\partial B_0}{\partial x} = \left. \frac{\partial B}{\partial x} \right|_{\mathbf{r}=\mathbf{r}_0}. \tag{1.59}$$

Next, we evaluate the equation of motion of a particle with first-order accuracy in the small parameter, $\epsilon = r_c/\ell$. In this approximation the magnetic field vector changes very little during one gyration period of the particle. We may consider the motion of the particle in the vicinity of our reference point, $\mathbf{r}_0$, as the superposition of a standard helical motion in the constant magnetic field, $\mathbf{B}_0$, and a first-order velocity perturbation, $\delta \mathbf{v}$, due to the slow variation of the magnetic field. Assuming that at $t = 0$ the guiding center of the particle is located at the reference point and neglecting second- and higher-order terms, one can express the magnetic field vector and the particle velocity along the perturbed particle trajectory as

$$\mathbf{B} = \begin{pmatrix} 0 \\ 0 \\ B_0 + \frac{v_\perp \sin \Omega_c t}{\Omega_c}\frac{\partial B_0}{\partial x} \end{pmatrix} \tag{1.60}$$

and

$$\mathbf{v} = \begin{pmatrix} v_\perp \cos \Omega_c t \\ \mp v_\perp \sin \Omega_c t \\ v_\parallel \end{pmatrix} + \delta \mathbf{v} = \mathbf{v}_{\text{helix}} + \delta \mathbf{v}, \tag{1.61}$$

where $\Omega_c = |q| B_0/m$ is the unperturbed gyrofrequency of the particle and $|\delta \mathbf{v}| \ll |\mathbf{v}_{\text{helix}}|$.

The Lorentz force acting on the particle can now be expressed with first-order accuracy:

$$\left(\frac{d\mathbf{v}}{dt}\right)_{\text{grad}} = \frac{q}{m}(\mathbf{v} \times \mathbf{B})$$

$$= \frac{q}{m}(\delta\mathbf{v} \times \mathbf{B}_0) + \begin{pmatrix} -v_\perp \Omega_c \sin \Omega_c t - \frac{v_\perp^2}{B_0} \sin^2 \Omega_c t \frac{\partial B_0}{\partial x} \\ \mp v_\perp \Omega_c \cos \Omega_c t \mp \frac{v_\perp^2}{B_0} \sin \Omega_c t \cos \Omega_c t \frac{\partial B_0}{\partial x} \\ 0 \end{pmatrix} .$$

$$\tag{1.62}$$

Next, we calculate the average acceleration of the particle over a gyroperiod due to the gradient of the magnetic field in the perpendicular direction. It can be shown that $\langle \delta\mathbf{v} \rangle_\Phi = 0$, and therefore one obtains

$$\mathbf{a}_{\text{grad}} = \left\langle \frac{d\mathbf{v}}{dt}_{\text{grad}} \right\rangle_\Phi = -\begin{pmatrix} \frac{v_\perp^2}{2B_0} \frac{\partial B_0}{\partial x} \\ 0 \\ 0 \end{pmatrix} = -\frac{v_\perp^2}{2B_0}\nabla B_0. \tag{1.63}$$

As mentioned earlier, a force acting on the particle will result in a slow drift of the guiding center. Because the average effect of the magnetic field gradient can be described in terms of an average acceleration, $\mathbf{a}_{\text{grad}}$, it will also result in a guiding center drift. The velocity of this guiding center drift, which is due to the gradient of the magnitude of the magnetic field, can be obtained by substituting Eq. (1.63) into (1.35):

$$\mathbf{v}_G = \frac{m}{q} \frac{v_\perp^2}{2B_0} \frac{\mathbf{B}_0 \times \nabla B_0}{B_0^2}. \tag{1.64}$$

The magnetic field given by Eq. (1.58) is divergenceless, but it is not *curl* free: $\nabla \times \mathbf{B} \neq 0$. This means that a current system is required to maintain the magnetic field configuration. In most cases such a current system is not available and the magnetic field is curl free. This means that the **B** vector cannot point to the z direction everywhere and there is a small magnetic field component in the (x, y) plane. For the sake of mathematical simplicity we assume that in the vicinity of the reference point $\mathbf{r}_0$, this small field component is in the x direction (this assumption results in the neglect of field line twists, which is a justifiable simplification, since the twisting of field lines does not cause first-order particle drifts). In this approximation the magnetic field vector can be expressed as

$$\mathbf{B} = \begin{pmatrix} (z - z_0)\frac{\partial B_0}{\partial x} \\ 0 \\ B_0 + (x - x_0)\frac{\partial B_0}{\partial x} \end{pmatrix} . \tag{1.65}$$

This magnetic field vector is curl free, and the field line has a local radius of curvature of

$$\frac{1}{R_0} = -\frac{1}{B_0}\frac{\partial B_0}{\partial x}. \tag{1.66}$$

This configuration is shown schematically in Figure 1.5.

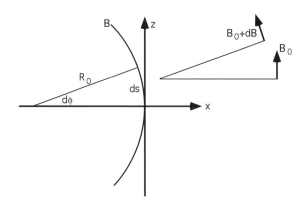

Figure 1.5 Curved magnetic field line.

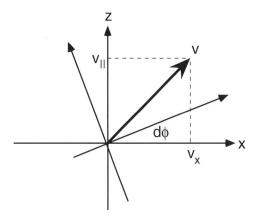

Figure 1.6 Velocity vector rotation due to magnetic field line curvature.

As the guiding center of the particle moves along the magnetic field line, the direction of the magnetic field slowly changes due to the curvature of the field line. Thus the velocity components also change, because the parallel and perpendicular velocity components are projected to slowly rotating directions (see Figure 1.6). The change of the velocity vector due to the field line curvature is the following:

$$
d\mathbf{v}_{\text{curv}} = \begin{pmatrix} v_x \cos d\phi + v_\parallel \sin d\phi \\ v_y \\ v_\parallel \cos d\phi - v_x \sin d\phi \end{pmatrix} - \begin{pmatrix} v_x \\ v_y \\ v_\parallel \end{pmatrix}
$$

$$
= \begin{pmatrix} v_x(\cos d\phi - 1) + v_\parallel \sin d\phi \\ 0 \\ v_\parallel(\cos d\phi - 1) - v_x \sin d\phi \end{pmatrix}. \tag{1.67}
$$

Assuming that $|d\phi| \ll 1$, Eq. (1.67) can be approximated with first-order accuracy in $d\phi$:

$$
d\mathbf{v}_{\text{curv}} = \begin{pmatrix} v_\parallel d\phi \\ 0 \\ -v_x d\phi \end{pmatrix}. \tag{1.68}
$$

The rate of change of the velocity vector averaged over a gyration period can be readily obtained:

$$\mathbf{a}_{\text{curv}} = \left\langle \frac{d\mathbf{v}_{\text{curv}}}{dt} \right\rangle_{\Phi} = \begin{pmatrix} v_{\parallel} \frac{d\phi}{dt} \\ 0 \\ 0 \end{pmatrix} \qquad (1.69)$$

because $\langle v_x \rangle_{\Phi} = 0$. However, we know that $d\phi = ds/R_0 = v_{\parallel} dt/R_0$ (see Figure 1.5), so therefore one obtains

$$\frac{d\phi}{dt} = \frac{v_{\parallel}}{R_0} = -\frac{v_{\parallel}}{B_0} \frac{\partial B_0}{\partial x}, \qquad (1.70)$$

and consequently the acceleration due to the field line curvature is

$$\mathbf{a}_{\text{curv}} = -\frac{v_{\parallel}^2}{B_0} \nabla_{\perp} B_0, \qquad (1.71)$$

where $\nabla_{\perp} B$ is the component of the gradient of the magnetic field magnitude, which is perpendicular to the magnetic field vector.

The guiding center drift due to this acceleration can be easily obtained by substituting Eq. (1.71) into the general expression for particle drift velocity, Eq. (1.35):

$$\mathbf{v}_C = \frac{m}{q} \frac{v_{\parallel}^2}{B_0} \frac{\mathbf{B}_0 \times \nabla B_0}{B_0^2}. \qquad (1.72)$$

In vacuum (where there is no electric current density) both the gradient and curvature effects are present and the particle experiences the combined drift, usually referred to as *gradient–curvature drift*:

$$\mathbf{v}_{\text{GC}} = \frac{m}{q} \frac{v_{\parallel}^2 + \frac{1}{2} v_{\perp}^2}{B_0} \frac{\mathbf{B}_0 \times \nabla B_0}{B_0^2}. \qquad (1.73)$$

It is important to note that the direction of the gradient–curvature drift depends on the sign of the charge of the particle: Electrons and ions drift in different directions. This effect leads to the formation of the *ring current*, to be discussed in the next section.

1.4 Adiabatic Invariants

We learned in classical mechanics that periodic motion can be characterized with conserved quantities. Such quantities can be obtained by the general action integral:

$$I = \oint \mathcal{P} \, d\mathcal{Q}, \qquad (1.74)$$

where $\mathcal{P}$ and $\mathcal{Q}$ represent the generalized momentum and coordinate associated with the periodic motion. It can be shown that the constant of motion, I, is invariant to slow

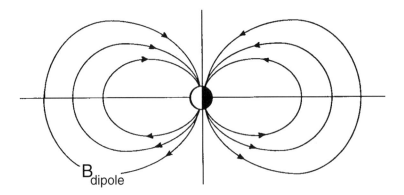

Figure 1.7 Magnetic field lines of a dipole.

changes in the system to higher-order accuracy (slow change means that the characteristic time of the change is very large compared to the period of the periodic motion).

Adiabatic invariants play a very important role in space physics, and they represent a simple but powerful tool in explaining a variety of phenomena in the space
environment. It is very important to point out that adiabatic invariants are approximately conserved even when "traditional" conserved quantities (such as momentum
or energy) are not. For instance, in a slowly varying magnetic field, the energy of the
particle changes, while the adiabatic invariants remain nearly constant. Under such
conditions the values of adiabatic invariants can well characterize particle populations.

The conservation of the first adiabatic invariant is widely used in all kinds of
magnetic geometries to understand particle transport. However, the second and third
adiabatic invariants are primarily used in dipolelike magnetic fields (such as closed
planetary magnetic field lines). For the sake of mathematical simplicity, we will introduce the second and third adiabatic invariants for dipole fields, so that we can
concentrate to the underlying physics and still underscore important physical impli-
cations for our space environment. Magnetic field lines of a dipole field are shown in
Figure 1.7.

The magnetic field of the dipole shown in Figure 1.7 can be expressed in the
following form using spherical coordinates (r, Φ, λ):

$$\mathbf{B} = \frac{\mu_0}{4\pi} \frac{M}{r^3} (-2\sin\lambda \, \mathbf{e}_r + \cos\lambda \, \mathbf{e}_\lambda), \tag{1.75}$$

where M is the strength of the magnetic moment (the magnetic moment vector is
aligned with the z axis in the $-z$ direction), r is radial distance from the center of the
sphere, while λ and Φ are magnetic latitude and longitude. ($\mathbf{e}_r$, $\mathbf{e}_\Phi$, and $\mathbf{e}_\lambda$ form a
right-handed coordinate system.) Note that there is no magnetic field component in
the azimuthal direction. The magnitude of the magnetic field at a given point is

$$B = \frac{\mu_0}{4\pi} \frac{M}{r^3} \sqrt{1 + 3\sin^2\lambda}. \tag{1.76}$$

The equation of a magnetic field line is given by

$$r = L\, R_e\, \cos^2 \lambda \tag{1.77}$$

where R_e is the average radius of the central body and L is the equatorial crossing distance of the magnetic field line normalized to the radius of the central body.

In many derivations it is useful to express the element of arc length along the field line. This can be expressed as

$$ds^2 = dr^2 + r^2 d\lambda^2. \tag{1.78}$$

One can now express $ds/d\lambda$ and ds/dr with the help of Eq. (1.77):

$$\frac{ds}{d\lambda} = L R_e \, \cos \lambda \sqrt{1 + 3 \sin^2 \lambda}, \tag{1.79}$$

$$\frac{ds}{dr} = \frac{\sqrt{1 + 3 \sin^2 \lambda}}{2 \sin \lambda}. \tag{1.80}$$

1.4.1   The First Adiabatic Invariant

The first adiabatic invariant is related to the cyclotron motion of the particle around the magnetic field line. In this case the generalized periodic momentum and coordinate are the perpendicular momentum and pathlength along the gyration path, $\mathcal{P}_1 = m v_\perp$ and $d\mathcal{Q} = r_c d\phi$ (where ϕ is the angle of the gyromotion). Substituting these quantities into Eq. (1.74) yields the first adiabatic invariant:

$$I_1 = \int_0^{2\pi} m v_\perp r_c \, d\phi = 2\pi m v_\perp r_c = 2\pi \frac{m^2 v_\perp^2}{|q| B} = 4\pi \frac{m}{|q|} \mu_m. \tag{1.81}$$

Thus the first adiabatic invariant, I_1, is proportional to the magnetic moment of the gyrating particle, μ_m. We have shown earlier that in slowly varying magnetic fields μ_m is conserved along the particle trajectory. It can also be shown that I_1 (and consequently μ_m) is conserved to a very high order in the smallness parameter (which is the ratio of the gyration period to the characteristic time of variation of the system).

1.4.2   The Second Adiabatic Invariant

For particles trapped between two magnetic mirrors, the second adiabatic invariant is related to the periodic (bouncing) motion along the magnetic field line. For this periodic motion the generalized momentum and coordinate are $\mathcal{P}_2 = m v_\parallel$ and $d\mathcal{Q} = ds$, where

ds is the element of pathlength of the guiding center along the field line. Then

$$I_2 = J = \oint m v_\parallel \, ds = 2m v \int_{s_{\min,1}}^{s_{\min,2}} \sqrt{1 - \frac{B}{B_0} \sin^2 \Theta_0} \, ds, \qquad (1.82)$$

where B_0 and Θ_0 represent the magnetic field magnitude and the particle pitch angle at a reference point, while $s_{\min,1}$ and $s_{\min,2}$ are the locations of the mirror points along the magnetic field line. When deriving Eq. (1.82) we used relation (1.55). The factor of 2 comes from the fact that the particle travels back and forth along the field line during a full bounce.

It is interesting to note that the quantity

$$I = \frac{J}{2mv} = \int_{s_{\min,1}}^{s_{\min,2}} \sqrt{1 - \frac{B}{B_0} \sin^2 \Theta_0} \, ds \qquad (1.83)$$

is independent of the magnitude of the particle velocity and it characterizes the magnetic field geometry. This expression can be evaluated for dipole field lines. We will use magnetic latitude, λ, as the integration variable. If Θ_0 and B_0 represent the particle pitch angle and the magnetic field magnitude at the magnetic equator, the following expression is obtained for I:

$$I = L R_e \int_{-\lambda_m}^{\lambda_m} \cos \lambda \sqrt{1 + 3 \sin^2 \lambda} \sqrt{1 - \sin^2 \Theta_0 \frac{\sqrt{1 + 3 \sin^2 \lambda}}{\cos^6 \lambda}} \, d\lambda. \qquad (1.84)$$

Here R_e is the radius of the sphere (planet). The magnetic latitude of the mirror point, λ_m, can be expressed with the help of the equatorial pitch angle (see Eq. 1.56):

$$\sin^2 \Theta_0 = \frac{B_0}{B_m} = \frac{\cos^6 \lambda_m}{\sqrt{1 + 3 \sin^2 \lambda_m}}. \qquad (1.85)$$

The travel time from one mirror point to the other and back is called the *bounce period*. In the case of a dipole field the bounce period can be obtained by the following integral:

$$\tau_B = 4 \int_0^{\lambda_m} \frac{ds}{v_\parallel} = \frac{4}{v} \int_0^{\lambda_m} \frac{1}{\sqrt{1 - \frac{B}{B_0} \sin^2 \Theta_0}} \, ds. \qquad (1.86)$$

The integral in Eq. (1.86) can be expressed in terms of the magnetic latitude:

$$\mathcal{S}(\Theta_0) = \frac{v \tau_B}{4 L R_e} = \int_0^{\lambda_m} \frac{\cos \lambda \sqrt{1 + 3 \sin^2 \lambda}}{\sqrt{1 - \sin^2 \Theta_0 \frac{\sqrt{1 + 3 \sin^2 \lambda}}{\cos^6 \lambda}}} \, d\lambda. \qquad (1.87)$$

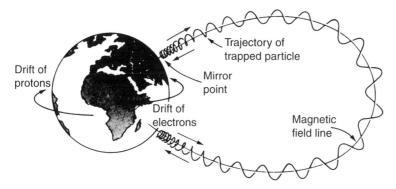

Figure 1.8 The motion of a trapped charged particle in Earth's magnetic field.

A very good approximation for $\mathcal{S}(\Theta_0)$ is the following (*Lyons & Willams 1984*):

$$\mathcal{S}(\Theta_0) \approx 1.38 - 0.32(\sin \Theta_0 + \sin^{1/2} \Theta_0). \tag{1.88}$$

1.4.3   The Third Adiabatic Invariant

We have seen that the direction of the gradient–curvature drift is determined by $\mathbf{B} \times \nabla B$ (cf. Eq. 1.73). In the case of a dipole field both $\mathbf{B}$ and ∇B are in the same meridional plane; consequently the gradient–curvature drift velocity always points to the azimuthal direction. Furthermore, since the direction of drift depends on the sign of the particle's charge, electrons and positive ions drift in opposite directions. The current generated by this charge-dependent drift is called the *ring current*. Figure 1.8 shows the drift patterns for particles trapped in the nearly dipolar magnetic field of the Earth. The slow gradient–curvature drift of the guiding center is the third quasi-periodic motion of trapped particles.

In the case of an axisymmetric magnetic field (such as the dipole field), the generalized coordinate is the azimuth angle, Φ, while the azimuthal component of the generalized momentum is the sum of the angular momentum with respect to the axis of symmetry and the magnetic flux through the portion of the spherical surface that extends from the particle's location (r and λ) to the "north pole" of the dipole [cf. *Rossi & Olbert (1970)*]:

$$\mathcal{P}_3 = m\, r^2 \cos^2 \lambda \frac{d\Phi}{dt} + \frac{q}{2\pi} \Phi_{\mathrm{B}}, \tag{1.89}$$

where

$$\Phi_{\mathrm{B}}(r, \lambda) = 2\pi r^2 \int_{\lambda}^{\pi/2} B_r(r, \lambda') \cos \lambda' d\lambda'. \tag{1.90}$$

The first term in Eq. (1.89) is the azimuthal component of the angular momentum of the drift motion; the second term is the magnetic flux, which explains the choice of the symbol Φ_B. Substituting Eq. (1.75) into expression (1.90) yields the function $\Phi_B(r, \lambda)$ for a magnetic dipole:

$$\Phi_B^{(\text{dipole})}(r, \lambda) = -2\pi \frac{\mu_0}{4\pi} \frac{M}{2r} \cos^2 \lambda. \tag{1.91}$$

In this case the generalized momentum component becomes

$$\mathcal{P}_3 = m\, r^2 \cos^2\!\lambda \left(\frac{d\Phi}{dt} - \frac{q}{m} \frac{\mu_0}{4\pi} \frac{M}{r^3} \right). \tag{1.92}$$

In axisymmetric magnetic fields $\mathcal{P}_3$ remains constant as the particle moves along its orbit ($\mathcal{P}_3$ is the well-known Størmer integral[1]). One can now express $d\Phi/dt$ with the help of the conserved quantity, $\mathcal{P}_3$:

$$\frac{d\Phi}{dt} = \frac{1}{r^2 \cos^2\!\lambda} \left(\frac{\mathcal{P}_3}{m} + \frac{q}{m} \frac{\mu_0}{4\pi} \frac{M}{r} \cos^2 \lambda \right). \tag{1.93}$$

The differential arc length for the three-dimensional motion of the particle is

$$ds^2 = dr^2 + r^2 d\lambda^2 + r^2 \cos^2 \lambda\, d\Phi^2. \tag{1.94}$$

With the help of this expression one obtains the following for $d\Phi/dt$:

$$\begin{aligned}
\left(\frac{d\Phi}{dt} \right)^2 &= \left(\frac{d\Phi}{ds} \right)^2 \left(\frac{ds}{dt} \right)^2 = v^2 \left(\frac{d\Phi}{ds} \right)^2 \\
&= \frac{v^2}{r^2 \cos^2 \lambda} \left[1 - \left(\frac{dr}{ds} \right)^2 - r^2 \left(\frac{d\lambda}{ds} \right)^2 \right].
\end{aligned} \tag{1.95}$$

This can be rearranged to obtain

$$\left(\frac{dr}{ds} \right)^2 + r^2 \left(\frac{d\lambda}{ds} \right)^2 = 1 - \left(\frac{r_s}{r} \frac{2\gamma_3}{\cos \lambda} + \frac{r_s^2}{r^2} \cos \lambda \right)^2, \tag{1.96}$$

where we introduced the following two quantities:

$$r_s^2 = \frac{q}{mv} \frac{\mu_0}{4\pi} M, \tag{1.97}$$

$$\gamma_3 = \frac{\mathcal{P}_3}{2mvr_s}. \tag{1.98}$$

Since the left-hand side of Eq. (1.96) is always positive or zero, the right-hand side must also be positive or zero for all physically meaningful particle orbits. Mathematically

[1] Størmer, C., "Sur les trajectoires des corpuscules électrisés dans l'espace sous l'action du magnétisme terrestre avec application aux aurores boréales," *Arch. Sci. Phys. et Nat. (Genève)*, **24**, 5–18, 113–58, 221–47, 317–54, 1907.

speaking, the

$$\left(\frac{r_s}{r} \frac{2\gamma_3}{\cos\lambda} + \frac{r_s^2}{r^2} \cos\lambda \right)^2 \leq 1 \tag{1.99}$$

condition must be met for all orbits. Depending on the value of γ_3, this inequality is satisfied in the following regions:

If $\gamma_3 \geq 0$, then the particle can move in the region where

$$r \geq r_s \frac{\gamma_3 + \sqrt{\gamma_3^2 + \cos^3\lambda}}{\cos\lambda}. \tag{1.100}$$

If $0 \geq \gamma_3 \geq -1$, then the particle can move in a complicated region. If $\cos^3\lambda \leq \gamma_3^2$, then there are two accessible radial regions given by

$$r_s \frac{\gamma_3 + \sqrt{\gamma_3^2 + \cos^3\lambda}}{\cos\lambda} \leq r \leq r_s \frac{-\gamma_3 - \sqrt{\gamma_3^2 - \cos^3\lambda}}{\cos\lambda} \tag{1.101}$$

and

$$r \geq r_s \frac{-\gamma_3 + \sqrt{\gamma_3^2 - \cos^3\lambda}}{\cos\lambda}. \tag{1.102}$$

If $\cos^3\lambda > \gamma_3^2$, then the accessible radial region is

$$r \geq r_s \frac{\gamma_3 + \sqrt{\gamma_3^2 + \cos^3\lambda}}{\cos\lambda}. \tag{1.103}$$

The most interesting situation is when $\gamma_3 < -1$. In this case there are two distinct regions accessible for the particle. The outer region is given by

$$r \geq r_s \frac{-\gamma_3 + \sqrt{\gamma_3^2 - \cos^3\lambda}}{\cos\lambda} \tag{1.104}$$

while there is a second, closed region where the particle can move:

$$r_s \frac{\gamma_3 + \sqrt{\gamma_3^2 + \cos^3\lambda}}{\cos\lambda} \leq r \leq r_s \frac{-\gamma_3 - \sqrt{\gamma_3^2 - \cos^3\lambda}}{\cos\lambda}. \tag{1.105}$$

If a particle gets into this region, it becomes "trapped," and it is confined to the narrow region defined by Eq. (1.105). This physical phenomenon explains the existence of trapped radiation in the magnetosphere, the plasmasphere, and the Van Allen radiation belts. The region of trapped radiation for $\gamma_3 = -1.5$ is shown in Figure 1.9.

It can be shown that for large values of $|\gamma_3|$ the boundaries of the trapped region converge to

$$r = \frac{r_s}{2\,|\gamma_3|} \cos^2\lambda. \tag{1.106}$$

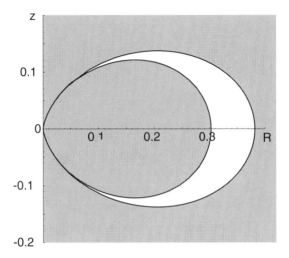

Figure 1.9 Example of a "trapped" trajectory in the meridional plane of a dipole for $\gamma_3 = -1.5$. All distances are measured in units of r_s.

This equation describes a dipole magnetic field line with equatorial radius of

$$r_0 = \frac{r_s}{2\,|\gamma_3|} = \left|\frac{q}{\mathcal{P}_3}\right| \frac{\mu_0}{4\pi} M. \tag{1.107}$$

The field line given by Eq. (1.106) is inside the trapped region for $\gamma_3 < -1$.

We have seen that in dipole-type magnetic fields gyration is the fastest periodic motion, followed by the field-aligned bounce, whereas the azimuthal drift is the slowest one. During each bounce the particle crosses the magnetic equator twice. At the equator ($\lambda = 0$) the $d\Phi/dt$ quantity becomes zero twice, as the particle gyrates around the field line. This means that for a trapped particle the conserved quantity, $\mathcal{P}_3$, can be expressed as (see Eq. 1.92)

$$\mathcal{P}_3 = -q\frac{\mu_0}{4\pi}\frac{M}{r} \approx -q\frac{\mu_0}{4\pi}\frac{M}{r_0} \tag{1.108}$$

because the radial distance of the particle when it crosses the magnetic equator is practically the equatorial radius of the magnetic field line containing the guiding center of the particle, r_0. With the help of Eq. (1.108) and the definition of the gyroradius ($r_c = mv/|q|B$) the dimensionless quantity γ_3 can be expressed as

$$\gamma_3 = -\frac{1}{2}\sqrt{\frac{r_0}{r_c}}. \tag{1.109}$$

Because particles are trapped as long as $\gamma_3 < -1$, we can conclude that charged particles become trapped in dipole magnetic fields as long as their gyroradius is much less than the equatorial distance of the field line.

For particles trapped in the terrestrial dipole field, the third adiabatic invariant becomes the following:

$$I_3 = \oint d\Phi \mathcal{P}_3 = -\oint d\Phi q \frac{\mu_0}{4\pi} \frac{M}{r} = -2\pi q \frac{\mu_0}{4\pi} \frac{M}{R_e} \frac{1}{L}, \tag{1.110}$$

because during a bounce the guiding center moves along the field line (given by Eq. 1.77). The third adiabatic invariant, in effect, characterizes the series of field lines the particle is drifting along. In the case of a dipole field the drift surface encompasses magnetic field lines with constant equatorial distances from the center (characterized by the dimensionless parameter, L). This drift surface is referred to as an *L-shell*. In summary, the third adiabatic invariant defines the L-shell of the particle drift.

One can calculate the drift period for a trapped particle. This will be done in two steps. First, we calculate the change in azimuth during a bounce period and then use this expression to obtain the time of a full revolution around the axis of symmetry. In a dipole field the gradient–curvature drift velocity is (see Eqs. 1.73 and 1.75–1.77):

$$\mathbf{v}_{GC} = -3 \frac{mv^2}{q} \frac{L^2 R_e^2}{M} \left(1 - \frac{1}{2}\sin^2\Theta_0 \frac{\sqrt{1+3\sin^2\lambda}}{\cos^6\lambda}\right) \frac{\cos^5\lambda(1+\sin^2\lambda)}{(1+3\sin^2\lambda)^2} \mathbf{e}_\Phi. \tag{1.111}$$

The change of the azimuth angle during a bounce period can be obtained by the following integral:

$$\Delta\Phi = 4 \int_0^{\lambda_m} \frac{d\Phi}{dt} \frac{dt}{ds} \frac{ds}{d\lambda'} d\lambda', \tag{1.112}$$

where $dt/ds = 1/v_\parallel$, $ds/d\lambda$ is given by Eq. (1.79), while

$$\frac{d\Phi}{dt} = \frac{|\mathbf{v}_{GC}|}{r\cos\lambda}. \tag{1.113}$$

Substituting these expressions into Eq. (1.112) yields

$$\Delta\Phi = 12 \frac{mv}{q} \frac{4\pi}{\mu_0} \frac{L^2 R_e^2}{M} \mathcal{Q}(\Theta_0), \tag{1.114}$$

where

$$\mathcal{Q}(\Theta_0) = \int_0^{\lambda_m} \frac{\cos^3\lambda'(1+\sin^2\lambda')}{(1+3\sin^2\lambda')^{3/2}} \frac{1 - \frac{1}{2}\sin^2\Theta_0 \frac{\sqrt{1+3\sin^2\lambda'}}{\cos^6\lambda'}}{1 - \sin^2\Theta_0 \frac{\sqrt{1+3\sin^2\lambda'}}{\cos^6\lambda'}} d\lambda'. \tag{1.115}$$

Using this expression, one can now easily obtain the bounce-averaged drift period:

$$\tau_D = \frac{2\pi}{\Delta\Phi} \tau_B = \frac{2\pi}{3} \frac{q}{mv^2} \frac{\mu_0}{4\pi} \frac{M}{LR_e} \frac{\mathcal{S}(\Theta_0)}{\mathcal{Q}(\Theta_0)}. \tag{1.116}$$

The Q/S ratio can be evaluated in a very good approximation as (*Lyons & Williams 1984*):

$$\frac{Q(\Theta_0)}{S(\Theta_0)} = 0.35 + 0.15 \, \sin \Theta_0. \tag{1.117}$$

One then obtains the following approximate expression for the bounce-averaged drift period:

$$\tau_D = \frac{2\pi}{3} \frac{q}{mv^2} \frac{\mu_0}{4\pi} \frac{M}{L R_e} \frac{1}{0.35 + 0.15 \, \sin \Theta_0}. \tag{1.118}$$

1.5   Problems

Problem 1.1 Show that the particle's relativistic energy, ε, can be expressed as stated in Eq. (1.13):

$$\frac{\varepsilon^2}{c^2} - \mathbf{p} \cdot \mathbf{p} = (mc)^2. \tag{1.119}$$

Problem 1.2 Derive the gyroradius, gyrofrequency, and magnetic moment of a relativistic particle.

Problem 1.3 Calculate the length of the $L = 5$ field line in the terrestrial dipole field.

Problem 1.4 Consider a planet with no atmosphere, but with a strong intrinsic dipole moment located at the center. An electron gun is placed at a distance of $4R_p$ from the center at the magnetic equator (R_p is the radius of the planet). The electron gun emits 1 keV particles in a way that all pitch angles are equally probable. There is an electron detector next to the electron gun that detects every electron returning toward the electron gun. What is the ratio of the emitted to detected electrons?

Problem 1.5 Consider a 1 eV proton in the terrestrial ionosphere with its velocity perpendicular to the magnetic field. What is the ratio of the magnitudes of the gravitational and Lorentz forces acting on this particle? (Neglect the ionospheric electric field.)

Problem 1.6 Approximate the Earth's intrinsic magnetic field with a dipole moment located at the center of the planet.

1. Consider magnetic field lines characterized by $L = 1.5, 2.0, 3.0$, and 4.0. Calculate the magnetic latitude where the field lines intersect the surface.
2. Plot the magnitude of the magnetic field for $L = 1.5, 2.0, 3.0$, and 4.0 field lines as a function of magnetic latitude for the portion of the field line that is above the surface.

3. Plot the gyroradius and gyrofrequency for electrons and single ionized hydrogen and oxygen ions as a function of geomagnetic latitude for $L = 1.5, 2.0, 3.0$, and 4.0 field lines. Assume that the charged particle population is in thermodynamic equilibrium at a temperature of $T = 2{,}000$ K and use the average thermal speed for each species to calculate r_c and Ω_c. (The average thermal speed is $\bar{v} = \sqrt{8kT/m\pi}$, where k is the Boltzmann constant).

4. A charged particle is lost to the dense neutral atmosphere if its altitude is less than 200 km above the surface. What is the minimum equatorial pitch angle of trapped electron and H^+ and O^+ ions on $L = 1.5, 2.0, 3.0$, and 4.0 field lines? (Trapped particles bounce many times along the field line.) What is the bounce period of these particles?

5. What is the equatorial drift period for $\Theta_0 = 45°$ electrons, protons, and O^+ ions on $L = 1.5, 2.0, 3.0$, and 4.0 field lines?

Chapter 2

Kinetic Theory

Maxwell's velocity distribution function together with Clausius's mean free path concept were the key elements that made it possible to develop connections between the motions of microscopic molecules (molecules are defined in a very broad sense here, referring to any neutral or charged particle composing the gas) and the macroscopic properties of gases (observable by the methods of classical physics).

In this chapter we introduce some fundamental concepts of kinetic theory. For more details we refer the reader to books on classical kinetic theory (cf. *Gombosi 1994*).

2.1 Collisions

First, we introduce several statistical quantities such as the mean free path, collision frequency, collision rate, collision cross section, and differential cross section. We shall also see that these quantities are closely related to each other and to other fundamental molecular quantities. We will use very simple physical models to emphasize the basic concepts behind these new statistical quantities.

2.1.1 Mean Free Path

The free path is the distance traveled by a molecule between two successive collisions. The mean free path is the average distance between two successive collisions of a single molecule.

Consider a single molecule with velocity $\mathbf{v}$. Assume that this particle suffers a collision with another molecule at a distance of $s = 0$. Let $P_{\text{path}}(s)$ denote the probability that this molecule survives a distance s without suffering a second collision.

If we neglect simultaneous collisions between three or more particles, then obviously $P_{\text{path}}(0) = 1$. Let $\alpha\, ds$ denote the probability that a given molecule suffers a collision between distance s and $s + ds$. The quantity α is the collision probability per unit distance for an individual particle (it is also called collisional probability density). It is assumed that the collisional probability density is independent of the past collisional history of the particle. In general, however, α may depend on the speed of the particles (faster particles travel larger distances and therefore encounter more molecules in the same time interval).

With the help of the parameter α, it is possible to derive the functional form of the survival probability, $P_{\text{path}}(s)$. The probability that the particle survives an interval $s + ds$ without suffering a collision must be equal to the product of the probabilities that no collision occurs in the subsequent intervals from 0 to s and from s to $s + ds$:

$$P_{\text{path}}(s + ds) = P_{\text{path}}(s)(1 - \alpha\, ds), \tag{2.1}$$

where we made use of the fact that $P_{\text{path}}(ds) = (1 - \alpha\, ds)$. The left-hand side of Eq. (2.1) can be expanded into a Taylor series, and one can stop after the linear term because ds is a very small quantity. This linear expansion yields the following differential equation for $P_{\text{path}}(s)$:

$$\frac{d P_{\text{path}}}{ds} = -\alpha P_{\text{path}}(s). \tag{2.2}$$

The solution of this differential equation (taking into account the $P_{\text{path}}(0) = 1$ initial condition) is the following:

$$P_{\text{path}}(s) = e^{-\alpha s}. \tag{2.3}$$

Equation (2.3) gives the probability that no collision occurs on the distance interval s. Based on this probability, one can readily derive the probability that the free path between two successive collisions of a given molecule is between s and $s + ds$. Let us denote this probability by $f_{\text{path}}(s)\, ds$. It is obvious that $f_{\text{path}}(s)\, ds$ must be equal to the product of the probabilities that no collision occurs in the interval from 0 to s, $P_{\text{path}}(s)$, and that a collision does occur between s to $s + ds$, $\alpha\, ds$:

$$f_{\text{path}}(s)\, ds = e^{-\alpha s}\alpha\, ds. \tag{2.4}$$

With the help of the probability density $f_{\text{path}}(s)$ (probability per unit length), one can derive the average distance traveled by a particle between two consecutive collisions, the *mean free path*, λ:

$$\lambda = \int_0^\infty ds\, s f_{\text{path}}(s) = \int_0^\infty ds\, \alpha s e^{-\alpha s} = \frac{1}{\alpha}. \tag{2.5}$$

Finally, one can express the distribution of distances traveled by the molecule between two consecutive collisions with the help of the mean free path:

$$f_{\text{path}}(s) = \frac{1}{\lambda} \exp\left(-\frac{s}{\lambda}\right). \tag{2.6}$$

2.1.2   Collision Cross Section

Collisions between molecules can be described in terms of the scattering cross section, σ. In the case of a single-species gas composed of hard sphere molecules the cross section is simply the projected area of the sphere of influence to a plane perpendicular to the relative velocity vector of the particles, $\sigma = \pi d_m^2$ (where d_m is the diameter of the hard sphere molecule).

A more general definition can be given the following way. Consider a beam of particles moving with a fixed velocity, $\mathbf{v}_1$, incident upon a single target molecule (moving with velocity $\mathbf{v}_2$) that deflects the beam molecules. In the frame of reference of the target molecule the beam is moving with the relative velocity, $\mathbf{g} = \mathbf{v}_1 - \mathbf{v}_2$. In this frame of reference the beam intensity (number of particles intersecting in unit time a unit area perpendicular to the beam) is denoted by I_0. As a result of collisions with the target molecule, a fraction of the beam, dI, will be scattered so that the direction of the scattered molecules is in a small solid angle element, $d\Omega'$. The number of particles scattered into the solid angle element in unit time, dN, is proportional to the intensity of the incident beam, I_0, and to the solid angle element, $d\Omega'$. One can express the scattering rate per solid-angle element as follows:

$$dN = SI_0\, d\Omega', \tag{2.7}$$

where S is a factor of proportionality, called the differential cross section. In general the differential cross section is a function of the relative velocity vectors before and after the collision, $S = S(\mathbf{g}, \mathbf{g}')$. The differential cross section can be calculated by using classical or quantum mechanical methods. It can be easily seen that the differential cross section has the dimensions of an area. The physical interpretation is that $S = S(\mathbf{g}, \mathbf{g}')$ is the perpendicular area of the molecule (in a plane perpendicular to the pre-collision relative velocity vector, $\mathbf{g}$) which scatters incident molecules with relative velocity $\mathbf{g}$ to an infinitesimal velocity space solid angle around the postcollision relative velocity vector, $\mathbf{g}'$.

The total rate of particles scattered in all directions, N, can be obtained by integrating Eq. (2.7) over all possible directions of the scattered relative velocity vector, $\mathbf{g}'$:

$$N = \sigma I_0 = I_0 \int_{4\pi} d\Omega'\, S(\mathbf{g}, \mathbf{g}'), \tag{2.8}$$

where the proportionality factor, σ, is called the total cross section and is given by the integral

$$\sigma(\mathbf{g}) = \int_{4\pi} d\Omega'\, S(\mathbf{g}, \mathbf{g}'). \tag{2.9}$$

2.1.3   Collision Frequency, Collision Rate

Consider a gas consisting of only a single kind of molecule. Let us take one single molecule (molecule 1) and calculate the number of its collisions per unit time. We denote the average concentration of molecules (the average number of molecules per unit volume) by n and assume that all molecules move with the same speed (speed here refers to the magnitude of the velocity vector), $\bar{v}$. The mean relative speed of the other molecules with respect to our molecule 1 is denoted by $\bar{g}$. The fixed molecule represents an effective target area of σ. This cross section is the total scattering cross section of the molecule. Consider the number of collisions of molecule 1 in time dt. At $t = 0$ those molecules that collide with molecule 1 in time dt have their centers within a circular cylinder of length $\bar{g}\,dt$ and volume $\sigma\bar{g}\,dt$. Now the number of collisions of molecule 1 in time dt is the volume of the impact cylinder times the number density of the impinging molecules, $n\sigma\bar{g}\,dt$.

The collision frequency, ν, is defined as the number of collisions of a single molecule per unit time. This means that

$$\nu = n\sigma\bar{g}. \tag{2.10}$$

Next, we estimate the relative speed, $\bar{g}$. Let us consider the relative velocity between our molecule 1 and another arbitrary molecule, molecule 2. It was assumed that all molecules move with the same speed, $\bar{v}$. Therefore the square of the relative speed can be written as

$$g_{12}^2 = v_1^2 + v_2^2 - 2v_1v_2\cos\Theta = 2\bar{v}^2 - 2\bar{v}^2\cos\Theta, \tag{2.11}$$

where we denoted the angle between the two velocity vectors by Θ. The average relative speed square can be obtained by taking the angular average of Eq. (2.11):

$$\overline{g^2} = \langle g_{12}^2 \rangle_\Theta = 2\bar{v}^2. \tag{2.12}$$

Approximating $\bar{g}^2$ by $\overline{g^2}$, one can express the collision frequency (number of collisions of a single molecule in unit time) in terms of well-understood fundamental parameters of the gas:

$$\nu = \sqrt{2}n\sigma\bar{v}. \tag{2.13}$$

It is obvious from the definitions of collision frequency and collision mean free path that $\lambda\nu = \bar{v}$. Therefore

$$\lambda = \frac{1}{\sqrt{2}n\sigma}. \tag{2.14}$$

This very important result tells us that the mean free path of the molecules depends only on the gas density and the scattering cross section of the molecules and is independent of the average molecular speed.

The collision rate is defined as the number of collisions per unit volume per unit time. Thus the collision rate and the collision frequency (number of collisions of a single molecule per unit time) are closely related quantities. The collision rate between molecules of types 1 and 2, N_{12}, is therefore the product of the collision frequency of a single molecule 1, ν_{12}, and the number density of molecules 1, n_1:

$$N_{12} = n_1\nu_{12}. \tag{2.15}$$

Here v_{12} is given by

$$v_{12} = n_2 \sigma_{12} \langle \bar{g}_{12} \rangle_\Theta \approx n_2 \sigma_{12} \sqrt{v_1^2 + v_2^2}. \tag{2.16}$$

2.2   The Boltzmann Equation

2.2.1   Phase-Space Distribution Function

Consider a single nonrelativistic particle moving in one dimension under the influence of a well-defined force field. This system can be completely described in terms of the particle spatial coordinate, x, and its velocity, v_x. This description is complete, because the laws of classical mechanics are such that the knowledge of x and v_x at any instant permits prediction of the values of x and v_x at any other time. The particle can be represented by a single point in the Cartesian coordinate system (x, v_x). Specification of location and velocity is equivalent to specifying a point in this two-dimensional space, called *phase space*. As the location and the velocity of the particle change in time, the point representing the particle moves around this two-dimensional phase space.

The three-dimensional motion of individual particles is governed by the equations of motion which is a set of three second-order differential equations. In each dimension we have two integration constants. These six constants are independent of each other, and they determine the particle orbit. Each particle can be represented by a single point in the *six-dimensional phase space* (x, y, z, v_x, v_y, v_z). In vector form the six-dimensional phase space (sometimes called μ-*space*) can be denoted as $(\mathbf{r}, \mathbf{v})$, where $\mathbf{r}$ is called the *configuration space* location, and $\mathbf{v}$ is the three-dimensional *velocity space* location. Just as in our previous example, the particle is represented by a single point in this six-dimensional phase space. The vectors $\mathbf{r}$ and $\mathbf{v}$ are the Eulerian coordinates of the phase space. As the location and the velocity of the particle change in time, the point representing the particle moves around this six-dimensional phase space.

Let us consider a gas that is nonuniformly distributed in a container. Let $\Delta^3 N$ be the number of molecules contained in the small configuration space volume element, $\Delta V = \Delta^3 r$, located between x and $x + \Delta x$, y and $y + \Delta y$, z and $z + \Delta z$. The superscript 3 refers to the three-dimensional nature of the small volume element. The mean density of the gas in the small volume element is $\Delta^3 N / \Delta^3 r$. Next, we take the limit that the volume element becomes macroscopically infinitesimal (in other words $\Delta^3 r$ approaches zero, but it still contains a statistically meaningful number of particles). This way we define the particle number density (or particle concentration):

$$n(\mathbf{r}) = \lim_{\Delta^3 r \to 0} \frac{\Delta^3 N}{\Delta^3 r}. \tag{2.17}$$

The local number density, $n(\mathbf{r})$, gives the number of molecules per unit volume in the infinitesimal vicinity of location $\mathbf{r}$. It is thus the measure of the distribution of molecules in configuration space; in short it is the configuration-space distribution function. If the configuration-space distribution function is known at a particular location, $\mathbf{r}$, the number of molecules, d^3N, in the macroscopically infinitesimal volume element, d^3r, is given by

$$d^3N = n(\mathbf{r})\,dx\,dy\,dz = n(\mathbf{r})\,d^3r. \tag{2.18}$$

It should be noted that the configuration-space distribution function characterizes the distribution of particles in space, regardless of their velocity.

At any instant, each of the gas molecules will have a velocity, which can be specified by a velocity space point, $\mathbf{v}$. One could define the velocity-space concentration function of all molecules in a macroscopic container. The velocity-space concentration defines the number of molecules in an infinitesimal velocity-space volume element around a given velocity vector independently of their spatial location in the container. However, such a velocity-space concentration is not a particularly useful physical quantity. A statistically useful physical quantity should characterize the distribution of molecular velocities at any given configuration-space location.

One can generalize the concept of a configuration-space particle distribution function to the six-dimensional phase space. The phase-space distribution function, $F(\mathbf{r}, \mathbf{v})$, characterizes the number of particles in a six-dimensional infinitesimal volume element, $d^3r\,d^3v$. If the phase-space distribution function, F, is known at a particular phase-space location, $(\mathbf{r}, \mathbf{v})$, the number of molecules, d^6N, to be found in the configuration space volume element between x and $x + dx$, y and $y + dy$, z and $z + dz$ with velocity components between v_x and $v_x + dv_x$, v_y and $v_y + dv_y$, v_z and $v_z + dv_z$ is given by

$$d^6N = F(\mathbf{r}, \mathbf{v})\,dx\,dy\,dz\,dv_x\,dv_y\,dv_z = F(\mathbf{r}, \mathbf{v})d^3r\,d^3v. \tag{2.19}$$

Equation (2.19) shows that the phase-space distribution function, $F(\mathbf{r}, \mathbf{v})$, is the density of particles in the $d^3r\,d^3v$ phase-space volume element around the phase-space point $(\mathbf{r}, \mathbf{v})$.

It is obvious that the functions $F(\mathbf{r}, \mathbf{v})$ and $n(\mathbf{r})$ must be closely related. The configuration-space distribution function, $n(\mathbf{r})$, describes the number of particles in a small volume element, d^3r, around r, regardless of their velocity. This means that if we integrate the phase-space distribution function, $F(\mathbf{r}, \mathbf{v})$, over all possible velocity values, we must get the configuration space distribution function, $n(\mathbf{r})$:

$$n(\mathbf{r}) = \int_{-\infty}^{+\infty} dv_x \int_{-\infty}^{+\infty} dv_y \int_{-\infty}^{+\infty} dv_z F(\mathbf{r}, \mathbf{v}) = \iiint_{\infty} F(\mathbf{r}, \mathbf{v})\,d^3v. \tag{2.20}$$

Equation (2.20) shows that the phase-space distribution function, $F(\mathbf{r}, \mathbf{v})$, is normalized to the particle number density, $n(\mathbf{r})$. One can also introduce the normalized phase-space distribution function, $f(\mathbf{r}, \mathbf{v})$:

$$f(\mathbf{r}, \mathbf{v}) = \frac{F(\mathbf{r}, \mathbf{v})}{n(\mathbf{r})}. \tag{2.21}$$

This definition means that $f(\mathbf{r}, \mathbf{v}) \, d^3r \, d^3v$ is the fraction of all particles in the spatial volume element d^3r around $\mathbf{r}$, which have their velocity vectors between $\mathbf{v}$ and $\mathbf{v} + d^3v$. In other words, $f(\mathbf{r}, \mathbf{v}) \, d^3r \, d^3v$ is the conditional probability of finding a particle in the velocity-space volume element d^3v around $\mathbf{v}$, provided that the particle is located in the configuration-space volume element d^3r around $\mathbf{r}$. The normalized phase-space distribution function, $f(\mathbf{r}, \mathbf{v})$, is the probability density of finding a particle at the velocity space point $\mathbf{v}$ at the configuration space location $\mathbf{r}$. It is easy to see that $f(\mathbf{r}, \mathbf{v})$ is normalized to 1:

$$\iiint_{\infty} f(\mathbf{r}, \mathbf{v}) \, d^3v = \frac{1}{n(\mathbf{r})} \iiint_{\infty} F(\mathbf{r}, \mathbf{v}) \, d^3v = 1. \tag{2.22}$$

In kinetic theory macroscopic physical quantities usually characterize the gas at a given location. Such quantities are the concentration of molecules, mass density, temperature, pressure, flow velocity, etc. These macroscopic quantities might vary with time or configuration-space location, but they do not reflect the distribution of molecular velocities at the given time and location. These macroscopic parameters are averages of quantities that depend on molecular velocities.

Let $Q(\mathbf{v})$ be any molecular quantity that is only a function of the velocity of the given molecule. Examples of such quantities include the molecular mass (which is independent of the molecular velocity in our nonrelativistic approximation), momentum, or energy. The average value of Q for molecules in a macroscopically infinitesimal configuration-space volume element, d^3r, located at position $\mathbf{r}$ is given by

$$\langle Q \rangle = \iiint_{\infty} Q(\mathbf{v}) f(\mathbf{r}, \mathbf{v}) \, d^3v. \tag{2.23}$$

For instance, the averages of the molecular mass, m, momentum, $m\mathbf{v}$, and kinetic energy, $mv^2/2$, are the following:

$$\langle m \rangle = \iiint_{\infty} m f(\mathbf{r}, \mathbf{v}) \, d^3v = m, \tag{2.24}$$

$$\langle m\mathbf{v} \rangle = \iiint_{\infty} m\mathbf{v} f(\mathbf{r}, \mathbf{v}) \, d^3v = m\mathbf{u}, \tag{2.25}$$

$$\left\langle \frac{1}{2} mv^2 \right\rangle = \iiint_{\infty} \frac{1}{2} mv^2 f(\mathbf{r}, \mathbf{v}) \, d^3v = E. \tag{2.26}$$

2.2.2 Evolution of the Phase-Space Distribution Function

The state of a gas or a gas mixture at a particular instant is completely specified for the purposes of kinetic theory if the distribution function of the molecular velocities and positions is known throughout the gas. Observable properties of the gas are obtained as appropriate averages over the phase-space distribution function.

The gas may be nonuniform (and consequently nonequilibrium), the distribution function may depend explicitly on time, and external fields of force (such as gravity) may also be present. These externally imposed force fields result in acceleration of the individual molecules. This acceleration, $\mathbf{a}$, is a function of the seven independent variables, time, t, configuration space location, $\mathbf{r}$, and velocity space location, $\mathbf{v}$, as well as of some intrinsic properties of the molecules themselves (such as mass). In kinetic theory it is usually assumed that the acceleration of a molecule, $\mathbf{a}(t, \mathbf{r}, \mathbf{v})$, is divergenceless in velocity space, $\nabla_v \cdot \mathbf{a} = 0$. Most external force fields satisfy this condition (all velocity-independent forces, such as gravity, automatically satisfy it, as does the Lorentz force acting on moving charged particles in electromagnetic fields).

Consider first the case of a gas composed of a single molecular species. Let the phase-space distribution function be denoted by $F(\mathbf{r}, \mathbf{v})$, so that the number of molecules in a d^3r volume element around the configuration space location, $\mathbf{r}$, with velocity vectors between $\mathbf{v}$ and $\mathbf{v} + d^3v$ is given by

$$d^6N = F(t, \mathbf{r}, \mathbf{v}) \, d^3r \, d^3v. \tag{2.27}$$

At time t, these particles are located very close to each other and they move with nearly identical velocities. If there were no collisions, then dt time later all these particles would be found in a volume element d^3r' near the spatial location $\mathbf{r}' = \mathbf{r} + \mathbf{v}dt$ with velocities between $\mathbf{v}' = \mathbf{v} + \mathbf{a}(t, \mathbf{r}, \mathbf{v}) \, dt$ and $\mathbf{v}' + d^3v'$. The number of particles in the modified phase-space element around $(\mathbf{r}', \mathbf{v}')$ at time $t + dt$ is given by

$$d^6N' = F(t + dt, \mathbf{r} + \mathbf{v}dt, \mathbf{v} + \mathbf{a}dt) \, d^3r' \, d^3v'. \tag{2.28}$$

The relation between the modified phase-space volume element, $d^3r'd^3v'$, and the initial one, $d^3r \, d^3v$, is given by the determinant of the Jacobian matrix of the transformation:

$$d^3r' \, d^3v' = |J| d^3r \, d^3v, \tag{2.29}$$

where

$$J = \frac{\partial(\mathbf{r}', \mathbf{v}')}{\partial(\mathbf{r}, \mathbf{v})}. \tag{2.30}$$

Substituting our expressions for $\mathbf{r}'$ and $\mathbf{v}'$ yields the following 6×6 Jacobian matrix:

$$J = \begin{bmatrix} 1 & 0 & 0 & \frac{\partial a_1}{\partial x_1}dt & \frac{\partial a_1}{\partial x_2}dt & \frac{\partial a_1}{\partial x_3}dt \\ 0 & 1 & 0 & \frac{\partial a_2}{\partial x_1}dt & \frac{\partial a_2}{\partial x_2}dt & \frac{\partial a_2}{\partial x_3}dt \\ 0 & 0 & 1 & \frac{\partial a_3}{\partial x_1}dt & \frac{\partial a_3}{\partial x_2}dt & \frac{\partial a_3}{\partial x_3}dt \\ dt & 0 & 0 & 1 + \frac{\partial a_1}{\partial v_1}dt & \frac{\partial a_1}{\partial v_2}dt & \frac{\partial a_1}{\partial v_3}dt \\ 0 & dt & 0 & \frac{\partial a_2}{\partial v_1}dt & 1 + \frac{\partial a_2}{\partial v_2}dt & \frac{\partial a_2}{\partial v_3}dt \\ 0 & 0 & dt & \frac{\partial a_3}{\partial v_1}dt & \frac{\partial a_3}{\partial v_2}dt & 1 + \frac{\partial a_3}{\partial v_3}dt \end{bmatrix}. \tag{2.31}$$

The determinant of the Jacobian can be evaluated as a power series of the infinitesimal time interval, dt. Because the time interval is very small, we can truncate after the linear term:

$$|J| = 1 + (\nabla_v \cdot \mathbf{a}) \, dt + \cdots. \tag{2.32}$$

At this point we recall our earlier assumption that the particle acceleration is divergenceless in the velocity coordinates. Therefore, we conclude that in the first-order approximation $|J| = 1$. This means that $d^3 r' \, d^3 v' = d^3 r \, d^3 v$.

If there were no scattering, all particles that were located inside the phase-space volume element $d^3 r \, d^3 v$ at the initial time, t, would be found inside $d^3 r' \, d^3 v'$ at the later time, $t + dt$. In other words, in the absence of particle scattering $d^6 N' - d^6 N = 0$. However, gas molecules do interact with each other. The scattering process changes the velocities of the colliding particles. Therefore, the total number of molecules inside our infinitesimal phase-space volume element may change during the short time interval, dt:

$$d^6 N' - d^6 N = d^6 N_{\text{coll}}, \tag{2.33}$$

where $d^6 N_{\text{coll}}$ denotes the net rate of change (change per unit time) of the number of particles in the phase-space volume element due to collisions. Equation (2.33) can be rewritten with the help of Eqs. (2.27) and (2.28):

$$[F(t + dt, \mathbf{r} + \mathbf{v}dt, \mathbf{v} + \mathbf{a} \, dt) - F(t, \mathbf{r}, \mathbf{v})] d^3 r \, d^3 v = \frac{\delta F(t, \mathbf{r}, \mathbf{v})}{\delta t} d^3 r \, d^3 v \, dt, \tag{2.34}$$

where $\delta F / \delta t$ is the rate of change of the phase-space distribution function. Here $\delta F / \delta t$ is only formally introduced; for a detailed discussion of this so-called collision term see *Gombosi* (*1994*).

In the next step we apply a linear Taylor expansion in the small time interval, dt, to evaluate the change in the total number of particles inside our phase-space volume element. The Taylor expansion is centered around time, t:

$$\left\{ \frac{\partial F(t, \mathbf{r}, \mathbf{v})}{\partial t} + (\mathbf{v} \cdot \nabla) F(t, \mathbf{r}, \mathbf{v}) + [\mathbf{a}(t, \mathbf{r}, \mathbf{v}) \cdot \nabla_v] F(t, \mathbf{r}, \mathbf{v}) \right\} d^3 r \, d^3 v \, dt$$
$$= \frac{\delta F(t, \mathbf{r}, \mathbf{v})}{\delta t} d^3 r \, d^3 v \, dt. \tag{2.35}$$

The expression in the curly braces is the total time derivative of the phase-space distribution function, dF/dt. Because Eq. (2.35) is valid for arbitrary phase-space volume elements and for arbitrary time intervals, one obtains the following equation:

$$\frac{\partial F(t, \mathbf{r}, \mathbf{v})}{\partial t} + (\mathbf{v} \cdot \nabla) F(t, \mathbf{r}, \mathbf{v}) + (\mathbf{a} \cdot \nabla_v) F(t, \mathbf{r}, \mathbf{v}) = \frac{\delta F(t, \mathbf{r}, \mathbf{v})}{\delta t}. \tag{2.36}$$

This is the Boltzmann equation, which describes the evolution of the phase-space distribution function, $F(t, \mathbf{r}, \mathbf{v})$. It should be emphasized that in the derivation of the Boltzmann equation we never assumed equilibrium conditions, and therefore this equation is equally valid for equilibrium and nonequilibrium gases. It is also important to note again that t, $\mathbf{r}$, and $\mathbf{v}$ are assumed to be independent variables.

The Boltzmann equation can also be expressed in terms of the random velocity (sometimes also called peculiar velocity or thermal velocity). The random velocity of a particle is its velocity with respect to the bulk gas flow at time, t, and location, $\mathbf{r}$:

$$\mathbf{c} = \mathbf{v} - \mathbf{u}(t, \mathbf{r}). \tag{2.37}$$

Here the average (or bulk) velocity, $\mathbf{u}(t, \mathbf{r})$, is a function of time and configuration space location and is defined the following way:

$$\mathbf{u}(t, \mathbf{r}) = \frac{\iiint\limits_{\infty} \mathbf{v} F(t, \mathbf{r}, \mathbf{v}) \, d^3 v}{\iiint\limits_{\infty} F(t, \mathbf{r}, \mathbf{v}) \, d^3 v}. \tag{2.38}$$

It is important to note that, unlike $\mathbf{v}$, the random velocity, $\mathbf{c}$, is not independent of t and $\mathbf{r}$.

In some cases it is more convenient to express the evolution of the phase-space distribution, F, in terms of t, $\mathbf{r}$, and $\mathbf{c}$ instead of the independent variables t, $\mathbf{r}$, and $\mathbf{v}$. This means that one has to introduce a "mixed" phase-space, where the configuration-space coordinate system is inertial, but the velocity-space coordinate system is an accelerating system because it is tied to the instantaneous local bulk velocity of the gas. In spite of the inherent difficulty of dealing with accelerating systems, it is often advantageous to use this mixed phase space.

Due to the accelerating nature of the veocity-space coordinate system, we want to transfer the Boltzmann equation into our mixed phase-space coordinate system. When replacing the independent variable, $\mathbf{v}$, with the random velocity, $\mathbf{c}$, one has to take into account the t and $\mathbf{r}$ dependence of the new variable and use the following substitutions:

$$\frac{\partial F(t, \mathbf{r}, \mathbf{v})}{\partial t} \Longrightarrow \frac{\partial F(t, \mathbf{r}, \mathbf{c})}{\partial t} + \left(\frac{\partial \mathbf{c}(t, \mathbf{r})}{\partial t} \cdot \nabla_c \right) F(t, \mathbf{r}, \mathbf{c})$$

$$= \frac{\partial F(t, \mathbf{r}, \mathbf{c})}{\partial t} - \left(\frac{\partial \mathbf{u}(t, \mathbf{r})}{\partial t} \cdot \nabla_c \right) F(t, \mathbf{r}, \mathbf{c}), \tag{2.39}$$

$$\nabla F(t, \mathbf{r}, \mathbf{v}) \Longrightarrow \nabla F(t, \mathbf{r}, \mathbf{c}) + (\nabla \mathbf{c}(t, \mathbf{r})) \cdot \nabla_c F(t, \mathbf{r}, \mathbf{v})$$

$$= \nabla F(t, \mathbf{r}, \mathbf{c}) - (\nabla \mathbf{u}(t, \mathbf{r})) \cdot \nabla_c F(t, \mathbf{r}, \mathbf{v}), \tag{2.40}$$

$$\nabla_v F(t, \mathbf{r}, \mathbf{v}) \Longrightarrow \nabla_c F(t, \mathbf{r}, \mathbf{c}). \tag{2.41}$$

Using relations (2.39) through (2.41), one can obtain the Boltzmann equation in the mixed phase-space coordinate system:

$$\frac{\partial F(t, \mathbf{r}, \mathbf{c})}{\partial t} + (\mathbf{c} \cdot \nabla) F(t, \mathbf{r}, \mathbf{c}) + (\mathbf{u}(t, \mathbf{r}) \cdot \nabla) F(t, \mathbf{r}, \mathbf{c})$$

$$- \left[\frac{\partial \mathbf{u}(t, \mathbf{r})}{\partial t} + (\mathbf{u}(t, \mathbf{r}) \cdot \nabla) \mathbf{u}(t, \mathbf{r}) + (\mathbf{c} \cdot \nabla) \mathbf{u}(t, \mathbf{r}) \right.$$

$$\left. - \mathbf{a}(t, \mathbf{r}, \mathbf{c}) \right] \nabla_c F(t, \mathbf{r}, \mathbf{c}) = \frac{\delta F(t, \mathbf{r}, \mathbf{c})}{\delta t}. \tag{2.42}$$

It should be emphasized that the two forms of the Boltzmann equation, given by (2.36) and (2.42), are both physically and mathematically equivalent. However, in specific problems it might be more convenient to use one or the other formulation. In general, the mixed form (Eq. 2.42) leads to so-called nonconservative formulations of transport problems, whereas the inertial form (Eq. 2.36) results in "conservative" forms of transport equations.

2.2.3   Local Equilibrium Distribution

The left-hand side of the Boltzmann equation is basically the full derivative of the phase-space distribution function. In dynamic equilibrium the distribution function does not change, that is, the full derivative must be zero and consequently the collision term must vanish. It can be shown that there is only an equilibrium solution of the Boltzmann equation, the so-called *Maxwell–Boltzmann distribution*. Also, in strict equilibrium there are no spatial gradients in the distribution function. In other words, the equilibrium distribution can only be a function of the particle velocity vector [this is shown in *Gombosi (1994)*].

When the gas is highly collisional, the assumption of *local therodynamic equilibrium* (LTE) is quite justified. In this approximation the gas is not strictly in equilibrium, so that spatial gradients may exist in the distribution function, but locally the distribution of molecular velocities can still be described by an equilibrium distribution. In other words, the macroscopic properties of the gas can slowly vary with time and location, but the local distribution of molecular velocities is an equilibrium distribution.

It can be shown that when the gas is in local dynamic equilibrium the logarithm of the distribution function of colliding molecules is a summation-invariant quantity (which was first shown by Boltzmann[1]). The quantity, $\mathcal{A}(\mathbf{v})$ is summation invariant if $\mathcal{A}(\mathbf{v}_1) + \mathcal{A}(\mathbf{v}_2)$ is conserved for all collisions of molecules 1 and 2.

We know that in the case of molecules without internal degrees of freedom there are four, and only four, independent summation-invariant quantities. These quantities are the three components of the momentum vector and the kinetic energy of the particles. A fifth summation-invariant quantity, the molecular mass, is independent of the molecular velocities. Any other summation invariant, such as log F, can be expressed as a linear combination of the five independent summation invariants:

$$\log F(t, \mathbf{r}, \mathbf{v}) = \alpha_0(t, \mathbf{r}) + \alpha_1(t, \mathbf{r})mv_x$$

$$+ \alpha_2(t, \mathbf{r})mv_y + \alpha_3(t, \mathbf{r})mv_z + \alpha_4(t, \mathbf{r})\frac{1}{2}mv^2. \tag{2.43}$$

Here α_0 through α_4 are velocity-independent coefficients. A more familiar form can be obtained for the distribution function by introducing another set of five independent

---

[1] Boltzmann, L.,"Über die Natur der Gasmolecüle," *Sitz. Math.-Naturwiss. Cl. Acad. Wiss., Wien II.*, **74**, 55, 1877.

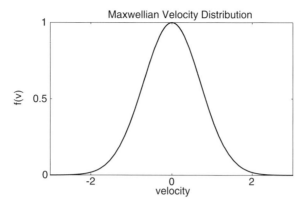

Figure 2.1 The distribution of molecular velocities in an equilibrium gas.

coefficients:

$$T = \frac{1}{k\alpha_4}, \quad u_1 = \frac{\alpha_1}{\alpha_4}, \quad u_2 = \frac{\alpha_2}{\alpha_4}, \quad u_3 = \frac{\alpha_3}{\alpha_4},$$
$$n = \left(\frac{2\pi}{m\alpha_4}\right)^{3/2} \exp\left(\alpha_0 + \frac{\alpha_1^2 + \alpha_2^2 + \alpha_3^2}{\alpha_4}\right), \tag{2.44}$$

where n is the particle number density, m is the molecular mass of the gas molecules, k is the Boltzmann constant ($k = 1.3807 \times 10^{-23}$ J/K), T is the gas temperature, and finally $\mathbf{u}$ is the bulk velocity of the gas. With these substitutions the distribution function can be written in the following form:

$$F_{\mathrm{M}}(\mathbf{r}, \mathbf{v}) = n\left(\frac{m}{2\pi kT}\right)^{3/2} \exp\left[-\frac{m(\mathbf{v} - \mathbf{u})^2}{2kT}\right], \tag{2.45}$$

which is the Maxwell–Boltzmann distribution of molecular velocities in an equilibrium gas.

This is a very important and far-reaching result. It tells us that in equilibrium the distribution of molecular velocities is always a Maxwellian. It also tells us that there is a one-to-one correspondence between equilibrium and Maxwellian velocity distribution: If and only if the distribution is Maxwellian, the gas is in equilibrium. This result also gives us a powerful tool to describe near-equilibrium conditions, because these distributions must be close to Maxwellians.

Equation (2.45) describes the equilibrium distribution of molecular velocities for nonrelativistic particles. The result can also be generalized for relativistic situations: In this case the relativistic Maxwell–Boltzmann distribution must be used.[2]

Figure 2.1 shows a schematic representation of the distribution of one velocity component in a stationary equilibrium gas. It is easy to see that this is a Gaussian

[2] Jüttner, F., "Das Maxwellsche Gesetz der Geschwindigkeitsverteilung in der Relativtheorie," *Annalen der Physik, IV*, Folge 34, 856, 1911.

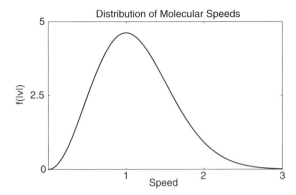

Figure 2.2 The distribution of molecular speeds in an equilibrium gas.

distribution, where the peak is located at the bulk velocity and the width is determined by the temperature.

The distribution of molecular speeds (speed is the magnitude of the velocity vector) for stationary ($\mathbf{u} = 0$) equilibrium gases can be easily obtained from Eq. (2.45). In this case one recognizes that

$$F_{\mathrm{M}}(\mathbf{r}, \mathbf{v}) \, d^3 v = 4\pi v^2 F_{\mathrm{M}}(\mathbf{r}, v) \, dv. \tag{2.46}$$

Therefore,

$$F_{\mathrm{M}}(\mathbf{r}, v) \, d^3 v = 4\pi v^2 n \left(\frac{m}{2\pi kT} \right)^{3/2} \exp\left[-\frac{mv^2}{2kT} \right] dv. \tag{2.47}$$

The distribution of molecular speeds in an equilibrium gas is illustrated in Figure 2.2.

With the help of the Maxwell–Boltzmann distribution, one can calculate the velocity moments for equilibrium gases. Some lower-order moments of particular interest are:

$$\iiint_{\infty} m F_{\mathrm{M}}(t, \mathbf{r}, \mathbf{v}) \, d^3 v = mn(t, \mathbf{r}), \tag{2.48}$$

$$\iiint_{\infty} m\mathbf{v} F_{\mathrm{M}}(t, \mathbf{r}, \mathbf{v}) \, d^3 v = mn(t, \mathbf{r})\mathbf{u}(t, \mathbf{r}), \tag{2.49}$$

$$\iiint_{\infty} m|\mathbf{v} - \mathbf{u}(t, \mathbf{r})| F_{\mathrm{M}}(t, \mathbf{r}, \mathbf{v}) \, d^3 v = mn(t, \mathbf{r})\sqrt{\frac{8kT}{\pi m}} = mn(t, \mathbf{r})\bar{v}(t, \mathbf{r}), \tag{2.50}$$

and

$$\iiint_{\infty} \frac{1}{2}mv^2 F_{\mathrm{M}}(t, \mathbf{r}, \mathbf{v}) \, d^3 v = \frac{1}{2}mn(t, \mathbf{r})u^2(t, \mathbf{r}) + \frac{3}{2}n(t, \mathbf{r})kT(t, \mathbf{r}). \tag{2.51}$$

Here $\bar{v} = \sqrt{8kT/\pi m}$ is the average random molecular speed in the gas (the average of the velocity magnitude).

2.2.4   Collision Terms

In the general sense a collision is any process resulting in a change of the particle's momentum vector. In a simple approximation we consider the effects of two different types of collisional processes resulting in changes of the phase-space distribution function:

1. short-range interactions between particles leading to large-angle collisions,
2. small-angle collisions caused by long-range interactions between charged particles or by wave–particle interactions.

In this section we examine the rate of change of the phase-space distribution function due to these two collisional processes.

Short-Range Interactions (Binary Collisions)   Neutral particles interact with other particles via short-range forces. This means that the vast majority of neutral particles travel a very large distance between two consecutive collisions compared to the characteristic size of the localized region where the collision takes place. This also means that the probability of three or more particles interacting at the same time via short-range forces is practically negligible. In other words, short-range particle interactions can be considered primarily two-particle (binary) collisions. The rate of change of the phase-space distribution function due to binary collisions was derived by Boltzmann [the derivation can be found in kinetic theory textbooks, cf. *Gombosi (1994)*]. This so-called Boltzmann collision integral is a complicated nonlinear, multidimensional integral of the distribution function that also depends on the differential cross section of the interaction. The rate of change described by the Boltzmann collision integral drives the solution of the Boltzmann equation toward equilibrium in an irreversible way.[3] It can also be shown that when the distribution function is the equilibrium distribution (Maxwell–Boltzmann distribution), the rate of change becomes zero.

There are several simplified approximations to the Boltzmann collision integral. The simplest and most widely used one is the so-called relaxation time approximation (sometimes also called the BGK approximation[4]). The relaxation time approximation assumes that collisions drive the phase-space distribution function toward equilibrium with a time constant, τ_{BGK}. This means that the actual phase-space distribution, F, is gradually replaced by an equilibrium (Maxwellian) distribution function, F_0:

$$\frac{\delta F(t, \mathbf{r}, \mathbf{v})}{\delta t} = -\frac{F(t, \mathbf{r}, \mathbf{v}) - F_0(t, \mathbf{r}, \mathbf{v})}{\tau_{BGK}}. \tag{2.52}$$

The five parameters of the Maxwellian, F_0, are chosen in such a way that mass,

---

[3] Boltzmann, L., "Über die Natur der Gasmolecüle," *Sitz. Math.-Naturwiss. Cl. Acad. Wiss. Wien*, **74**, 55, 1877.

[4] Bhatnagar, P. L., Gross, E. P., and Krook, M., "A model for collision processes in gases, I.," *Phys. Rev.*, **94**, 511, 1954.

momentum, and energy are all conserved while the gas is driven toward equilibrium:

$$n_0 = \iiint_\infty F(t, \mathbf{r}, \mathbf{v}) \, d^3 v, \tag{2.53}$$

$$\mathbf{u}_0 = \frac{1}{n_0} \iiint_\infty \mathbf{v} F(t, \mathbf{r}, \mathbf{v}) \, d^3 v, \tag{2.54}$$

$$T_0 = \frac{m}{3 n_0 k} \iiint_\infty \left(v^2 - u_0^2 \right) F(t, \mathbf{r}, \mathbf{v}) \, d^3 v. \tag{2.55}$$

Long-Range Interactions (Fokker–Planck Approximation) Small-angle collisions (when the particles are deflected only by an infinitesimal angle in every individual collision) represent a random process, and therefore the cumulative effects of these collisions must be treated with statistical methods. A widely used statistical technique leads to the so-called Fokker–Planck approximation of the collision term. The Fokker–Planck approximation can be derived for any phenomenon that at least approximately can be considered to be a Markov process. A Markov process is when the value of a physical parameter at the next measuring time depends only on its value at the present time and not on its value at any previous measuring time. (This means that the system has a one time step (or less) "memory").

A trivial example is flipping a coin. The next toss is not dependent on the present result, much less on earlier ones. A better (but not trivial) example is the Brownian motion. On the time scale of molecular collisions this process is not Markovian, because the molecule needs several collisions to "forget" its earlier velocity. However, on a time scale of many collision times this process is very nearly Markovian.

Let us consider a gas where interactions (for instance Coulomb collisions or wave–particle interactions) can produce collisions that deflect the particles only slightly. These interaction processes result in a random sequence of very small changes of the particle velocity vector. Thus if we measure the particle velocity vector at t, $t + \Delta t$, $t + 2\Delta t$, etc. (where Δt is much larger than the mean free time between individual collisions) then the values of the particle's velocity vector follow a Markov process.

Let $Q_p(\mathbf{v} - \Delta\mathbf{v}, \Delta\mathbf{v}, \Delta t)$ denote the transition probability that, in time Δt, the velocity $\mathbf{v} - \Delta\mathbf{v}$ changes by $\Delta\mathbf{v}$ to $\mathbf{v}$. We know that the normalized particle velocity function, $f(t, \mathbf{r}, \mathbf{v})$, is defined as the probability of finding a particle at location $\mathbf{r}$ and in the infinitesimal vicinity of velocity $\mathbf{v}$ at time t. The probability of finding a particle at location $\mathbf{r}$ and in the infinitesimal vicinity of velocity $\mathbf{v}$ at time t can also be expressed as

$$f(t, \mathbf{r}, \mathbf{v}) = \iiint_\infty Q_p(\mathbf{v} - \mathbf{w}, \mathbf{w}, \Delta t) f(t - \Delta t, \mathbf{r}, \mathbf{v} - \mathbf{w}) \, d^3 w. \tag{2.56}$$

The meaning of this expression is fairly obvious: To obtain the probability of finding a particle in the infinitesimal vicinity of $\mathbf{v}$ at time t, one has to integrate over all possible transitions resulting in that particular velocity. This integral relation is called the Chapman–Kolmogorov or Schmoluchowsky equation.

The integrand in Eq. (2.56) can be expanded into a Taylor series around the t and $\mathbf{v}$ point. Expanding to first-order accuracy in Δt and second-order accuracy in $\mathbf{w}$ yields the following:

$$Q_p(\mathbf{v} - \mathbf{w}, \mathbf{w}, \Delta t) f(t - \Delta t, \mathbf{r}, \mathbf{v} - \mathbf{w})$$

$$= Q_p(\mathbf{v}, \mathbf{w}, \Delta t) f(t, \mathbf{r}, \mathbf{v}) - \Delta t \, Q_p(\mathbf{v}, \mathbf{w}, \Delta t) \frac{df(t, \mathbf{r}, \mathbf{v})}{dt}$$

$$- w_i \frac{\partial[Q_p(\mathbf{v}, \mathbf{w}, \Delta t) f(t, \mathbf{r}, \mathbf{v})]}{\partial v_i}$$

$$+ \frac{1}{2} w_i w_j \frac{\partial^2[Q_p(\mathbf{v}, \mathbf{w}, \Delta t) f(t, \mathbf{r}, \mathbf{v})]}{\partial^2 v_i v_j} + \cdots . \tag{2.57}$$

In the Fokker–Planck approximation Δt is assumed to be small (but large compared to the collision mean free time); therefore, one can use first-order accuracy in Δt. However, one needs to keep the second-order term in the velocity transition expansion, because $\mathbf{w}$ is the result of many small random changes that can add up to somewhat larger values. At the same time the transition probabilities of large $|\mathbf{w}|$ values are assumed to be negligible. Let us examine the individual terms of the Schmoluchowsky integral using our present expansion. The first term is the following:

$$\iiint_\infty Q_p(\mathbf{v}, \mathbf{w}, \Delta t) f(t, \mathbf{r}, \mathbf{v}) \, d^3 w$$

$$= f(t, \mathbf{r}, \mathbf{v}) \iiint_\infty Q_p(\mathbf{v}, \mathbf{w}, \Delta t) \, d^3 w = f(t, \mathbf{r}, \mathbf{v}), \tag{2.58}$$

because the $Q_p(\mathbf{v}, \mathbf{w}, \Delta t)$ probability distribution is normalized to 1. Similarly, the next term is

$$\iiint_\infty \Delta t \, Q_p(\mathbf{v}, \mathbf{w}, \Delta t) \frac{df(t, \mathbf{r}, \mathbf{v})}{dt} \, d^3 w$$

$$= \Delta t \frac{df(t, \mathbf{r}, \mathbf{v})}{dt} \iiint_\infty Q_p(\mathbf{v}, \mathbf{w}, \Delta t) \, d^3 w = \Delta t \frac{df(t, \mathbf{r}, \mathbf{v})}{dt}. \tag{2.59}$$

The integral of the third term in the Taylor expansion is

$$\iiint_\infty w_i \frac{\partial[Q_p(\mathbf{v}, \mathbf{w}, \Delta t) f(t, \mathbf{r}, \mathbf{v})]}{\partial v_i} \, d^3 w$$

$$= \frac{\partial}{\partial v_i} \left[f(t, \mathbf{r}, \mathbf{v}) \iiint_\infty Q_p(\mathbf{v}, \mathbf{w}, \Delta t) w_i d^3 w \right]$$

$$= \Delta t \frac{\partial}{\partial v_i} \left[\left\langle \frac{\Delta v_i}{\Delta t} \right\rangle f(t, \mathbf{r}, \mathbf{v}) \right], \tag{2.60}$$

where we introduced the following notation:

$$\left\langle \frac{\Delta v_i}{\Delta t} \right\rangle = \frac{1}{\Delta t} \iiint_\infty Q_p(\mathbf{v}, \mathbf{w}, \Delta t) w_i \, d^3 w. \tag{2.61}$$

The integral of the last term in the Taylor expansion can be written as

$$\iiint_\infty \frac{1}{2} w_i w_j \frac{\partial^2 [Q_p(\mathbf{v}, \mathbf{w}, \Delta t) f(t, \mathbf{r}, \mathbf{v})]}{\partial v_i \partial v_j} \, d^3 w$$

$$= \frac{1}{2} \frac{\partial^2}{\partial v_i \partial v_j} \left[f(t, \mathbf{r}, \mathbf{v}) \iiint_\infty Q_p(\mathbf{v}, \mathbf{w}, \Delta t) w_i w_j \, d^3 w \right]$$

$$= \Delta t \frac{1}{2} \frac{\partial^2}{\partial v_i \partial v_j} \left[\left\langle \frac{\Delta v_i \Delta v_j}{\Delta t} \right\rangle f(t, \mathbf{r}, \mathbf{v}) \right], \tag{2.62}$$

where

$$\left\langle \frac{\Delta v_i \Delta v_j}{\Delta t} \right\rangle = \frac{1}{\Delta t} \iiint_\infty Q_p(\mathbf{v}, \mathbf{w}, \Delta t) w_i w_j \, d^3 w. \tag{2.63}$$

Next, we substitute all these integrals into the Schmoluchowky equation (2.56):

$$f(t, \mathbf{r}, \mathbf{v}) = f(t, \mathbf{r}, \mathbf{v}) - \Delta t \left\{ \frac{df(t, \mathbf{r}, \mathbf{v})}{dt} + \frac{\partial}{\partial v_i} \left[\left\langle \frac{\Delta v_i}{\Delta t} \right\rangle f(t, \mathbf{r}, \mathbf{v}) \right] \right.$$

$$\left. - \frac{1}{2} \frac{\partial^2}{\partial v_i \partial v_j} \left[\left\langle \frac{\Delta v_i \Delta v_j}{\Delta t} \right\rangle f(t, \mathbf{r}, \mathbf{v}) \right] \right\}. \tag{2.64}$$

This equation can also be written as

$$\frac{df(t, \mathbf{r}, \mathbf{v})}{dt} = -\frac{\partial}{\partial v_i} \left[\left\langle \frac{\Delta v_i}{\Delta t} \right\rangle f(t, \mathbf{r}, \mathbf{v}) \right] + \frac{1}{2} \frac{\partial^2}{\partial v_i \partial v_j} \left[\left\langle \frac{\Delta v_i \Delta v_j}{\Delta t} \right\rangle f(t, \mathbf{r}, \mathbf{v}) \right]. \tag{2.65}$$

One has to realize that in this approximation all changes in the distribution function occur as a result of collisions; therefore, the right-hand side of Eq. (2.65) is the rate of change of the normalized distribution function due to long-range collisions. This is the Fokker–Planck collision term. The relation between the full phase-space distribution function, F, and the normalized function, f, is $F(t, \mathbf{r}, \mathbf{v}) = n(t, \mathbf{r}) f(t, \mathbf{r}, \mathbf{v})$, where n is the particle number density, which is independent of velocity. This means that the Fokker–Planck collision term for the full distribution function can be written as

$$\left(\frac{dF(t, \mathbf{r}, \mathbf{v})}{dt} \right)_{FP} = -\frac{\partial}{\partial v_i} \left[\left\langle \frac{\Delta v_i}{\Delta t} \right\rangle F(t, \mathbf{r}, \mathbf{v}) \right]$$

$$+ \frac{1}{2} \frac{\partial^2}{\partial v_i \partial v_j} \left[\left\langle \frac{\Delta v_i \Delta v_j}{\Delta t} \right\rangle F(t, \mathbf{r}, \mathbf{v}) \right]. \tag{2.66}$$

The first term in the Fokker–Planck collision term describes a phenomenon similar to dynamical friction. This term is proportional to $\langle \Delta \mathbf{v} \rangle$, which is essentially the average velocity of our particles with respect to the scattering medium. The second term describes diffusion in velocity space. This term is proportional to $\langle \Delta \mathbf{v} \Delta \mathbf{v} \rangle$ and results in the broadening of the velocity distribution function.

It should be repeated again that the Fokker–Planck collision term is valid for any system in which individual collisions produce only small changes in the velocity of particles, with large changes occurring only as a result of many small changes. Another important point to be made is that the scattering process is not necessarily constrained to intermolecular collisions. For instance, small-angle scattering of charged particles can also be caused by fluctuating external fields (for instance magnetic field fluctuations in interplanetary space).

Finally, let us examine the evolution of the distribution function due to long-range interactions. To demonstrate the fundamental physics, let us consider a greatly simplified situation in which there is no bulk flow ($\mathbf{u} = \mathbf{0}$), there are no spatial gradients ($\nabla F = 0$), and there are no external force fields ($\mathbf{a} = \mathbf{0}$). Also, we neglect dynamical friction. In this case the Boltzmann equation simplifies to

$$\frac{\partial F(t, \mathbf{v})}{\partial t} = \frac{\partial^2}{\partial v_i \partial v_j} \left[\frac{1}{2} \left\langle \frac{\Delta v_i \Delta v_j}{\Delta t} \right\rangle F(t, \mathbf{v}) \right] = \frac{\partial^2}{\partial v_i \partial v_j} [D_{ij} F(t, \mathbf{v})], \quad (2.67)$$

where we introduced the tensor $\mathbf{D}$ as

$$D_{ij} = \frac{1}{2} \left\langle \frac{\Delta v_i \Delta v_j}{\Delta t} \right\rangle. \quad (2.68)$$

Additional simplification can be achieved by considering a one-dimensional problem when only one component of $\mathbf{D}$ is nonzero. For instance, let us consider the case when $D_{xx} = const. \neq 0$, while all other components are zero. Then Eq. (2.67) becomes

$$\frac{\partial F(t, v_x)}{\partial t} = D_{xx} \frac{\partial^2 F(t, v_x)}{\partial^2 v_x}. \quad (2.69)$$

Equation (2.69) can be readily solved for the case when at $t = 0$ the particle distribution function is a narrow Gaussian around $v_x = 0$:

$$F(t, v_x) = \frac{N}{\sqrt{4\pi (\sigma_{xx} + D_{xx} t)}} \exp\left(-\frac{v_x^2}{4(\sigma_{xx} + D_{xx} t)} \right), \quad (2.70)$$

where N is the total number of particles in the system and σ_{xx} characterizes the width of the distribution at $t = 0$. The solution described by Eq. (2.70) is a Gaussian with a gradually broadening width. Equivalently, the distribution can be considered to be a Maxwellian with increasing temperature. The temperature linearly increases with time as

$$T = \frac{2m(\sigma_{xx} + D_{xx} t)}{k}. \quad (2.71)$$

The interpretation of this solution is that the Fokker–Planck collision term results in diffusion in velocity space. As time evolves, the solution becomes broader and broader. In the $t \to \infty$ limit the velocity distribution becomes flat (isotropic), but the exponential goes to zero. Thus as $t \to \infty$ the probability of finding a particle with a given velocity is becoming extremely small, but it is the same for all velocities. This diffusion in velocity space occurs as a natural consequence of the large number of random small collisions. In effect, we have a random walk in velocity space.

2.3   Multispecies Gases

So far we have considered only the case when all molecules in the gas are identical. When more than one molecular species occupies the same volume (each species is characterized by a different molecule) the phase-space distribution functions of the various species might be quite different from each other. Also, chemical reactions, ionization, or recombination can change the nature of molecules: These changes must be reflected by appropriate changes of the distribution functions.

One of the fundamental assumptions of kinetic theory is that the physical sizes of the individual molecules are much smaller than the average distance between the particles. This assumption means that the presence of more than one species only negligibly reduces the volume available for a given gas component. Consequently, the concentrations of the individual species remain unmodified. With this assumption the total (not normalized) distribution function of the gas mixture can be written in the following form:

$$F(t, \mathbf{r}, \mathbf{v}) = \sum_s F_s(t, \mathbf{r}, \mathbf{v}_s), \tag{2.72}$$

where F_s denotes the distribution function of particles of type s. It should be noted that the summation rule for the normalized distribution function is more complicated. It can be obtained by dividing Eq. (2.72) by the total particle density, $n = \sum n_s$:

$$\frac{F(t, \mathbf{r}, \mathbf{v})}{n(t, \mathbf{r})} = f(t, \mathbf{r}, \mathbf{v}) = \sum_s \frac{n_s(t, \mathbf{r})}{n(t, \mathbf{r})} f_s(t, \mathbf{r}, \mathbf{v}_s). \tag{2.73}$$

In the case of multicomponent gases, each gas component is described by a separate Boltzmann equation:

$$\frac{\partial F_s(t, \mathbf{r}, \mathbf{v}_s)}{\partial t} + (\mathbf{v}_s \cdot \nabla) F_s(t, \mathbf{r}, \mathbf{v}_s) + [\mathbf{a}_s(t, \mathbf{r}, \mathbf{v}_s) \cdot \nabla_{vs}] F_s(t, \mathbf{r}, \mathbf{v}_s)$$

$$= \frac{\delta F_s(t, \mathbf{r}, \mathbf{v}_s)}{\delta t}, \tag{2.74}$$

where the subscript s refers to the given particle species. The interaction between the various gas components is described by the collision term, $\delta F_s / \delta t$. In the mixed phase-space coordinate system the same equation becomes

$$
\frac{\partial F_s(t, \mathbf{r}, \mathbf{c}_s)}{\partial t} + (\mathbf{c}_s \cdot \nabla) F_s(t, \mathbf{r}, \mathbf{c}_s) + (\mathbf{u}_s(t, \mathbf{r}) \cdot \nabla) F_s(t, \mathbf{r}, \mathbf{c}_s)
$$

$$
- \left[\frac{\partial \mathbf{u}_s}{\partial t} + (\mathbf{u}_s(t, \mathbf{r}) \cdot \nabla) \mathbf{u}_s(t, \mathbf{r}) + (\mathbf{c}_s \cdot \nabla) \mathbf{u}_s(t, \mathbf{r}) \right.
$$

$$
\left. - \mathbf{a}_s(t, \mathbf{r}, \mathbf{c}_s) \right] \nabla_{cs} F_s(t, \mathbf{r}, \mathbf{c}_s) = \frac{\delta F_s(t, \mathbf{r}, \mathbf{c}_s)}{\delta t}. \tag{2.75}
$$

The most interesting term in these equations is the collision term, $\delta F_s / \delta t$. This term describes the effects of collisions, ionization, recombination, chemical reactions, etc. In what follows we briefly examine the collision term for several important processes. For the sake of mathematical simplicity, we will consider the collision term for the mixed phase-space system. The collision terms for other phase-space systems can be obtained from these results in a relatively straightforward manner.

2.4 Elastic Binary Collisions

The effects of elastic collisions between two particles can be approximated by using a generalized form of the relaxation-time (BGK) approximation (Eq. 2.52):

$$
\left(\frac{\delta F_s(t, \mathbf{r}, \mathbf{c}_s)}{\delta t} \right)_{\text{elastic}} = -\sum_t \frac{F_s(t, \mathbf{r}, \mathbf{c}_s) - F_{0s(st)}(t, \mathbf{r}, \mathbf{c}_s)}{\tau_{st}}, \tag{2.76}
$$

where the subscript t refers to all species present in the gas mixture (including species s), τ_{st} is a velocity-independent average collision time between particles s and t, and $F_{0s(st)}$ is an equilibrium distribution function (Maxwellian). The subscripts in $F_{0s(st)}$ indicate that the distribution of s particles would asymptotically approach this distribution as a result of collisions between s and t particles. It is important to note that the parameters of $F_{0s(st)}$ also depend on the actual distribution of species t; therefore, $F_{0s(st)}$ changes as the distributions of both s and t particles evolve with time. The parameters of $F_{0s(st)}$ can be obtained by requiring that the total mass, momentum, and energy of the s and t gases be conserved (cf. *Gombosi 1994*):

$$
F_{0s(st)} = n_s \left(\frac{m_s}{2\pi \, k T_{s(st)}} \right)^{3/2} \exp \left\{ -\frac{m_s}{2k T_{s(st)}} \left[\mathbf{c}_s - \frac{m_t(\mathbf{u}_t - \mathbf{u}_s)}{m_s + m_t} \right]^2 \right\}, \tag{2.77}
$$

where

$$
T_{s(st)} = T_s + \frac{m_s m_t}{(m_s + m_t)^2} \left[2(T_t - T_s) + \frac{m_t}{3k} (\mathbf{u}_t - \mathbf{u}_s)^2 \right]. \tag{2.78}
$$

In general, $T_{s(st)} \neq T_{t(ts)}$. This expression tells us that the gas components, s and t, reach equilibrium through different intermediate states. The interpretation of the different terms in expression (2.78) is quite straightforward: The temperature of a gas component changes due to heat exchange between the constituents (the $T_t - T_s$ term) and due to friction as the two gases move through each other (the velocity difference term).

To illuminate this result let us consider a very simple example. Consider a volume of gas composed of two different molecules of identical masses, $m_1 = m_2 = m$ (for instance CO and N_2). Furthermore, let us assume that there are no external forces ($\mathbf{a} = \mathbf{0}$), the gas is spatially homogeneous (there are no spatial gradients), that none of the gas components have any bulk motion ($\mathbf{u}_1 = \mathbf{u}_2 = \mathbf{0}$), and that the two components are each near equilibrium at different temperatures (initially at T_{1i} and T_{2i}, respectively). In this case the transport equations simplify to the following:

$$\frac{\partial F_1(t, \mathbf{c}_1)}{\partial t} = -\frac{F_1(t, \mathbf{c}_1) - F_{01(11)}(t, \mathbf{c}_1)}{\tau_{11}} - \frac{F_1(t, \mathbf{c}_1) - F_{01(12)}(t, \mathbf{c}_1)}{\tau_{12}} \quad (2.79)$$

and

$$\frac{\partial F_2(t, \mathbf{c}_2)}{\partial t} = -\frac{F_2(t, \mathbf{c}_2) - F_{02(21)}(t, \mathbf{c}_2)}{\tau_{21}} - \frac{F_2(t, \mathbf{c}_2) - F_{02(22)}(t, \mathbf{c}_2)}{\tau_{22}}, \quad (2.80)$$

where

$$F_{01(11)} = n_1 \left(\frac{m}{2\pi k T_1} \right)^{3/2} \exp\left(-\frac{m}{2k T_1} c_1^2 \right),$$

$$F_{01(12)} = n_1 \left(\frac{m}{\pi k (T_1 + T_2)} \right)^{3/2} \exp\left(-\frac{m}{k (T_1 + T_2)} c_1^2 \right),$$

$$F_{02(21)} = n_2 \left(\frac{m}{\pi k (T_1 + T_2)} \right)^{3/2} \exp\left(-\frac{m}{k (T_1 + T_2)} c_2^2 \right),$$

and

$$F_{02(22)} = n_2 \left(\frac{m}{2\pi k T_2} \right)^{3/2} \exp\left(-\frac{m}{2k T_2} c_2^2 \right). \quad (2.81)$$

To emphasize the physical meaning of individual terms, let us make a few more simplifying assumptions. First of all, let us assume that the collision times of self-collisions (τ_{11} and τ_{22}) is much smaller than the collision time between different molecules. For the sake of additional simplicity, we also assume that $\tau_{12} = \tau_{21} = \tau$. Thus the distribution functions, F_1 and F_2, remain nearly Maxwellian as they evolve in time. In other words, the temperatures, T_1 and T_2, change with time, but the sum of the two temperatures remains constant and the distribution functions remain nondrifting Maxwellians with total number densities of n_1 and n_2. However, this assumption also means that at any given time $F_1 \approx F_{01(11)}$ and $F_2 \approx F_{02(22)}$. One can also see that

$$F_{01(12)}(t, \mathbf{c}_1) = n_1 f_0(\mathbf{c}_1),$$
$$F_{02(21)}(t, \mathbf{c}_2) = n_2 f_0(\mathbf{c}_2), \quad (2.82)$$

where

$$f_0(\mathbf{c}) = \left(\frac{m}{\pi k (T_{1i} + T_{2i})} \right)^{3/2} \exp\left(-\frac{m}{k (T_{1i} + T_{2i})} c^2 \right). \quad (2.83)$$

In this approximation Eqs. (2.79) and (2.80) become

$$\frac{\partial f_1(t, \mathbf{c}_1)}{\partial t} = -\frac{f_1(t, \mathbf{c}_1) - f_0(\mathbf{c}_1)}{\tau},$$

$$\frac{\partial f_2(t, \mathbf{c}_2)}{\partial t} = -\frac{f_2(t, \mathbf{c}_2) - f_0(\mathbf{c}_2)}{\tau}. \tag{2.84}$$

Equation (2.84) can be readily solved to obtain

$$f_1(t, \mathbf{c}_1) = f_0(\mathbf{c}_1) + \left[f_{1i}(\mathbf{c}_1) - f_0(\mathbf{c}_1) \right] \exp\left(-\frac{t}{\tau}\right),$$

$$f_2(t, \mathbf{c}_1) = f_0(\mathbf{c}_2) + \left[f_{2i}(\mathbf{c}_2) - f_0(\mathbf{c}_2) \right] \exp\left(-\frac{t}{\tau}\right), \tag{2.85}$$

where

$$f_{1i}(\mathbf{c}_1) = \left(\frac{m}{2\pi k T_{1i}}\right)^{3/2} \exp\left(-\frac{mc_1^2}{2k T_{1i}}\right),$$

$$f_{2i}(\mathbf{c}_1) = \left(\frac{m}{2\pi k T_{2i}}\right)^{3/2} \exp\left(-\frac{mc_2^2}{2k T_{2i}}\right). \tag{2.86}$$

This result shows that the initial distribution functions exponentially decay (with a characteristic time τ) and they are gradually replaced by the new equilibrium distributions with common temperature $(T_{1i} + T_{2i})/2$.

2.5 Problems

Problem 2.1 At room temperature $(T = 300 \text{ K})$ the total scattering cross section for electrons on air molecules is $\sigma_t = 10^{-19} \text{ m}^2$. At what gas density will 90% of the electrons emitted from a pointlike cathode reach a pointlike anode 0.2 m away? (Assume that scattered electrons cannot reach the anode.)

Problem 2.2 A beam of cold electrons (mass m_e) is impacting on stationary protons (mass m_p) with velocity $\mathbf{u}$. The densities are n_e and n_p, respectively, and $n_e \neq n_p$. The normalized distribution functions are given by

$$f_e = \delta^3(\mathbf{v}_e - \mathbf{u})$$

and

$$f_p = \delta^3(\mathbf{v}_p).$$

1. What is the normalized distribution of the electron–proton mixture?
2. What is the average velocity of the particles in the mixture?
3. What is the average particle energy in the mixture?

Problem 2.3 In the previous problem assume that $n_e = 10^{12} \text{ m}^{-3}$, $n_p = 10^{10} \text{ m}^{-3}$, $u = 1{,}000 \text{ km/s}$, and the electron–proton collision cross section is $\sigma = 10^{-14} \text{ m}^2$.

1. What is the electron–proton collision rate?
2. What is the electron–proton collision frequency?
3. What is the proton–electron collision frequency?

Problem 2.4 A stationary volume of gas is in thermodynamic equilibrium at temperature T. Calculate the following quantities: $\langle mv^4 \rangle$, $\langle mv_z^2 \rangle$, and $\langle m/v \rangle$.

Problem 2.5 In a two-dimensional world a stationary volume of gas is in thermodynamic equilibrium at temperature T. Calculate the average kinetic energy.

Problem 2.6 Consider a volume of gas (consisting of a single species) with no spatial gradients and no external forces, and use the relaxation-time approximation for the collision term. Assume that at $t = 0$ the distribution function is the following:

$$
F_{in} = \begin{cases} \frac{3n_0}{4\pi v_0^3} & |\mathbf{v}| \le v_0, \\ 0 & |\mathbf{v}| > v_0. \end{cases}
$$

What is the time evolution of the distribution function?

Chapter 3

Basic Plasma Phenomena

Most *space plasmas*[1] are quasi-neutral statistical systems containing mobile charged particles. On the average, the potential energy of a mobile particle due to its nearest neighbor is much smaller than its kinetic energy. This definition excludes high density plasmas (such as solid states or stellar interiors), but the description of these forms of matter goes far beyond the scope of this book.

Owing to the long-range nature of electromagnetic forces, each charged particle in the plasma interacts simultaneously with a large number of other charged particles. This process results in collective behavior of the plasma particles: In some respect even a low density space plasma behaves as a continuous medium.

In gaseous, nonrelativistic plasmas the motion of individual particles is governed by electromagnetic fields, which are a combination of internally generated (due to the presence and motion of charged particles) and externally imposed fields. The motion and interaction of plasma particles can be described by nonrelativistic classical mechanics and by electrodynamics, quantum mechanical effects are usually neglected.

The interaction of charged particles with electromagnetic fields is described by the force (Eq. 1.16), whereas the electromagnetic field itself obeys Maxwell's equations (Eqs. 1.1, 1.2, 1.3, and 1.4). It should be noted that one must include the contributions of plasma particles to the charge and electric current densities. The individual particles are thus self-conﬁstly coupled to the electromagnetic fields.

In this chapter we examine some of the basic plasma phenomena and outline some of the fundamental elements of theoretical descriptions of space plasmas.

---

[1] The word *plasma* comes from the Greek $\pi\lambda\acute{\alpha}\sigma\mu\alpha$, meaning "something formed or molded."

3.1   Debye Shielding

Let us start by considering the electric field generated by a stationary pointlike charged particle (charge q_0) located at the origin of the coordinate system. If there are no other charged particles in the system, the central charge produces an electrostatic field given by the Poisson equation (Eq. 1.1):

$$\nabla \cdot \mathbf{E} = \frac{q_0}{\varepsilon_0} \delta^3(\mathbf{r}), \tag{3.1}$$

where $\delta(x)$ is the Dirac delta function (see Appendix D). The solution of this well-known problem is the electric field of a single charge:

$$\mathbf{E}_0(\mathbf{r}) = \frac{q_0}{4\pi \varepsilon_0} \frac{\mathbf{r}}{r^3}. \tag{3.2}$$

It can be easily shown that $\nabla \times \mathbf{E}_0 = \mathbf{0}$. Therefore, the electric field vector can be expressed with the help of a scalar potential as $\mathbf{E}_0 = -\nabla \psi_0$. The scalar potential can now be expressed as

$$\psi_0(\mathbf{r}) = \frac{q_0}{4\pi \varepsilon_0} \frac{1}{r}. \tag{3.3}$$

Equation (3.3) is the so-called *Coulomb potential*, which is the electric potential field generated by a stationary charged particle in vacuum.

   In a plasma there are many particles moving around a central test charge. Particles with opposite charges are attracted toward the central charge, whereas particles with similar charge (positive or negative) are repelled. Thus, around the central test charge an oppositely charged shielding cloud is formed, which tends to cancel the electric field generated by the central charge. This effect is called *Debye shielding*. In what follows we will give a simple mathematical description of Debye shielding.

   Let us place our stationary central test particle into a quasi-neutral gas containing mobile charged particles in thermodynamic equilibrium. As discussed earlier for equilibrium gases, the total time derivative of the distribution function vanishes, and therefore we obtain the following equation for the distribution of electrons or ions (see Eq. 2.74):

$$(\mathbf{v}_s \cdot \nabla) F_s(\mathbf{r}, \mathbf{v}_s) + [\mathbf{a}_s(\mathbf{r}, \mathbf{v}_s) \cdot \nabla_{vs}] F_s(\mathbf{r}, \mathbf{v}_s) = 0, \tag{3.4}$$

where the subscript s refers to electrons or a particular type of ions, and $\mathbf{a}_s$ is acceleration due to a conservative external force. In this case the acceleration can also be written as

$$\mathbf{a}_s(\mathbf{r}) = -\frac{1}{m_s} \nabla U_s(\mathbf{r}), \tag{3.5}$$

where $U(\mathbf{r})$ is the potential of the external force field acting on the particle.

   We also discussed in the previous chapter that the velocity distribution function of equilibrium gases can be described by a Maxwellian distribution. Here we assume that the gas is very near to equilibrium, and the distribution function locally can be approximated by a Maxwellian, but the macroscopic parameters of the Maxwellian (density, bulk velocity, and temperature) slowly vary with location. This approximation is called *local thermodynamic equilibrium*. In this case we know that F_s can be written as (see Eq. 2.45):

$$F_s(\mathbf{r}, \mathbf{v}_s) = n_s(\mathbf{r}) \left(\frac{m_s}{2\pi k T_s} \right)^{3/2} \exp\left[-\frac{m_s v_s^2}{2k T_s} \right]. \tag{3.6}$$

For the sake of simplicity, we assumed constant temperature. Substituting Eqs. (3.5) and (3.6) into Eq. (3.4), we obtain the following equation for the number density distribution of particles s:

$$\left[(\mathbf{v}_s \cdot \nabla) n_s + \frac{\mathbf{v}_s \cdot \nabla U}{k T_s} n_s \right] \left(\frac{m_s}{2\pi k T_s} \right)^{3/2} \exp\left[-\frac{m_s v_s^2}{2k T_s} \right] = 0. \tag{3.7}$$

This equation can be satisfied if

$$(\mathbf{v}_s \cdot \nabla) n_s + \frac{\mathbf{v}_s \cdot \nabla U}{k T_s} n_s = 0. \tag{3.8}$$

The solution is the following:

$$n_s(\mathbf{r}) = n_{s0} \exp\left[-\frac{U_s(\mathbf{r})}{k T_s} \right]. \tag{3.9}$$

Finally, the phase-space distribution function becomes

$$F_s(\mathbf{r}, \mathbf{v}_s) = n_{s0} \left(\frac{m_s}{2\pi k T_s} \right)^{3/2} \exp\left[-\frac{m_s v_s^2 + 2U_s(\mathbf{r})}{2k T_s} \right]. \tag{3.10}$$

Next, we consider the distribution of charged particles around a stationary central test charge, q_0. Because the presence of other charged particles modifies the potential field, the electrostatic potential will be different from ψ_0.

For the sake of simplicity, we neglect the magnetic field generated by the random motion of the charged particles. This means that $\nabla \times \mathbf{E} = \mathbf{0}$, and consequently the electric field can be characterized by a scalar potential, $\psi(\mathbf{r})$. In this case the potential energy of a particle with charge q_s becomes

$$U_s(\mathbf{r}) = q_s \psi(\mathbf{r}). \tag{3.11}$$

To simplify further, we constrain ourselves to a single kind of single ionized ions. In this case $q_e = -e$ and $q_{ion} = e$, where e is the elementary charge ($e = 1.6022 \times 10^{-19}$ coulomb). Also, the gas is assumed to be quasi-neutral, that is, $n_{e0} = n_{ion0} = n_0$. Now the charge density becomes

$$\rho = q_0 \delta^3(\mathbf{r}) - e n_0 \left[\exp\left(\frac{e\psi}{k T_e} \right) - \exp\left(\frac{-e\psi}{k T_{ion}} \right) \right]. \tag{3.12}$$

The potential is determined by Poisson's equation (Eq. 1.1):

$$\nabla^2 \psi(\mathbf{r}) = -\frac{\rho}{\varepsilon_0} = -\frac{q_0}{\varepsilon_0}\delta^3(\mathbf{r}) + \frac{en_0}{\varepsilon_0}\left[\exp\left(\frac{e\psi}{kT_e}\right) - \exp\left(\frac{-e\psi}{kT_{ion}}\right)\right]. \qquad (3.13)$$

It was mentioned at the beginning of this chapter that we concentrate on gaseous plasmas where the average particle potential energy is small compared to the average kinetic energy. This means that we limit ourselves to $r > 0$ (since for our case the charges cannot be very close to each other) and to $|e\psi/kT| \ll 1$ for both electrons and ions. In this case the exponentials can be expanded to first order, yielding

$$\nabla^2 \psi(\mathbf{r}) = -\frac{\rho}{\varepsilon_0} = \frac{en_0}{\varepsilon_0}\left[1 + \frac{e\psi}{kT_e} - 1 + \frac{e\psi}{kT_{ion}}\right] = \frac{e^2 n_0}{\varepsilon_0 kT^*}\psi, \qquad (3.14)$$

where the effective plasma temperature, T^*, is given by

$$\frac{1}{T^*} = \frac{1}{T_e} + \frac{1}{T_{ion}}. \qquad (3.15)$$

One can introduce a new length parameter:

$$\lambda_D^2 = \frac{\varepsilon_0 kT^*}{e^2 n_0} \qquad (3.16)$$

The parameter, λ_D, has the dimension of length, and it is called the *Debye length*.

The solution of Eq. (3.14) gives the *Debye potential*:

$$\psi(\mathbf{r}) = \frac{q_0}{4\pi\varepsilon_0}\frac{1}{r}\exp\left(-\frac{r}{\lambda_D}\right) = \psi_0(\mathbf{r})\exp\left(-\frac{r}{\lambda_D}\right), \qquad (3.17)$$

where ψ_0 is the potential generated by the central charge in vacuum. The physical meaning of this result is clear: For a distance much smaller than the Debye length, the electrostatic potential field is essentially the same as for a single charge in vacuum (Coulomb potential), whereas for distances larger than λ_D the potential vanishes exponentially. Far from the central charge, the long-range electrostatic force field is effectively shielded by the induced space-charge field due to the surrounding ions and electrons. This *Debye shielding* is a fundamental feature of gaseous plasmas.

The definition of Debye length can be extended to multispecies plasmas:

$$\frac{1}{\lambda_D^2} = \sum_s \frac{1}{\lambda_{Ds}^2} \qquad (3.18)$$

where $\lambda_{Ds}^2 = \varepsilon_0 kT_s/q_s^2 n_s$ is the Debye length of the individual species s.

3.2   Plasma Parameter

We mentioned that in most plasmas the average kinetic energy of the particles is much larger that the average potential energy due to the presence of neighboring charged particles. In this section we will examine this ratio using the Debye potential determined above. One can express the charge density (for $r > 0$) with the help of Poisson's equation

$$\rho = -\varepsilon_0 \nabla^2 \psi. \tag{3.19}$$

In the case of the Debye potential field (given by 3.17), this yields

$$\rho = -\frac{\varepsilon_0}{\lambda_D^2}\psi = -\frac{q_0}{4\pi\lambda_D^2}\frac{1}{r}\exp\left(-\frac{r}{\lambda_D}\right). \tag{3.20}$$

Let us consider a plasma composed of electrons and one kind of single charged ion. The potential energy density of the plasma particles due to the electric potential is given by $\rho\psi$. In this case the potential energy density becomes

$$w = \rho\psi = -\frac{e^2}{(4\pi)^2\varepsilon_0\lambda_D^2}\frac{1}{r^2}\exp\left(-2\frac{r}{\lambda_D}\right). \tag{3.21}$$

The average potential energy (energy per particle) can now be determined by integrating this energy density over the entire volume:

$$
\begin{aligned}
\overline{W} &= \iiint_\infty d^3r\, w(r) = \iiint_\infty d^3r\, \rho(r)\psi(r) \\
&= -\iiint_\infty d^3r\, \frac{e^2}{(4\pi)^2\varepsilon_0\lambda_D^2}\frac{1}{r^2}\exp\left(-2\frac{r}{\lambda_D}\right) = -\frac{e^2}{8\pi\varepsilon_0\lambda_D}.
\end{aligned}
\tag{3.22}
$$

The average kinetic energy of a plasma particle is given by

$$\overline{E}_s = \frac{3}{2}kT_s. \tag{3.23}$$

Now the ratio of the average potential and kinetic energies is

$$g_p = \frac{|\overline{W}|}{\overline{E}_s} = \frac{e^2}{12\pi\varepsilon_0\lambda_D kT_s}. \tag{3.24}$$

Because there are equal numbers of ions and electrons in the plasma (the electron density is n_0), the average energy ratio per plasma particle can be written as

$$g_p = \frac{|\overline{W}|}{\overline{E}} = \frac{1}{2}\left(\frac{|\overline{W}|}{\overline{E}_e} + \frac{|\overline{W}|}{\overline{E}_{ion}}\right) = \frac{e^2}{12\pi\varepsilon_0\lambda_D k}\left(\frac{1}{T_e} + \frac{1}{T_{ion}}\right) = \frac{1}{12\pi n_0\lambda_D^3}. \tag{3.25}$$

The total number of plasma particles inside a sphere with radius equal to the Debye length (Debye sphere) is $N_D = 8\pi n_0\lambda_D^3/3$ (because the total particle density is $2n_0$). This means that the energy ratio (also called the *plasma parameter*) can be written as

$$g_p = \frac{|\overline{W}|}{\overline{E}} = \frac{1}{12\pi n_0\lambda_D^3} = \frac{2}{9N_D}. \tag{3.26}$$

This result means that the plasma parameter is proportional to $1/N_D$. Usually a very large number of particles reside inside the Debye sphere, so consequently the plasma parameter is very small. Table 3.1 summarizes the Debye length and plasma parameter values for some space plasmas. Inspection of Table 3.1 shows that $N_D \gg 1$ (or $g_p \ll 1$) for gaseous plasmas, which are the subjects of classical plasma physics.

3.3   Plasma Frequency

Consider a slab of plasma of thickness L (see Figure 3.1). For the sake of simplicity, let us assume that the plasma is quasi-neutral (with electron density of n_0) and contains only one type of singly ionized ion. We also assume that the thermal motion of all particles can be neglected. In this approximation all electrons and all ions move together, but the ion and electron slabs can move with respect to each other.

At $t = 0$, the electron slab is displaced by a distance of $\delta_e \ll L$ and the ion slab is displaced to the opposite direction by a distance of $\delta_i \ll L$. The total distance between

Table 3.1. *Typical characteristics of some plasmas.*

Plasma	n (m^{-3})	T^* (K)	λ_D (m)	N_D
Interstellar gas	10^6	10^4	7×10^0	1×10^9
Gaseous nebula	10^9	10^4	2×10^{-1}	4×10^7
Solar atmosphere	10^{20}	10^4	7×10^{-7}	1×10^2
Solar corona	10^{15}	10^6	2×10^{-3}	4×10^7
Solar wind (1 AU)	10^7	10^5	7×10^0	1×10^{10}
Ionosphere (F layer)	10^{12}	10^3	2×10^{-3}	4×10^4
Tenuous laboratory plasma	10^{17}	10^4	2×10^{-5}	4×10^3
Dense laboratory plasma	10^{21}	10^5	7×10^{-7}	1×10^3
Thermonuclear plasma	10^{22}	10^8	7×10^{-6}	1×10^7
Laser plasma	10^{26}	10^6	7×10^{-9}	1×10^2
Metal	10^{29}	10^2	2×10^{-12}	4×10^{-6}
Stellar interior	10^{33}	10^7	7×10^{-12}	1×10^0

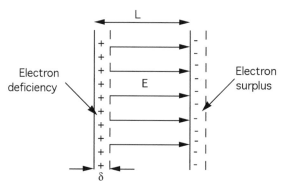

Figure 3.1 Schematic of a plasma slab.

the two slabs is now $\delta = \delta_e - \delta_i$. According to Poisson's equation (Eq. 1.1), the displacements result in a uniform electric field in the direction of the net displacement (except near the edges of the slab):

$$E = \frac{e n_0}{\varepsilon_0} \delta \tag{3.27}$$

The equations of motion of the electron and ion slabs are:

$$m_e \frac{d^2 \delta_e}{dt^2} = -eE, \tag{3.28}$$

$$m_{ion} \frac{d^2 \delta_i}{dt^2} = eE. \tag{3.29}$$

One can combine the electron and ion equations of motion to get

$$\frac{d^2 \delta}{dt^2} = \frac{d^2 \delta_e}{dt^2} - \frac{d^2 \delta_i}{dt^2} = -\left(\frac{e^2 n_0}{\varepsilon_0 m_e} + \frac{e^2 n_0}{\varepsilon_0 m_{ion}} \right) \delta. \tag{3.30}$$

Equation (3.30) describes an oscillating relative motion between the electron and ion slabs. The frequency of this oscillation is called the *plasma frequency*:

$$\omega_p^2 = \omega_{pe}^2 + \omega_{pi}^2, \tag{3.31}$$

where the electron and ion plasma frequencies are given by

$$\omega_{pe}^2 = \frac{e^2 n_0}{\varepsilon_0 m_e} \tag{3.32}$$

$$\omega_{pi}^2 = \frac{e^2 n_0}{\varepsilon_0 m_{ion}}. \tag{3.33}$$

It can be easily seen that because of the mass difference $\omega_{pe} \gg \omega_{pi}$. Therefore, $\omega_p \approx \omega_{pe}$.

Thus the plasma frequency is basically determined by the electron density. In a good approximation one gets

$$\omega_p \approx \omega_{pe} = 56.41 \sqrt{n_0} \quad \text{rad/s}, \tag{3.34}$$

where n_0 is measured in units of m^{-3}. This is the natural frequency of a plasma.

In the case of multispecies plasmas the plasma frequency becomes

$$\omega_p^2 = \frac{e^2 n_0}{\varepsilon_0 m_e} + \sum_{s=ions} \frac{e^2 n_0}{\varepsilon_0 m_s}. \tag{3.35}$$

Finally, it is interesting to point out that for a given species the product of the Debye length and the plasma frequency is the characteristic thermal speed, $v_{th} = (kT/m)^{1/2}$:

$$\omega_{ps} \lambda_{Ds} = \sqrt{\frac{e^2 n_s}{\varepsilon_0 m_s}} \sqrt{\frac{\varepsilon_0 k T_s}{e^2 n_s}} = \sqrt{\frac{k T_s}{m_s}} = v_{th,s}. \tag{3.36}$$

3.4   Problems

Problem 3.1 A distant galaxy is composed solely of protons and antiprotons (an antiproton is a negatively charged proton), each with density $n = 10^6$ m^{-3} and temperature 100 K. What is

1. the Debye length, and
2. the plasma frequency?

Problem 3.2 Show that expression (3.17) is the solution of Eq. (3.14).

Problem 3.3 The induced charge density around a central test charge, q_0, is given by Eq. (3.20). Calculate the total induced charge

$$Q_0 = \iiint_\infty d^3r \rho(r).$$

Problem 3.4 Calculate the Debye length and the number of particles in the Debye sphere for:

1. the solar wind at Saturn, where the average energy of an electron is 10 eV and the average electron density is 10^5 m^{-3};
2. the Earth's lower inosphere ($\sim$100 km), where the average energy of an electron is 0.01 eV and the average electron density is 10^{11} m^{-3};
3. the magnetospheric plasma sheet, where the average energy of an electron is 1 keV and the average electron density is 10^6 m^{-3}.

Chapter 4

Fluid and MHD Theory

In its most general form the Boltzmann equation is a seven-dimensional nonlinear integro-differential equation. The solutions of the Boltzmann equation provide a full description of the phase-space distribution function at all times. In most cases, however, it is next to impossible to solve the full Boltzmann equation and one has to resort to various approximate methods to describe the spatial and temporal evolution of macroscopic quantities characterizing the gas.

Transport equations for macroscopic molecular averages are obtained by taking velocity moments of the Boltzmann equation. This seemingly straightforward technique runs into considerable difficulties because the governing equations for the components of the n-th velocity moment also depend on components of the $(n + 1)$-th moment. In order to get a closed transport equation system, one has to use closing relations (expressing a higher-order velocity moment of the distribution function in terms of the components of lower moments) and thus make implicit assumptions about the distribution function.

4.1 Moment Equations

4.1.1 Velocity Moments

We start by examining the physical interpretation of the various velocity moments of the phase-space distribution function.

Macroscopic variables, such as number density, average flow velocity, kinetic pressure, and so on, can be considered as average values of molecular properties. These macroscopic variables are related to the velocity moments of the phase-space distribution function. Next, we introduce a consistent hierarchy of velocity moments. This hierarchy is based on the "order" of a given velocity moment, that is, the sum of the powers of velocity components in the moment integral. For instance, the integral

$$M_6 = \iiint_\infty v^2 v_1^3 v_2 F(t, \mathbf{r}, \mathbf{v}) \, d^3 v \qquad (4.1)$$

is a sixth-order velocity moment of the distribution function, because the sum of the velocity powers of the integrand is equal to six.

Let us start with the lowest velocity moments of the phase-space distribution function. We know that the normalization of the distribution function gives the particle number density:

$$n(t, \mathbf{r}) = \iiint_\infty F(t, \mathbf{r}, \mathbf{v}) \, d^3 v. \qquad (4.2)$$

Obviously, the number density is the zeroth velocity moment of the phase-space distribution function, because the distribution function is not multiplied by any velocity components.

We know that the components of the average molecular velocity are defined the following way:

$$\mathbf{u}(t, \mathbf{r}) = \frac{1}{n(t, \mathbf{r})} \iiint_\infty \mathbf{v} F(t, \mathbf{r}, \mathbf{v}) \, d^3 v. \qquad (4.3)$$

In other words, this is the first velocity moment of the distribution function. With the help of the average molecular velocity, one can introduce the random velocity vector, $\mathbf{c}$:

$$\mathbf{c}(t, \mathbf{r}) = \mathbf{v} - \mathbf{u}(t, \mathbf{r}). \qquad (4.4)$$

Note that although t, $\mathbf{r}$, and $\mathbf{v}$ are independent variables, the new velocity variable, $\mathbf{c}(t, \mathbf{r})$, is not independent of t and $\mathbf{r}$. This fact has to be taken into account when deriving transport equations for the distribution of random velocities.

The zeroth moment of the distribution of random velocities is still the particle number density, because in our nonrelativistic approximation the number of particles in a given configuration space volume is independent of the velocity of the coordinate system. However, it follows from the definition of the random velocity that the first moment of this distribution must vanish:

$$\iiint_\infty \mathbf{c} F(t, \mathbf{r}, \mathbf{c}) \, d^3 c = 0. \qquad (4.5)$$

The pressure of a gas is usually defined as the force per unit area exerted by the gas molecules through collisions with the wall. This force is equal to the rate of transfer of molecular momentum to the wall. Next, we generalize this definition. Pressure is now defined as the net rate of transport of molecular momentum per unit area, that is,

the net flux of momentum across unit area due to the random particle motion. This
definition makes it obvious that the pressure in the molecular sense is the following
tensor quantity:

$$P_{ij}(t, \mathbf{r}) = m \iiint_{\infty} c_i c_j F(t, \mathbf{r}, \mathbf{c}) \, d^3 c. \tag{4.6}$$

With the help of the general pressure tensor, P_{ij}, one can define a kinetic scalar
pressure (and temperature), as well as a stress tensor. The definition of the scalar
pressure is related to the trace of the pressure tensor:

$$p = \frac{1}{3} P_{ii} = \frac{1}{3}(P_{xx} + P_{yy} + P_{zz}) = m \iiint_{\infty} c^2 F(t, \mathbf{r}, \mathbf{c}) \, d^3 c. \tag{4.7}$$

The temperature of the particles, T, is a measure of the mean kinetic energy of the
random particle motion. Because we know that the scalar pressure, p, can be expressed
as $p = nkT$, one obtains for the temperature:

$$T = \frac{p}{nk} = \frac{m}{k} \frac{1}{n(t, \mathbf{r})} \iiint_{\infty} c^2 F(t, \mathbf{r}, \mathbf{c}) \, d^3 c. \tag{4.8}$$

There is another frequently used pressure-related quantity in addition to the scalar
pressure, the stress tensor. Following the definition most frequently used in fluid dynamics, we define the stress tensor, as

$$\tau_{ij} = p\delta_{ij} - P_{ij}. \tag{4.9}$$

In the case of a Maxwellian velocity distribution (with number density n, and temperature T), the pressure tensor simply becomes $P_{ij} = nkT\delta_{ij}$. The stress tensor, therefore,
expresses the deviation of the pressure tensor from the equilibrium case (characterized
by a Maxwellian velocity distribution).

The third velocity moment of the phase-space distribution function describes the
flow of the random energy due to the thermal motion of the molecules. In other
words, this term describes heat conduction in the gas. Based on this definition, one
can introduce the heat flow vector in the gas:

$$\mathbf{h}(t, \mathbf{r}) = \iiint_{\infty} \left(\frac{1}{2} m c^2 \right) \mathbf{c} F(t, \mathbf{r}, \mathbf{c}) \, d^3 c. \tag{4.10}$$

With the help of the moments of the distribution function, we are now ready to
derive transport equations for macroscopic quantities.

4.1.2   The Euler and Navier–Stokes Equations

For the sake of simplicity, let us first examine a single species gas and take the first
few velocity moments of the distribution function. In this case the collision term is

approximated by the simple relaxation-time approximation given by Eq. (2.52):

$$
\frac{\partial F(t, \mathbf{r}, \mathbf{c})}{\partial t} + (\mathbf{c} \cdot \nabla) F(t, \mathbf{r}, \mathbf{c}) + (\mathbf{u} \cdot \nabla) F(t, \mathbf{r}, \mathbf{c})
$$

$$
- \left[\frac{\partial \mathbf{u}}{\partial t} + (\mathbf{u} \cdot \nabla)\mathbf{u} + (\mathbf{c} \cdot \nabla)\mathbf{u} - \mathbf{a} \right] \nabla_{\mathbf{c}} F(t, \mathbf{r}, \mathbf{c})
$$

$$
= -\nu_{\text{coll}} [F(t, \mathbf{r}, \mathbf{c}) - F_0(t, \mathbf{r}, \mathbf{c})], \tag{4.11}
$$

where ν_{coll} is the collision frequency and F_0 is given by

$$
F_0 = n \left(\frac{m}{2\pi kT} \right)^{3/2} \exp \left[-\frac{mc^2}{2kT} \right]. \tag{4.12}
$$

Here n and T can be obtained from the zeroth and second moments of the distribution function, $F(t, \mathbf{r}, \mathbf{c})$ (cf. Eqs. 4.2 and 4.8).

Let us calculate the zeroth velocity moment of Eq. (4.11). The zeroth moments of the individual terms are the following:

$$
\iiint_{\infty} \frac{\partial F(t, \mathbf{r}, \mathbf{c})}{\partial t} \, d^3c = \frac{\partial}{\partial t} \iiint_{\infty} F(t, \mathbf{r}, \mathbf{c}) \, d^3c = \frac{\partial n(t, \mathbf{r})}{\partial t}, \tag{4.13}
$$

$$
\iiint_{\infty} (\mathbf{c} \cdot \nabla) F(t, \mathbf{r}, \mathbf{c}) \, d^3c = \frac{\partial}{\partial x_i} \iiint_{\infty} c_i F(t, \mathbf{r}, \mathbf{c}) \, d^3c = 0, \tag{4.14}
$$

$$
\iiint_{\infty} (\mathbf{u} \cdot \nabla) F(t, \mathbf{r}, \mathbf{c}) \, d^3c = (\mathbf{u} \cdot \nabla) \iiint_{\infty} F(t, \mathbf{r}, \mathbf{c}) d^3c
$$

$$
= (\mathbf{u} \cdot \nabla) n(t, \mathbf{r}), \tag{4.15}
$$

$$
\iiint_{\infty} \left[\frac{\partial \mathbf{u}}{\partial t} + (\mathbf{u} \cdot \nabla)\mathbf{u} \right] \nabla_{\mathbf{c}} F(t, \mathbf{r}, \mathbf{c}) \, d^3c
$$

$$
= \left[\frac{\partial \mathbf{u}}{\partial t} + (\mathbf{u} \cdot \nabla)\mathbf{u} \right] \iiint_{\infty} \nabla_c F(t, \mathbf{r}, \mathbf{c}) \, d^3c = 0, \tag{4.16}
$$

because the distribution function vanishes for infinite velocities. Furthermore,

$$
\iiint_{\infty} (\mathbf{c} \cdot \nabla)(\mathbf{u} \cdot \nabla_c) F(t, \mathbf{r}, \mathbf{c}) \, d^3c
$$

$$
= \frac{\partial \mathbf{u}_j}{\partial x_i} \iiint_{\infty} c_i \frac{\partial F(t, \mathbf{r}, \mathbf{c})}{\partial c_j} \, d^3c
$$

$$
= \frac{\partial \mathbf{u}_j}{\partial x_i} \iiint_{\infty} \frac{\partial c_i F(t, \mathbf{r}, \mathbf{c})}{\partial c_j} \, d^3c - \frac{\partial \mathbf{u}_j}{\partial x_i} \iiint_{\infty} \delta_{ij} F(t, \mathbf{r}, \mathbf{c}) \, d^3c
$$

$$
= -(\nabla \cdot \mathbf{u}) \iiint_{\infty} F(t, \mathbf{r}, \mathbf{c}) \, d^3c = -(\nabla \cdot \mathbf{u}) n(t, \mathbf{r}) \tag{4.17}
$$

and

$$\iiint_{\infty} [\mathbf{a}(t, \mathbf{r}, \mathbf{c}) \cdot \nabla_c] F(t, \mathbf{r}, \mathbf{c}) \, d^3 c$$

$$= \iiint_{\infty} \frac{\partial a_i(t, \mathbf{r}, \mathbf{c}) F(t, \mathbf{r}, \mathbf{c})}{\partial c_i} \, d^3 c - \iiint_{\infty} \frac{\partial a_i(t, \mathbf{r}, \mathbf{c})}{\partial c_i} F(t, \mathbf{r}, \mathbf{c}) \, d^3 c = 0,$$

(4.18)

because for most accelerations $\partial a_i / \partial c_i = 0$. Also,

$$\iiint_{\infty} F(t, \mathbf{r}, \mathbf{v}) \, d^3 v = \iiint_{\infty} F_0(t, \mathbf{r}, \mathbf{v}) \, d^3 v = n(t, \mathbf{r}). \tag{4.19}$$

One can put all these results together to obtain the zeroth moment of the Boltzmann equation:

$$\frac{\partial n}{\partial t} + (\mathbf{u} \cdot \nabla) n + n (\nabla \cdot \mathbf{u}) = \frac{\partial n}{\partial t} + \nabla \cdot (n \mathbf{u}) = 0. \tag{4.20}$$

This is the well-known *continuity equation*.

The first moment of the Boltzmann equation can be obtained by multiplying Eq. (4.11) by $m\mathbf{c}$ and integrating over the random velocity. This operation leads to the following result:

$$mn \frac{\partial \mathbf{u}}{\partial t} + mn (\mathbf{u} \cdot \nabla) \mathbf{u} + (\nabla \cdot P) - m \iiint_{\infty} \mathbf{a}(t, \mathbf{r}, \mathbf{c}) F(t, \mathbf{r}, \mathbf{c}) \, d^3 c = 0. \tag{4.21}$$

Let us next substitute expression (4.9) for the pressure tensor and express $\nabla \cdot P$ in terms of the scalar pressure and the traceless stress tensor:

$$\nabla \cdot P = \nabla p - \nabla \cdot \tau. \tag{4.22}$$

One can also substitute the specific form of the acceleration we usually encounter in space physics (cf. Eq. 1.17):

$$m \iiint_{\infty} \mathbf{a}(t, \mathbf{r}, \mathbf{c}) F(t, \mathbf{r}, \mathbf{c}) \, d^3 c$$

$$= \iiint_{\infty} [m\mathbf{g} + q(\mathbf{E} + \mathbf{v} \times \mathbf{B})] F(t, \mathbf{r}, \mathbf{c}) \, d^3 c$$

$$= mn\mathbf{g} + qn(\mathbf{E} + \mathbf{u} \times \mathbf{B}). \tag{4.23}$$

Now the first moment of the Boltzmann equation becomes

$$mn \frac{\partial \mathbf{u}}{\partial t} + mn (\mathbf{u} \cdot \nabla) \mathbf{u} + \nabla p - \nabla \cdot \tau - mn\mathbf{g} - qn(\mathbf{E} + \mathbf{u} \times \mathbf{B}) = \mathbf{0}. \tag{4.24}$$

This is the *momentum equation* of the fluid. It can be shown by direct substitution that for gases in equilibrium the stress tensor vanishes. In a first approximation one can neglect the contribution of the stress tensor and obtain the following momentum equation for a gas in local thermodynamic equilibrium:

$$mn\frac{\partial \mathbf{u}}{\partial t} + mn(\mathbf{u}\cdot\nabla)\mathbf{u} + \nabla p - mn\mathbf{g} - qn(\mathbf{E}+\mathbf{u}\times\mathbf{B}) = \mathbf{0}. \tag{4.25}$$

In a second approximation the components of the stress tensor can be expressed in terms of the velocity "shear" tensor (*Gombosi 1994*):

$$\tau_{ij} = \eta\left(\frac{\partial u_i}{\partial x_j} + \frac{\partial u_j}{\partial x_i} - \frac{2}{3}\delta_{ij}(\nabla\cdot\mathbf{u})\right), \tag{4.26}$$

where η is the *coefficient of viscosity*. In this case one obtains a more realistic approximation for the momentum equation:

$$mn\frac{\partial \mathbf{u}}{\partial t} + mn(\mathbf{u}\cdot\nabla)\mathbf{u} + \nabla p - mn\mathbf{g} - qn(\mathbf{E}+\mathbf{u}\times\mathbf{B})$$

$$= \nabla\cdot\left\{\eta\left[(\nabla\mathbf{u}) + (\nabla\mathbf{u})^T - \frac{2}{3}I(\nabla\cdot\mathbf{u})\right]\right\}, \tag{4.27}$$

where I is the unit tensor. The term on the right-hand side describes the viscous force acting on the gas.

Finally, one can obtain the second moment of the single-fluid Boltzmann equation by multiplying Eq. (4.11) by $mc^2/2$ and integrating over the random velocity. This operation leads to the following transport equation:

$$\frac{3}{2}\frac{\partial p}{\partial t} + \frac{3}{2}(\mathbf{u}\cdot\nabla)p + \frac{5}{2}p(\nabla\cdot\mathbf{u}) + (\nabla\cdot\mathbf{h}) = 0. \tag{4.28}$$

This is the *energy equation*. The energy equation contains the divergence of the third velocity moment of the distribution function, $\nabla\cdot\mathbf{h}$. It can be shown by direct substitution that for equilibrium gases $\mathbf{h} = \mathbf{0}$.

In a more realistic approximation the series of transport equations is closed by applying the widely used Fourier approximation for the heat flux:

$$\mathbf{h} = -\kappa\nabla T, \tag{4.29}$$

where κ is the heat conductivity.

Equation (4.28) implicitly assumes that the particles have no internal degrees of freedom. If we allow for internal degrees of freedom (such as rotational and vibrational excitations), we get the following result (cf. *Gombosi 1994*):

$$\frac{1}{\gamma-1}\frac{\partial p}{\partial t} + \frac{1}{\gamma-1}(\mathbf{u}\cdot\nabla)p + \frac{\gamma}{\gamma-1}p(\nabla\cdot\mathbf{u}) + (\nabla\cdot\mathbf{h}) = 0, \tag{4.30}$$

where γ is the specific heat ratio for the gas. If the heat flow vector can be neglected, Eqs. (4.20) and (4.30) can be combined to obtain

$$\frac{\partial}{\partial t}\left(\frac{p}{(mn)^\gamma}\right) + (\mathbf{u}\cdot\nabla)\left(\frac{p}{(mn)^\gamma}\right) = 0. \tag{4.31}$$

This means that in the absence of external heat sources the *entropy function*

$$\frac{p}{(mn)^{\gamma}} \tag{4.32}$$

remains constant as the fluid moves. This is the *adiabatic energy relation* (or *adiabatic equation of state*), which is valid for gases in local equilibrium.

In summary, the lowest order (local thermodynamic equilibrium or LTE) approximation of the transport equations can be obtained by neglecting the viscous force and the heat flow term. This approximation leads to the set of *Euler equations*:

$$\frac{\partial n}{\partial t} + \nabla \cdot (n\mathbf{u}) = 0,$$

$$mn\frac{\partial \mathbf{u}}{\partial t} + mn(\mathbf{u} \cdot \nabla)\mathbf{u} + \nabla p - mn\mathbf{g} - qn(\mathbf{E} + \mathbf{u} \times \mathbf{B}) = \mathbf{0}, \tag{4.33}$$

$$\frac{1}{\gamma - 1}\frac{\partial p}{\partial t} + \frac{1}{\gamma - 1}(\mathbf{u} \cdot \nabla)p + \frac{\gamma}{\gamma - 1}p(\nabla \cdot \mathbf{u}) = 0.$$

In the second approximation we keep the viscous force and the heat flow terms to obtain the more realistic set of *Navier–Stokes* equations:

$$\frac{\partial n}{\partial t} + \nabla \cdot (n\mathbf{u}) = 0,$$

$$mn\frac{\partial \mathbf{u}}{\partial t} + mn(\mathbf{u} \cdot \nabla)\mathbf{u} + \nabla p - mn\mathbf{g} - qn(\mathbf{E} + \mathbf{u} \times \mathbf{B})$$

$$= -\nabla \cdot \left\{ \eta \left[(\nabla\mathbf{u}) + (\nabla\mathbf{u})^T - \frac{2}{3}I(\nabla \cdot \mathbf{u}) \right] \right\}, \tag{4.34}$$

$$\frac{1}{\gamma - 1}\frac{\partial p}{\partial t} + \frac{1}{\gamma - 1}(\mathbf{u} \cdot \nabla)p + \frac{\gamma}{\gamma - 1}p(\nabla \cdot \mathbf{u}) = \nabla \cdot (\kappa \nabla T).$$

This result shows that for gases near equilibrium (when the stress tensor and heat flow vector are negligible) the Euler equations represent a reasonable first approximation of the transport equations describing the evolution of the fundamental macroscopic quantities (corresponding to low order velocity moments). A better (but more complicated) approximation is the Navier–Stokes equation system.

In most space physics and upper atmospheric science applications the viscous terms are neglected, while the heat conduction terms are kept in the Navier–Stokes equations. In this book we will generally use this approximation and will neglect the viscous terms.

4.2   Transport Equations for Multispecies Gases

In the case of multispecies gases the left-hand side of the Euler equations remain practically unchanged, but the right-hand side describes the interactions between the various species. Transport equations for multispecies gases are obtained by taking the zeroth, first, and second velocity moments of the multispecies Boltzmann equation (given by 2.75). This can be done by multiplying Eq. (2.75) by m_s, $m_s\mathbf{c}_s$, and $m_s c_s^2/2$

and integrating over the random velocity space. This process leads to the following continuity, momentum, and energy equations for species s:

$$\frac{\partial m_s n_s}{\partial t} + \nabla \cdot (m_s n_s \mathbf{u}_s) = m_s \frac{\delta n_s}{\delta t},$$

$$m_s n_s \frac{\partial \mathbf{u}_s}{\partial t} + m_s n_s (\mathbf{u}_s \cdot \nabla) \mathbf{u}_s + \nabla p_s$$

$$- m_s n_s \mathbf{g} - q_s n_s (\mathbf{E} + \mathbf{u}_s \times \mathbf{B}) = m_s n_s \frac{\delta \mathbf{u}_s}{\delta t}, \tag{4.35}$$

$$\frac{1}{\gamma - 1} \frac{\partial p_s}{\partial t} + \frac{1}{\gamma - 1} (\mathbf{u}_s \cdot \nabla) p_s$$

$$+ \frac{\gamma}{\gamma - 1} p_s (\nabla \cdot \mathbf{u}_s) - \nabla \cdot (\kappa_s \nabla T_s) = \frac{1}{\gamma - 1} \frac{\delta p_s}{\delta t}.$$

The interaction between the various species is described by the moments of the collision terms. These moments can be evaluated for the collision term describing the effects of elastic binary collisions. This collision term was discussed in the previous chapter.

In this case we take the velocity moments of the collision term given by Eq. (2.76):

$$\left(\frac{\delta F_s(t, \mathbf{r}, \mathbf{c}_s)}{\delta t} \right)_{\text{elastic}} = - \sum_t \nu_{st} [F_s(t, \mathbf{r}, \mathbf{c}_s) - F_{0s(st)}(t, \mathbf{r}, \mathbf{c}_s)], \tag{4.36}$$

where ν_{st} is the collision frequency of elastic collisions between species s and t. The integrals can be evaluated [for details see *Gombosi* (*1994*)] to yield

$$m_s \frac{\delta n_s}{\delta t} = 0,$$

$$m_s n_s \frac{\delta \mathbf{u}_s}{\delta t} = \sum_t m_s n_s \bar{\nu}_{st} (\mathbf{u}_t - \mathbf{u}_s), \tag{4.37}$$

$$\frac{\delta p_s}{\delta t} = \sum_t \frac{m_s n_s \bar{\nu}_{st}}{m_s + m_t} \left[2k(T_t - T_s) + \frac{2}{3} m_t (\mathbf{u}_t - \mathbf{u}_s)^2 \right],$$

where the momentum transfer collision frequency, $\bar{\nu}_{st}$, is defined as

$$\bar{\nu}_{st} = \frac{m_t}{m_s + m_t} \nu_{st}. \tag{4.38}$$

The physical interpretation of these terms is quite straightforward. First of all, elastic collisions do not create or destroy particles; therefore, there are no particle sources or sinks. The collision term for the momentum equation describes simple mechanical friction between the gas components exhibiting bulk motion relative to each other. Finally, the source term for the energy equation shows that friction transforms the energy of relative bulk motion to heat (energy of random motion). Moreover, the temperature difference between the two species heats the colder gas and cools the warmer one (if $T_t > T_s$, the temperature difference term is positive, and the species s is heated; otherwise it is cooled).

4.3   MHD Equations for Conducting Fluids

A mixture of neutral and ionized gases can also be considered as a conducting fluid, without specifying its various individual species. The transport equations, derived earlier in this chapter, describe the macroscopic behavior of each individual species (electrons, ions, neutral particles).

Next, we derive a new set of transport equations that describe the transport of the ionized gas (plasma) as a whole. This procedure will yield governing equations describing the evolution of the global macroscopic plasma parameters, such as mass density, charge and current densities, etc.

Let us start with a few definitions. The total mass density, ρ_m, is the mass per unit volume:

$$\rho_m = \sum_s m_s n_s. \tag{4.39}$$

The electric charge density, ρ, is the net electric charge per unit volume:

$$\rho = \sum_s q_s n_s, \tag{4.40}$$

where q_s is the charge of species s (in the case of neutral particles $q_s = 0$). The mean flow velocity of the fluid, $\mathbf{u}$, is defined such that the momentum density is the same as if each particle were moving with the same flow velocity:

$$\rho_m \mathbf{u} = \sum_s m_s n_s \mathbf{u}_s. \tag{4.41}$$

The electric current density, $\mathbf{j}$, is the following:

$$\mathbf{j} = \sum_s q_s n_s \mathbf{u}_s. \tag{4.42}$$

These quantities were easy to define for the multispecies fluid as a whole. It is somewhat more complicated to define the pressure tensor and the heat flow. We know that in a gas the pressure tensor is defined as the second random velocity moment (with respect to the bulk velocity of the entire fluid):

$$\begin{aligned}
P_{ij} &= \sum_s m_s \iiint_\infty (v_{si} - u_i)(v_{sj} - u_j) F_s(t, \mathbf{r}, \mathbf{c}_s)\, d^3 c_s \\
&= \sum_s m_s \iiint_\infty (c_{si} + u_{si} - u_i)(c_{sj} + u_{sj} - u_j) F_s(t, \mathbf{r}, \mathbf{c}_s)\, d^3 c_s \\
&= \sum_s P_{sij} + \sum_s m_s n_s u_{si} u_{sj} - \rho_m u_i u_j.
\end{aligned} \tag{4.43}$$

This equation means that in the multispecies gas both the thermal motion of the molecules and the relative bulk motion of the gases contribute to the pressure. The scalar pressure of the mixture is

$$p = \frac{1}{3} P_{ii} = \sum_s p_s + \frac{1}{3}\left(\sum_s m_s n_s u_s^2 - \rho_m u^2 \right). \tag{4.44}$$

In a similar manner one can also obtain the heat flow vector for a gas mixture:

$$\mathbf{h} = \frac{1}{2} \sum_s m_s \iiint_\infty (\mathbf{v}_s - \mathbf{u})^2 (\mathbf{v}_s - \mathbf{u}) F_s(t, \mathbf{r}, \mathbf{c}_s) \, d^3 c_s$$

$$= \sum_s \mathbf{h}_s + \sum_s (\mathbf{u}_s - \mathbf{u}) \left[\frac{5}{2} p_s + \frac{1}{2} m_s n_s (\mathbf{u}_s - \mathbf{u})^2 \right]. \tag{4.45}$$

4.3.1 Single-Fluid Equations

With the introduction of the global quantities, we can now derive single-fluid transport equations for conducting fluids. For the sake of simplicity, we neglect all reactive collisions (including ionization, charge exchange, recombination, etc.) so that only elastic collisions are taken into account. Also, we neglect the viscous terms. Furthermore, it is assumed that the constituent species have no internal degrees of freedom (atoms and ions are monatomic). In this approximation the continuity, momentum, and energy equations for the individual species are the following:

$$\frac{\partial m_s n_s}{\partial t} + \nabla \cdot (m_s n_s \mathbf{u}_s) = 0, \tag{4.46}$$

$$m_s n_s \frac{\partial \mathbf{u}_s}{\partial t} + m_s n_s (\mathbf{u}_s \cdot \nabla) \mathbf{u}_s + \nabla p_s - m_s n_s \mathbf{g} - q_s n_s (\mathbf{E} + \mathbf{u}_s \times \mathbf{B})$$

$$= \sum_t m_s n_s \bar{\nu}_{st} (\mathbf{u}_t - \mathbf{u}_s), \tag{4.47}$$

$$\frac{3}{2} \frac{\partial p_s}{\partial t} + \frac{3}{2} (\mathbf{u}_s \cdot \nabla) p_s + \frac{5}{2} p_s (\nabla \cdot \mathbf{u}_s) + \nabla \cdot \mathbf{h}_s$$

$$= \sum_t \frac{m_s n_s \bar{\nu}_{st}}{m_s + m_t} \left[3k(T_t - T_s) + m_t (\mathbf{u}_t - \mathbf{u}_s)^2 \right]. \tag{4.48}$$

First, we derive a continuity equation for the entire gas mixture. This can be obtained by summing the continuity equations for all species (neutral and charged):

$$\frac{\partial \sum_s m_s n_s}{\partial t} + \nabla \cdot \left(\sum_s m_s n_s \mathbf{u}_s \right) = 0. \tag{4.49}$$

Using Eqs. (4.40) and (4.41) this becomes

$$\frac{\partial \rho_m}{\partial t} + \nabla \cdot (\rho_m \mathbf{u}) = 0. \tag{4.50}$$

This is the continuity equation for the gas as a single conducting fluid.

The single-fluid momentum equation can be obtained by summing the momentum equations for all species:

$$\sum_s m_s n_s \frac{\partial \mathbf{u}_s}{\partial t} + \sum_s m_s n_s (\mathbf{u}_s \cdot \nabla) \mathbf{u}_s + \sum_s \nabla p_s - \sum_s m_s n_s \mathbf{g}$$

$$- \sum_s q_s n_s (\mathbf{E} + \mathbf{u}_s \times \mathbf{B}) = \sum_s \sum_t m_s n_s \bar{\nu}_{st} (\mathbf{u}_t - \mathbf{u}_s). \tag{4.51}$$

The right-hand side of the equation is zero, because for elastic collisions $m_s n_s \bar{v}_{st} = m_t n_t \bar{v}_{ts}$ (cf. *Gombosi 1994*) and the summations for both s and t run through all species. Also, one can use our definitions of single-fluid quantities to obtain

$$\sum_s m_s n_s \frac{\partial \mathbf{u}_s}{\partial t} + \sum_s m_s n_s (\mathbf{u}_s \cdot \nabla) \mathbf{u}_s$$
$$+ \nabla \sum_s p_s - \rho_m \mathbf{g} - \rho \mathbf{E} - \mathbf{j} \times \mathbf{B} = \mathbf{0}. \tag{4.52}$$

The equation can be further manipulated by examining the divergence of the single-fluid pressure tensor (Eq. 4.43):

$$\frac{\partial P_{ij}}{\partial x_i} = \sum_s \frac{\partial P_{sij}}{\partial x_i} + \frac{\partial \sum_s m_s n_s u_{si} u_{sj}}{\partial x_i} - \frac{\partial \rho_m u_i u_j}{\partial x_i}. \tag{4.53}$$

In our approximation $P_{sij} = p_s \delta_{ij}$ (where δ_{ij} is the Kronecker delta). Therefore, one can write

$$(\nabla \cdot P) = \sum_s \nabla p_s + \sum_s (\nabla \cdot m_s n_s \mathbf{u}_s) \mathbf{u}_s$$
$$+ \sum_s (m_s n_s \mathbf{u}_s \cdot \nabla) \mathbf{u}_s - (\nabla \cdot \rho_m \mathbf{u}) \mathbf{u} - (\rho_m \mathbf{u} \cdot \nabla) \mathbf{u}. \tag{4.54}$$

This can be further manipulated by using the multispecies and single-fluid continuity equations (4.46 and 4.50) to give

$$\sum_s \nabla p_s + \sum_s (m_s n_s \mathbf{u}_s \cdot \nabla) \mathbf{u}_s$$
$$= (\nabla \cdot P) + \sum_s \mathbf{u}_s \frac{\partial m_s n_s}{\partial t} - \mathbf{u} \frac{\partial \rho_m}{\partial t} + (\rho_m \mathbf{u} \cdot \nabla) \mathbf{u}. \tag{4.55}$$

Substituting Eq. (4.55) into Eq. (4.52) yields the single-fluid momentum equation:

$$\rho_m \frac{\partial \mathbf{u}}{\partial t} + \rho_m (\mathbf{u} \cdot \nabla) \mathbf{u} + (\nabla \cdot P) - \rho_m \mathbf{g} - \rho \mathbf{E} - \mathbf{j} \times \mathbf{B} = \mathbf{0}. \tag{4.56}$$

The single-fluid energy equation can be derived in a similar manner. This calculation is straightforward but a little lengthy. (We encourage the reader to carry out the derivation.) Eventually one obtains the following result:

$$\frac{3}{2} \frac{\partial p}{\partial t} + \frac{3}{2} \nabla \cdot (p\mathbf{u}) + (P \cdot \nabla) \cdot \mathbf{u} + (\nabla \cdot \mathbf{h}) = (\mathbf{j} - \rho \mathbf{u}) \cdot (\mathbf{E} + \mathbf{u} \times \mathbf{B}). \tag{4.57}$$

The terms on the left-hand side are the well-known terms of a single-fluid energy equation (except the $(P \cdot \nabla) \cdot \mathbf{u}$ term which is different from the standard $p(\nabla \cdot \mathbf{u})$ expression owing to the relative motion of the various species). The interesting new terms are on the right-hand side: These terms reflect the fact that electric charges and currents can be present in conducting fluids. It should be noted that in the frame of reference moving with the fluid the current density and electric field become $\mathbf{j}' = \mathbf{j} - \rho \mathbf{u}$

and $\mathbf{E}' = \mathbf{E} + \mathbf{u} \times \mathbf{B}$, respectively. This means that in the plasma frame the right-hand side of the MHD energy equation becomes

$$(\mathbf{j} - \rho \mathbf{u}) \cdot (\mathbf{E} + \mathbf{u} \times \mathbf{B}) = \mathbf{j}' \cdot \mathbf{E}'. \tag{4.58}$$

This source of thermal energy is generally referred to as Joule heating.

4.3.2 Generalized Ohm's Law

Conservation laws for electric charge and current densities can be obtained with the help of the multispecies transport equations.

The derivation of the charge conservation equation starts by multiplying Eq. (4.46) by q_s/m_s and summing over all species:

$$\frac{\partial \sum\limits_s q_s n_s}{\partial t} + \nabla \cdot \left(\sum_s q_s n_s \mathbf{u}_s \right) = 0. \tag{4.59}$$

Using the definition of the total charge and current densities (Eqs. 4.40 and 4.42), this equation yields a conservation equation for the electric charge density:

$$\frac{\partial \rho}{\partial t} + \nabla \cdot \mathbf{j} = 0. \tag{4.60}$$

This equation tells us that the time rate of change of the total charge is the source of currents. In most practical applications $\partial \rho / \partial t$ is negligible, and therefore the current density is divergenceless, $\nabla \cdot \mathbf{j} = 0$. In other words, all current systems must be closed. This fact, of course, is well-known in classical physics.

Next, we will derive a conservation law for the current density, $\mathbf{j}$. We multiply the multispecies momentum equations (4.47) by q_s/m_s and add them together to get

$$\sum_s q_s n_s \frac{\partial \mathbf{u}_s}{\partial t} + \sum_s q_s n_s (\mathbf{u}_s \cdot \nabla) \mathbf{u}_s + \sum_s \frac{q_s}{m_s} \nabla p_s - \sum_s q_s n_s \mathbf{g}$$
$$- \sum_s \frac{q_s^2}{m_s} n_s (\mathbf{E} + \mathbf{u}_s \times \mathbf{B}) = \sum_s q_s n_s \sum_t \overline{\nu}_{st} (\mathbf{u}_t - \mathbf{u}_s). \tag{4.61}$$

Next, we multiply the continuity equations (4.46) by $q_s \mathbf{u}_s / m_s$ and add them together. This yields

$$\sum_s \mathbf{u}_s \frac{\partial q_s n_s}{\partial t} + \sum_s \mathbf{u}_s \nabla \cdot (q_s n_s \mathbf{u}_s) = 0. \tag{4.62}$$

In the next step we add Eqs. (4.61) and (4.62) to obtain

$$\frac{\partial \left(\sum\limits_s q_s n_s \mathbf{u}_s \right)}{\partial t} + \nabla \cdot \left(\sum_s q_s n_s \mathbf{u}_s \mathbf{u}_s \right) + \sum_s \frac{q_s}{m_s} \nabla p_s - \sum_s q_s n_s \mathbf{g}$$
$$- \sum_s \frac{q_s^2}{m_s} n_s (\mathbf{E} + \mathbf{u}_s \times \mathbf{B}) = \sum_s q_s n_s \sum_t \overline{\nu}_{st} (\mathbf{u}_t - \mathbf{u}_s) \tag{4.63}$$

or

$$\frac{\partial \mathbf{j}}{\partial t} + \nabla \cdot \left(\sum_s q_s n_s \mathbf{u}_s \mathbf{u}_s \right) + \sum_s \frac{q_s}{m_s} \nabla p_s - \rho \mathbf{g}$$

$$- \sum_s \frac{q_s^2}{m_s} n_s (\mathbf{E} + \mathbf{u}_s \times \mathbf{B}) = \sum_s q_s n_s \sum_t \overline{\nu}_{st} (\mathbf{u}_t - \mathbf{u}_s). \tag{4.64}$$

Equation (4.64) is the desired conservation law for the current density. However, it is not in a particularly useful form. To simplify this equation we need to make a few more additional assumptions. These assumptions do not really restrict the validity of the result, but they help us to reveal the physical meaning of the equation.

Let us assume that the plasma is composed of a single ion (single ionized) and electrons. In this case we get

$$\rho = e(n_{\text{ion}} - n_{\text{e}}),$$

$$\mathbf{j} = e(n_{\text{ion}} \mathbf{u}_{\text{ion}} - n_{\text{e}} \mathbf{u}_{\text{e}}),$$

$$\rho_m = m_{\text{ion}} n_{\text{ion}} + m_{\text{e}} n_{\text{e}} \approx m_{\text{ion}} n_{\text{ion}}, \tag{4.65}$$

$$\mathbf{u} = \frac{1}{\rho_m} (m_{\text{ion}} n_{\text{ion}} \mathbf{u}_{\text{ion}} + m_{\text{e}} n_{\text{e}} \mathbf{u}_{\text{e}}).$$

One can now express the electron and ion velocities as

$$\mathbf{u}_{\text{ion}} = \frac{m_{\text{e}}}{m_{\text{e}} + m_{\text{ion}}} \frac{1}{n_{\text{ion}}} \left(\frac{\rho_m}{m_{\text{e}}} \mathbf{u} + \frac{\mathbf{j}}{e} \right) \approx \mathbf{u} + \frac{m_{\text{e}}}{m_{\text{ion}}} \frac{\mathbf{j}}{e n_{\text{ion}}},$$

$$\mathbf{u}_{\text{e}} = \frac{m_{\text{ion}}}{m_{\text{e}} + m_{\text{ion}}} \frac{1}{n_{\text{e}}} \left(\frac{\rho_m}{m_{\text{ion}}} \mathbf{u} - \frac{\mathbf{j}}{e} \right) \approx \frac{n_{\text{ion}}}{n_{\text{e}}} \mathbf{u} - \frac{\mathbf{j}}{e n_{\text{ion}}}. \tag{4.66}$$

Next, we substitute these expressions into Eq. (4.65). Additional simplification can be achieved by assuming quasi-neutrality, $n_{\text{ion}} = n_{\text{e}}$. This naturally also means that $\rho = 0$. Now we get

$$\frac{\partial \mathbf{j}}{\partial t} + \nabla \cdot \left(\mathbf{u} \mathbf{j} + \mathbf{j} \mathbf{u} - \frac{1}{e n_{\text{e}}} \mathbf{j} \mathbf{j} \right) + \frac{e}{m_{\text{ion}}} \nabla p_{\text{ion}} - \frac{e}{m_{\text{e}}} \nabla p_{\text{e}}$$

$$- \frac{e^2 n_{\text{e}}}{m_{\text{e}}} (\mathbf{E} + \mathbf{u} \times \mathbf{B}) + \frac{e}{m_{\text{e}}} \mathbf{j} \times \mathbf{B} = -(\overline{\nu}_{\text{ion,e}} + \overline{\nu}_{\text{e,ion}}) \mathbf{j}. \tag{4.67}$$

When deriving this equation, we made use of the fact that $m_{\text{e}} \ll m_{\text{ion}}$. Equation (4.67) can be further simplified. First of all, one must recognize that the ion and electron pressure gradients are usually of the same order of magnitude, but because of the mass difference the ion pressure gradient term can be neglected. Secondly, the electron–ion momentum transfer collision frequency is much larger than the ion–electron momentum transfer collision frequency, $\overline{\nu}_{\text{e,ion}} \gg \overline{\nu}_{\text{ion,e}}$. The reason is that the light electron very efficiently transfers its momentum to a much heavier ion during collisions, but a heavy ion cannot efficiently transfer momentum to an electron. This can also be seen by examining Eq. (4.38) which shows the presence of a mass factor of $m_t/(m_s + m_t)$ in the momentum transfer collision frequency. For electron–ion collisions this factor is nearly 1, whereas for ion–electron collisions it is smaller than 10^{-3}. With these simplifications Eq. (4.67) becomes

$$\frac{\partial \mathbf{j}}{\partial t} + \nabla \cdot \left(\mathbf{uj} + \mathbf{ju} - \frac{1}{en_e} \mathbf{jj} \right)$$

$$= \frac{e^2 n_e}{m_e} \left[\frac{1}{en_e} \nabla p_e + (\mathbf{E} + \mathbf{u} \times \mathbf{B}) - \frac{1}{en_e} \mathbf{j} \times \mathbf{B} - \frac{1}{\overline{\sigma}_0} \mathbf{j} \right], \qquad (4.68)$$

where we introduced the electrical conductivity of the ionized gas, $\overline{\sigma}_0$, which is defined as

$$\overline{\sigma}_0 = \frac{e^2 n_e}{\overline{\nu}_{e,ion} m_e}. \qquad (4.69)$$

Equation (4.68) is the generalized Ohm's law. At first glance it is very different than the Ohm's law we know from classical physics. However, if we consider a steady-state situation with no magnetic field, in which $\mathbf{u}$ and $\mathbf{j}$ are first-order perturbations (i.e., all quadratic terms containing these quantities can be neglected), and take the cold plasma limit (when the pressure gradient terms can be neglected), Eq. (4.68) yields the well-known form of Ohm's law:

$$\mathbf{j} = \overline{\sigma}_0 \mathbf{E}. \qquad (4.70)$$

The generalized Ohm's law is a fundamental element of magnetohydrodynamic (MHD) theory. In its most widely used form $\mathbf{u}$ and $\mathbf{j}$ are assumed to be first-order perturbations, and the time derivative of the current density is neglected (this is justified in most cases). This approximation yields the following expression for the current density:

$$\mathbf{j} = \overline{\sigma}_0 (\mathbf{E} + \mathbf{u} \times \mathbf{B}) - \frac{\overline{\sigma}_0}{en_e} \mathbf{j} \times \mathbf{B} + \frac{\overline{\sigma}_0}{en_e} \nabla p_e. \qquad (4.71)$$

The first term tells us that the current is proportional to the electric field as measured in the frame of reference moving with the plasma, the second term describes the so-called Hall effect (this term is frequently neglected), while the third term describes the contribution of the electron pressure gradient to the current density.

In Eq. (4.71) the electric current density appears in two terms. This can be also written in the form

$$\mathbf{A} \cdot \mathbf{j} = \overline{\sigma}_0 \left[(\mathbf{E} + \mathbf{u} \times \mathbf{B}) + \frac{1}{en_e} \nabla p_e \right], \qquad (4.72)$$

where the elements of tensor $\mathbf{A}$ can be expressed as

$$A_{ij} = \delta_{ij} + \frac{\overline{\sigma}_0}{en_e} \epsilon_{ijk} B_k, \qquad (4.73)$$

where ϵ_{ijk} is the antisymmetric permutation tensor ($\epsilon_{ijk} = 0$ if any two indices are the same; $\epsilon_{ijk} = 1$ if i, j, and k form an even permutation of 1, 2, and 3; and $\epsilon_{ijk} = -1$ if i, j, and k form an odd permutation of 1, 2, and 3). The tensor A can be inverted to obtain

$$(A^{-1})_{ij} = \frac{\overline{v}^2_{e,ion}}{\overline{v}^2_{e,ion} + \Omega_e^2} \left[\delta_{ij} + \left(\frac{\Omega_e}{\overline{v}_{e,ion}} \right)^2 b_i b_j - \frac{\Omega_e}{\overline{v}_{e,ion}} \epsilon_{ijk} b_k \right], \qquad (4.74)$$

where Ω_c is the electron gyrofrequency and $\mathbf{b} = \mathbf{B}/B$ is the unit vector along the magnetic field line. Now the current density can be expressed in the following form:

$$\mathbf{j} = \overline{\sigma} \cdot \left[(\mathbf{E} + \mathbf{u} \times \mathbf{B}) + \frac{1}{en_e} \nabla p_e \right], \qquad (4.75)$$

where we introduced the *conductivity tensor*, $\overline{\sigma}$, defined as

$$\overline{\sigma} = \overline{\sigma}_0 A^{-1} = \overline{\sigma}_P \delta_{ij} + (\overline{\sigma}_0 - \overline{\sigma}_P) b_i b_j - \overline{\sigma}_H \epsilon_{ijk} b_k, \qquad (4.76)$$

where the Hall and Pedersen conductivities ($\overline{\sigma}_H$ and $\overline{\sigma}_P$, respectively) are defined by

$$\overline{\sigma}_P = \frac{\overline{v}^2_{e,ion}}{\overline{v}^2_{e,ion} + \Omega_e^2} \overline{\sigma}_0,$$

$$\overline{\sigma}_H = \frac{\Omega_e \overline{v}_{e,ion}}{\overline{v}^2_{e,ion} + \Omega_e^2} \overline{\sigma}_0. \qquad (4.77)$$

In the case of a vertical magnetic field ($\mathbf{B} = B\mathbf{e}_z$), the conductivity tensor can be written as

$$\overline{\sigma} = \begin{pmatrix} \overline{\sigma}_P & -\overline{\sigma}_H & 0 \\ \overline{\sigma}_H & \overline{\sigma}_P & 0 \\ 0 & 0 & \overline{\sigma}_0 \end{pmatrix}. \qquad (4.78)$$

In a multicomponent ionospheric plasma the collision frequency, $\overline{v}^2_{e,ion}$, is replaced by the total electron collision frequency. This will be discussed in detail in connection with ionospheric conductivities.

The form of Ohm's law that is typically used in magnetohydrodynamics can be obtained from Eq. (4.71) by neglecting the Hall and the electron pressure gradient terms:

$$\mathbf{j} = \overline{\sigma}_0 (\mathbf{E} + \mathbf{u} \times \mathbf{B}). \qquad (4.79)$$

4.4   Ideal MHD

The macroscopic transport equations for the whole gas as a single conducting fluid together with Maxwell's equations constitute the equations of magnetohydrodynamics. These equations together form a complete set of partial differential equations which fully determine the fluid and field quantities.

In practice, usually a simplified set of MHD equations is used. These equations are obtained with the following simplifying assumptions:

1. The gas components are not far from local thermodynamic equilibrium. Therefore, the scalar pressure can be used (instead of the full pressure tensor).
2. Heat flow is neglected in the fluid.
3. Charge quasi-neutrality is assumed, $\rho \approx 0$.
4. The high-frequency component of the electric field is neglected (this means that the time derivative in Ampère's law is neglected, $\partial \mathbf{E}/\partial t = \mathbf{0}$).

In this approximation the transport equations become the following:

$$
\begin{aligned}
&\frac{\partial \rho_m}{\partial t} + \nabla \cdot (\rho_m \mathbf{u}) = 0, \\
&\rho_m \frac{\partial \mathbf{u}}{\partial t} + \rho_m (\mathbf{u} \cdot \nabla)\mathbf{u} + \nabla p - \rho_m \mathbf{g} - \mathbf{j} \times \mathbf{B} = 0, \\
&\frac{3}{2}\frac{\partial p}{\partial t} + \frac{3}{2}(\mathbf{u} \cdot \nabla)p + \frac{5}{2}p(\nabla \cdot \mathbf{u}) = \mathbf{j} \cdot (\mathbf{E} + \mathbf{u} \times \mathbf{B}).
\end{aligned}
\tag{4.80}
$$

These equations are supplemented by the generalized Ohm's law (neglecting the Hall term) and the simplified set of Maxwell's equations:

$$
\mathbf{j} = \overline{\sigma}_0 (\mathbf{E} + \mathbf{u} \times \mathbf{B}),
\tag{4.81}
$$

$$
\begin{aligned}
&\nabla \cdot \mathbf{E} = 0, \\
&\nabla \cdot \mathbf{B} = 0, \\
&\nabla \times \mathbf{E} = -\frac{\partial \mathbf{B}}{\partial t}, \\
&\nabla \times \mathbf{B} = \mu_0 \mathbf{j}.
\end{aligned}
\tag{4.82}
$$

These equations can be further simplified and the electric field and the electric current density eliminated from the equations. In order to do this, let us take the curl of Ohm's law (Eq. 4.81):

$$
\nabla \times \mathbf{j} = \overline{\sigma}_0 [\nabla \times \mathbf{E} + \nabla \times (\mathbf{u} \times \mathbf{B})].
\tag{4.83}
$$

One can use the simplified form of Maxwell's equations to express $\mathbf{j}$ and $\mathbf{E}$ and substitute them into Eq. (4.83) to get

$$
\nabla \times (\nabla \times \mathbf{B}) = \overline{\sigma}_0 \mu_0 \left[-\frac{\partial \mathbf{B}}{\partial t} + \nabla \times (\mathbf{u} \times \mathbf{B}) \right].
\tag{4.84}
$$

However, since

$$
\nabla \times (\nabla \times \mathbf{B}) = -\nabla^2 \mathbf{B}
\tag{4.85}
$$

Eq. (4.84) can be rewritten in the following form:

$$
\frac{\partial \mathbf{B}}{\partial t} = \nabla \times (\mathbf{u} \times \mathbf{B}) + \eta_m \nabla^2 \mathbf{B} = -\nabla \cdot (\mathbf{u}\mathbf{B} - \mathbf{B}\mathbf{u}) + \eta_m \nabla^2 \mathbf{B},
\tag{4.86}
$$

where we introduced the magnetic viscosity

$$\eta_m = \frac{1}{\overline{\sigma}_0 \mu_0}. \tag{4.87}$$

Equation (4.86) is the so-called *induction equation*.

On the right-hand side of Eq. (4.86) the first term describes convection of the magnetic field by the moving fluid, while the second term describes the diffusion of the magnetic field in the moving fluid.

The electric field and current density vectors can also be eliminated from the momentum and energy equations. This process leads to the following set of equations for ρ_m, $\mathbf{u}$, $\mathbf{B}$, and p:

$$\frac{\partial \rho_m}{\partial t} + \nabla \cdot (\rho_m \mathbf{u}) = 0,$$

$$\rho_m \frac{\partial \mathbf{u}}{\partial t} + \rho_m (\mathbf{u} \cdot \nabla) \mathbf{u} + \nabla \cdot \left[\left(p + \frac{B^2}{2\mu_0} \right) I - \frac{\mathbf{BB}}{\mu_0} \right] = \rho_m \mathbf{g},$$

$$\frac{\partial \mathbf{B}}{\partial t} + \nabla \cdot (\mathbf{uB} - \mathbf{Bu}) = \eta_m \nabla^2 \mathbf{B},$$

$$\frac{3}{2} \frac{\partial p}{\partial t} + \frac{3}{2} (\mathbf{u} \cdot \nabla) p + \frac{5}{2} p (\nabla \cdot \mathbf{u}) = \frac{1}{\sigma_0} j^2, \tag{4.88}$$

where I is the 3×3 unit matrix. In the limit when the plasma is perfectly conducting ($\sigma_0 \to \infty$ and consequently $\eta_m \to 0$), these equations reduce to the governing equations of ideal MHD:

$$\frac{\partial \rho_m}{\partial t} + \nabla \cdot (\rho_m \mathbf{u}) = 0,$$

$$\rho_m \frac{\partial \mathbf{u}}{\partial t} + \rho_m (\mathbf{u} \cdot \nabla) \mathbf{u} + \nabla \cdot \left[\left(p + \frac{B^2}{2\mu_0} \right) I - \frac{\mathbf{BB}}{\mu_0} \right] = \rho_m \mathbf{g},$$

$$\frac{\partial \mathbf{B}}{\partial t} + \nabla \cdot (\mathbf{uB} - \mathbf{Bu}) = \mathbf{0},$$

$$\frac{3}{2} \frac{\partial p}{\partial t} + \frac{3}{2} (\mathbf{u} \cdot \nabla) p + \frac{5}{2} p (\nabla \cdot \mathbf{u}) = 0. \tag{4.89}$$

The equations of ideal MHD can also be written in the so-called conservative form (when all spatial derivatives can be expressed as divergences of generalized fluxes):

$$\frac{\partial \rho_m}{\partial t} + \nabla \cdot (\rho_m \mathbf{u}) = 0,$$

$$\frac{\partial (\rho_m \mathbf{u})}{\partial t} + \nabla \cdot \left(\rho_m \mathbf{uu} + pI + \frac{B^2}{2\mu_0} I - \frac{\mathbf{BB}}{\mu_0} \right) = \rho_m \mathbf{g},$$

$$\frac{\partial \mathbf{B}}{\partial t} + \nabla \cdot (\mathbf{uB} - \mathbf{Bu}) = \mathbf{0}, \tag{4.90}$$

$$\frac{\partial}{\partial t} \left(\frac{1}{2} \rho_m u^2 + \frac{3}{2} p + \frac{B^2}{2\mu_0} \right)$$

$$+ \nabla \cdot \left(\frac{1}{2} \rho_m u^2 \mathbf{u} + \frac{5}{2} p \mathbf{u} + \frac{(\mathbf{B} \cdot \mathbf{B}) \mathbf{u} - \mathbf{B} (\mathbf{B} \cdot \mathbf{u})}{\mu_0} \right) = \rho_m (\mathbf{g} \cdot \mathbf{u}).$$

It can be seen that the conservative form of the ideal MHD equations represents conservation laws for the mass density, ρ_m, mass flux, $\rho_m \mathbf{u}$, and for the total energy density,

$$\varepsilon = \frac{1}{2}\rho_m u^2 + \frac{3}{2}p + \frac{B^2}{2\mu_0}. \tag{4.91}$$

4.5   Problems

Problem 4.1 Obtain the conservative form of the ideal MHD equations (4.90) from the nonconservative form (4.89).

Problem 4.2 Consider a "force-free" situation in the MHD limit (neglect the displacement current). In force-free fields the current density is parallel to the magnetic field, $\mathbf{j} = \alpha \mathbf{B}$.

1. Show that α is constant along magnetic field lines (the $\mathbf{B} \cdot \nabla \alpha = 0$ equation must be satisfied).
2. Show that in a stationary fluid α is constant and the force-free field decays as

$$\frac{\partial \mathbf{B}}{\partial t} = -\frac{\mu_0 \alpha^2}{\overline{\sigma}_0}\mathbf{B}. \tag{4.92}$$

Problem 4.3 Consider the simplified MHD equations for a steady-state hot plasma in the absence of gravity. A cylindrical column of hot plasma with radius a contains a coaxial external magnetic field, $\mathbf{B}_0 = (0, 0, B_0)$, and it has a radially varying pressure profile given by $p = p_0 \cos^2(\pi r/2a)$ (where r is the radial distance from the axis). To counteract the pressure gradient there must be an azimuthal diamagnetic current in this system.

1. What is the maximum possible value of the constant p_0?
2. Using the maximum value of p_0, calculate the diamagnetic current, $\mathbf{j}$, and the total magnetic field, $\mathbf{B}$.

Chapter 5

Waves and Oscillations

Nature produces many different kinds of waves and oscillations. In general, these periodic phenomena are very different from each other. However, certain physical and mathematical properties are common to small-amplitude waves almost irrespective of their nature. For instance, all small-amplitude waves can be characterized by dispersion relationhips and they transport physical quantities with the group velocity of the wave.

In this chapter we will examine some fundamental proprties of small-amplitude waves in neutral and conducting fluids. We will see that although neutral fluids exhibit a relatively small number of fundamental wave phenomena, conducting fluids (especially when they are magnetized) are a very fertile medium for the generation of a huge variety of plasma waves.

Here we shall concentrate on small-amplitude waves, when the wave equations can be linearized. This does not mean that nonlinear phenomena are unimportant – they are just too complicated for this introductory text. Also, we will limit our discussions to single species gases (or in the case of plasmas, to single ion plasmas). The results can be generalized to multispecies plasmas, when needed.

5.1   Linearized Fluid Equations

First of all, let us consider the linearized version of the ideal MHD equations. We choose to use the MHD equations, because mathematically the Euler equations represent a subset of these equations (one just has to set $\mathbf{B}$ to zero everywhere at all times). Let us assume that we have a solution of the full equation set, that is, ρ_{m0}, $\mathbf{u}_0$, p_0, and $\mathbf{B}_0$ represent a steady-state solution of Eqs. (4.89). We consider a small perturbative solution with respect to this known solution. For the sake of simplicity, we consider waves in stationary fluids (i.e., we assume that $\mathbf{u}_0 = \mathbf{0}$). In this case the perturbative solution can be written as

$$\rho_m = \rho_{m0} + \rho_{m1},$$
$$\mathbf{u} = \mathbf{u}_1,$$
$$p = p_0 + p_1,$$
$$\mathbf{B} = \mathbf{B}_0 + \mathbf{B}_1.$$

(5.1)

Here the subscript "1" refers to small perturbations.

Next, we can linearize the ideal MHD equations by neglecting all terms containing higher order perturbations (for instance terms like $\rho_{m1}\mathbf{u}_1$). This results in the following linearized set of MHD equations:

$$\frac{\partial \rho_{m1}}{\partial t} + \rho_{m0}(\nabla \cdot \mathbf{u}_1) + (\mathbf{u}_1 \cdot \nabla)\rho_{m0} = 0,$$
$$\rho_{m0}\frac{\partial \mathbf{u}_1}{\partial t} + \nabla(p_0 + p_1) - (\rho_{m0} + \rho_{m1})\mathbf{g}$$
$$+ \nabla\frac{\mathbf{B}_0 \cdot \mathbf{B}_1}{\mu_0} - \nabla \cdot \left(\frac{\mathbf{B}_0\mathbf{B}_1 + \mathbf{B}_1\mathbf{B}_0}{\mu_0}\right) = \mathbf{0},$$
$$\frac{\partial \mathbf{B}_1}{\partial t} + \nabla \cdot (\mathbf{u}_1\mathbf{B}_0 - \mathbf{B}_0\mathbf{u}_1) = \mathbf{0},$$
$$\frac{3}{2}\frac{\partial p_1}{\partial t} + \frac{3}{2}(\mathbf{u}_1 \cdot \nabla)p_0 + \frac{5}{2}p_0(\nabla \cdot \mathbf{u}_1) = 0.$$

(5.2)

Let us examine a few special cases that lead to some fundamental waves in gases and plasmas.

5.2 Sound Waves

The simplest wavelike phenomenon in compressible fluids is the sound wave. For the sake of simplicity, let us consider a compressible neutral gas characterized by the Euler equations. Let us neglect all external forces (including gravity) and assume that the gas is stationary and is in equilibrium with density ρ_{m0} and pressure p_0. We further assume that the steady-state equilibrium solution is also spatially uniform: $\nabla\rho_{m0} = \mathbf{0}$ and $\nabla p_0 = \mathbf{0}$. In this case the governing equations for small-amplitude perturbations become the following:

$$\frac{\partial \rho_{m1}}{\partial t} + \rho_{m0}(\nabla \cdot \mathbf{u}_1) = 0,$$
$$\rho_{m0}\frac{\partial \mathbf{u}_1}{\partial t} + \nabla p_1 = \mathbf{0},$$
$$\frac{3}{2}\frac{\partial p_1}{\partial t} + \frac{5}{2}p_0(\nabla \cdot \mathbf{u}_1) = 0.$$

(5.3)

The continuity and energy equations can be combined to yield the adiabatic relationship between the pressure and density perturbations (note that ρ_{m0} and p_0 are constants):

$$\frac{\partial}{\partial t}\left(p_1 - \frac{5}{3}\frac{p_0}{\rho_{m0}}\rho_{m1}\right) = 0. \tag{5.4}$$

Thus we obtain the following relationship between the density and pressure disturbances:

$$p_1 = \frac{5}{3}\frac{p_0}{\rho_{m0}}\rho_{m1}. \tag{5.5}$$

Equation (5.5) can be used to eliminate the energy equation from our system, giving,

$$\frac{\partial \rho_{m1}}{\partial t} + \rho_{m0}(\nabla \cdot \mathbf{u}_1) = 0,$$
$$\rho_{m0}\frac{\partial \mathbf{u}_1}{\partial t} + \frac{5}{3}\frac{p_0}{\rho_{m0}}\nabla\rho_{m1} = \mathbf{0}. \tag{5.6}$$

These equations can be combined to yield wave equations for the density and velocity
perturbations:

$$\frac{\partial^2 \rho_{m1}}{\partial t} = \frac{5}{3}\frac{p_0}{\rho_{m0}}\nabla^2\rho_{m1},$$
$$\frac{\partial^2 \mathbf{u}_1}{\partial t} = \frac{5}{3}\frac{p_0}{\rho_{m0}}\nabla^2\mathbf{u}_1. \tag{5.7}$$

These equations describe a compressive wave that propagates with the speed of

$$a_s = \sqrt{\frac{5}{3}\frac{p_0}{\rho_{m0}}}. \tag{5.8}$$

This wave is called the *acoustic wave* or *sound wave* and a_s is the *sound speed* in
the gas.

   It is very interesting to note that in ionized gases (plasmas) the sound speeds are
somewhat different than in neutral gases. In a quasineutral, single-ion plasma the
background pressure and density can be written as

$$\rho_{m0} = n(m_e + m_{ion}) \approx nm_{ion},$$
$$p_0 = p_e + p_{ion} = nk(T_e + T_{ion}), \tag{5.9}$$

where n is the concentration of a charged species (the electron and ion concentrations
are the same). Now the sound speed in this plasma becomes

$$a_{ion-acoustic} = \sqrt{\frac{5}{3}\frac{k(T_e + T_{ion})}{m_{ion}}}. \tag{5.10}$$

This wave is called the *ion–acoustic wave* and $a_{ion-acoustic}$ is the *ion–acoustic speed*.
Since in many space plasmas $T_e \gg T_{ion}$, the ion–acoustic speed is typically controlled
by the electron temperature and the ion mass.

5.3   Alfvén Waves

Next, we discuss one of the most fundamental waves in magnetized plasmas. Let us consider a cold magnetized plasma (a plasma is cold when the pressure term is negligible in the momentum equation, and therefore the energy equation is not necessary to describe the plasma) and neglect gravity. For the sake of mathematical simplicity, we also assume that the background solution corresponds to constant density and magnetic field ($\rho_{m0} = const.$ and $\mathbf{B}_0 = const.$). Finally, in this homogeneous magnetized medium we consider noncompressive perturbations. Noncompressive perturbations do not compress the medium, which mathematically can be written as $\rho = const.$ or $(\nabla \cdot \mathbf{u}) = 0$. In our particular case this means that $(\nabla \cdot \mathbf{u}_1) = 0$. This equation replaces the energy equation. With these simplifications the linearized MHD equations become the following:

$$\frac{\partial \rho_{m1}}{\partial t} = 0,$$

$$\rho_{m0}\frac{\partial \mathbf{u}_1}{\partial t} + \frac{\mathbf{B}_0}{\mu_0} \times (\nabla \times \mathbf{B}_1) = \mathbf{0}, \qquad (5.11)$$

$$\frac{\partial \mathbf{B}_1}{\partial t} - (\mathbf{B}_0 \cdot \nabla)\mathbf{u}_1 = \mathbf{0}.$$

For the sake of additional simplicity, we consider magnetic perturbations that are perpendicular to $\mathbf{B}_0$ and propagate along the ambient magnetic field (i.e., $\mathbf{B}_1$ varies only in the direction of $\mathbf{B}_0$). Therefore, one can write that

$$\rho_{m1} = const.,$$

$$\rho_{m0}\frac{\partial \mathbf{u}_1}{\partial t} = \left(\frac{\mathbf{B}_0}{\mu_0} \cdot \nabla\right)\mathbf{B}_1 = \frac{B_0}{\mu_0}\frac{\partial \mathbf{B}_1}{\partial z}, \qquad (5.12)$$

$$\frac{\partial \mathbf{B}_1}{\partial t} = (\mathbf{B}_0 \cdot \nabla)\mathbf{u}_1 = B_0\frac{\partial \mathbf{u}_1}{\partial z},$$

where the coordinate z is in the direction of $\mathbf{B}_0$. These two equations can be readily combined into two wave equations:

$$\frac{\partial^2 \mathbf{u}_1}{\partial t^2} = \frac{B_0^2}{\rho_{m0}\mu_0}\frac{\partial^2 \mathbf{u}_1}{\partial z^2},$$

$$\frac{\partial^2 \mathbf{B}_1}{\partial t^2} = \frac{B_0^2}{\rho_{m0}\mu_0}\frac{\partial^2 \mathbf{B}_1}{\partial z^2}. \qquad (5.13)$$

Equations (5.13) describe a noncompressive disturbance wave propagating along the ambient magnetic field line with the so-called *Alfvén speed*:

$$V_A = \frac{B_0}{\sqrt{\mu_0 \rho_{m0}}}. \qquad (5.14)$$

These waves are called *Alfvén waves*. With a simple analogy these waves can be considered waves propagating along a stretched string (the "strength" of the string is basically defined by the magnetic stress). Alfvén waves are very important in space physics.

5.4   Plane Waves

Periodic motions of fluids can be decomposed into a superposition of sinusoidal oscillations. The individual sinusoidal oscillations are characterized by their frequency, ω, and wave vector, $\mathbf{k}$. The magnitude of $\mathbf{k}$ is called the *wavenumber* and it is related to the wavelength, λ, by $k = 2\pi/\lambda$. The direction of the wave vector defines the direction of propagation of the sinusoidal oscillation.

In general, any sinusoidally oscillating quantity can be represented by a plane wave:

$$A = \tilde{A} \exp\left[i(\mathbf{k} \cdot \mathbf{r} - \omega t)\right], \tag{5.15}$$

where $\tilde{A}$ is the amplitude of the oscillation. By convention, the exponential notation means that the real part of the complex exponential represents a measurable quantity. If $\tilde{A}$ is real (for one quantity this can always be achieved by a suitable choice of the origin), then

$$\Re(A) = \tilde{A} \cos(\mathbf{k} \cdot \mathbf{r} - \omega t). \tag{5.16}$$

Expression (5.16) represents the value of the quantity A at a spatial location $\mathbf{r}$ at a given time t. It should be noted that for any given time the value of the physical quantity A is the same at all points in any plane perpendicular to $\mathbf{k}$. For this reason the wave given by expression (5.15) is called a *plane wave*.

Planes of constant phase on the wave move so that

$$\frac{d}{dt}(\mathbf{k} \cdot \mathbf{r} - \omega t) = 0 \tag{5.17}$$

or

$$\mathbf{k} \cdot \frac{d\mathbf{r}}{dt} = \omega. \tag{5.18}$$

Hence the velocity of constant phase on the wave in the direction of propagation ($\hat{\mathbf{k}} = \mathbf{k}/k$) can be expressed as

$$v_{\text{ph}} = \frac{\omega}{k}. \tag{5.19}$$

The quantity v_{ph} is called the *phase velocity*. It is important to note that the speed of propagation of constant phase in a given direction $\mathbf{e}$ is not $v_{\text{ph}}(\hat{\mathbf{k}} \cdot \mathbf{e})$ but $v_{\text{ph}}/(\hat{\mathbf{k}} \cdot \mathbf{e})$. This means that even though the phase velocity is associated with a direction of propagation, it is not a real vector quantity.

Another strange property of the phase velocity is the fact that it may exceed the speed of light. However, this does not violate the fundamental laws of physics, because no physical quantity or information propagates with the phase velocity. In other words, an infinitely long wave train of constant amplitude cannot carry information.

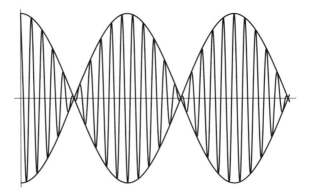

Figure 5.1 Spatial variation of the sum of two waves with a
small wavenumber and frequency difference.

Information is carried by modulations of wave trains, and modulations do not propagate
faster than the speed of light.

To illustrate this point let us consider a modulated wave formed by adding two
waves of nearly equal frequencies. Let us assume that both waves propagate in the
x direction and let us denote the observable physical properties of these two waves
by

$$\Re(A_1) = \tilde{A} \cos[(k + \Delta k)x - (\omega + \Delta\omega)t],$$
$$\Re(A_2) = \tilde{A} \cos[(k - \Delta k)x - (\omega - \Delta\omega)t].$$
(5.20)

Here A_1 and A_2 differ in wavenumber by $2\Delta k$ and in frequency by $2\Delta\omega$. Now the
sum of the two waves is

$$\Re(A_1 + A_2) = \tilde{A} \cos[(kx - \omega t) + (x\Delta k - t\Delta\omega)]$$
$$+ \tilde{A} \cos[(kx - \omega t) - (x\Delta k - t\Delta\omega)]$$
$$= 2\tilde{A} \cos(x\Delta k - t\Delta\omega) \cos(kx - \omega t).$$
(5.21)

This is a sinusoidally modulated wave (see Figure 5.1). It is the modulated amplitude
that carries information. In the limit when $\Delta k \to 0$ and $\Delta\omega \to 0$ the information
travels with the *group velocity*:

$$v_{\mathrm{g}} = \frac{d\omega}{dk}.$$
(5.22)

It is important to emphasize that the group velocity cannot exceed the velocity of light.

5.5   Internal Gravity Waves

Next, we will show how to find simple plane wave solutions to the linearized fluid
equations. Let us consider a stationary, single species, incompressible neutral fluid in
the presence of a constant gravitational field. For the sake of mathematical simplicity,

we assume that the background pressure gradient compensates the gravitational force: $\rho_{m0}\mathbf{g} = \nabla p_0$ (in effect, this assumption translates to hydrostatic equilibrium). The assumption of incompressibility means that $\nabla \cdot \mathbf{u}_1 = 0$. In incompressible fluids the density and pressure are independent of each other (but the density still may depend on the temperature) and the condition of incompressibilty can be used to replace the energy equation.

We have the freedom to choose the z axis to be along the direction of gravity in a way that $\mathbf{g} = (0, 0, -g)$. It is also assumed that both the background pressure and density vary only in the z direction. In this case the linearized continuity and momentum equations become the following:

$$\frac{\partial \rho_{m1}}{\partial t} - N^2 \frac{\rho_{m0}}{g} u_{1z} = 0,$$
$$\rho_{m0} \frac{\partial \mathbf{u}_1}{\partial t} + \nabla p_1 - \rho_{m1}\mathbf{g} = \mathbf{0}, \tag{5.23}$$

where we introduced the *buoyancy frequency* or *Brunt–Väisälä* frequency,[1,2] N, as

$$N^2 = -\frac{g}{\rho_{m0}} \frac{d\rho_{m0}}{dz}. \tag{5.24}$$

Equations (5.23) are supplemented with the condition of incompressibility:

$$\nabla \cdot \mathbf{u}_1 = 0. \tag{5.25}$$

One can now substitute the plane wave solution into our linearized sytem given by Eqs. (5.23). Now the perturbed quantities are approximated as

$$\rho_m = \rho_{m0} + \tilde{\rho} \exp[i(\mathbf{k} \cdot \mathbf{r} - \omega t)],$$
$$\mathbf{u} = \tilde{\mathbf{u}} \exp[i(\mathbf{k} \cdot \mathbf{r} - \omega t)], \tag{5.26}$$
$$p = p_0 + \tilde{p} \exp[i(\mathbf{k} \cdot \mathbf{r} - \omega t)].$$

With the help of these assumptions, we obtain the following set of algebraic equations for the wave amplitudes:

$$-i\omega\tilde{\rho}_m - N^2 \frac{\rho_{m0}}{g} \tilde{u}_z = 0,$$
$$-i\omega\rho_{m0}\tilde{\mathbf{u}} + i\mathbf{k}\tilde{p} - \tilde{\rho}\mathbf{g} = \mathbf{0}, \tag{5.27}$$
$$\mathbf{k} \cdot \tilde{\mathbf{u}} = 0.$$

This equation can also be written in matrix form:

$$\mathcal{A} \cdot \mathcal{W} = \mathbf{0}, \tag{5.28}$$

[1] Brunt, D., "The period of simple vertical oscillations in the atmosphere," *Q. J. R. Meteorol. Soc.*, **53**, 30, 1927.

[2] Väisälä, V., "Über die Wirkung der Windschwankungen auf die Pilotbeobachtungen," *Soc. Sci. Fenn. Commentat. Phys–Math.*, **2**(19), 19, 1925.

where the matrix $\mathcal{A}$ and the vector $\mathcal{W}$ are given by

$$
\mathcal{A} = \begin{pmatrix}
-i\omega & 0 & 0 & -N^2\frac{\rho_{m0}}{g} & 0 \\
0 & -i\omega\rho_{m0} & 0 & 0 & ik_x \\
0 & 0 & -i\omega\rho_{m0} & 0 & ik_y \\
g & 0 & 0 & -i\omega\rho_{m0} & ik_z \\
0 & k_x & k_y & k_z & 0
\end{pmatrix}
\tag{5.29}
$$

and

$$
\mathcal{W} = \begin{pmatrix}
\tilde{\rho} \\
\tilde{u}_x \\
\tilde{u}_y \\
\tilde{u}_z \\
\tilde{p}
\end{pmatrix}.
\tag{5.30}
$$

Equation (5.28) has nontrivial (not identically zero) solutions only if the determinant of the characteristic matrix, $\mathcal{A}$, vanishes. This requirement leads to the following condition:

$$
\omega^2 = \frac{k_x^2 + k_y^2}{k^2}N^2 = N^2 \sin^2 \vartheta,
\tag{5.31}
$$

where ϑ is the angle between the z axis and the wave vector $\mathbf{k}$.

Equation (5.31) is the *dispersion relation* for *internal gravity waves*. It shows that these waves can have any frequency between zero and a maximum value of N. Equation (5.31) also reveals that the frequency of internal gravity waves depends on the direction of propagation with respect to the vertical direction (defined by the direction of gravity). It can be seen that gravity waves cannot exist with strictly vertical wave vectors, because for these waves $k_x = k_y = 0$ and consequently $\omega = 0$. In the horizontal direction, however, internal gravity waves oscillate at the maximum frequency, N.

Later in the book we will discuss more complicated gravity waves in compressible media (atmospheric and solar gravity waves).

5.6   Waves in Field-Free Plasmas

Simple small-amplitude waves can be represented by the plane waves discussed above. A plane wave is characterized by four quantities given by the frequency (ω) and the three components of the wave vector ($\mathbf{k}$). However, in fluids these four quantities are related by a *dispersion relation*. Arbitrary (real or complex) values can be assigned to three quantities and the dispersion relation then determines the fourth one.

In order to derive dispersion relations for small-amplitude waves in neutral and conducting fluids, we assume that the temporal and spatial variation of a perturbed quantity can be approximated by a plane wave (the amplitudes of these waves might also contain nonzero imaginary parts). This is a reasonable assumption, since more complex periodic motions can be described as a sum of plane waves.

For the sake of simplicity, we restrict our consideration to *cold fluids*. In cold fluids the pressure term can be neglected in the momentum equation and therefore the energy equation can be neglected. Later in the book we will consider situations when the pressure term is important and the cold fluid approximation cannot be used. However, this simplification helps us to emphasize the basic physical concepts.

In this section we consider dispersion relations for cold single-ion plasmas. Later in the book we will also apply the concept to the upper atmosphere and the solar interior. However, as a first application, we consider simple plasma waves.

Let us start by considering a mixture of stationary cold electron and ion gases and neglect the effects of collisions and all external forces. The physical quantities are given in the form of

$$n_e = n_0 + \tilde{n}_e \exp\left[i(\mathbf{k} \cdot \mathbf{r} - \omega t)\right],$$

$$n_{\text{ion}} = n_0 + \tilde{n}_{\text{ion}} \exp\left[i(\mathbf{k} \cdot \mathbf{r} - \omega t)\right],$$

$$\mathbf{u}_e = \tilde{\mathbf{u}}_e \exp\left[i(\mathbf{k} \cdot \mathbf{r} - \omega t)\right],$$

$$\mathbf{u}_{\text{ion}} = \tilde{\mathbf{u}}_{\text{ion}} \exp\left[i(\mathbf{k} \cdot \mathbf{r} - \omega t)\right], \tag{5.32}$$

$$\mathbf{B} = \mathbf{B}_0 + \tilde{\mathbf{B}} \exp\left[i(\mathbf{k} \cdot \mathbf{r} - \omega t)\right],$$

$$\mathbf{E} = \mathbf{E}_0 + \tilde{\mathbf{E}} \exp\left[i(\mathbf{k} \cdot \mathbf{r} - \omega t)\right].$$

Here all quantities denoted by ~ represent first-order small complex amplitudes. The stationary equilibrium solution is assumed to be characterized by a constant electron (and ion) concentration, n_0, and by background magnetic and electric fields of $\mathbf{B}_0$ and $\mathbf{E}_0$, respectively.

First, we consider the simplest case, when there are no background electric and magnetic fields, $\mathbf{B}_0 = \mathbf{0}$ and $\mathbf{E}_0 = \mathbf{0}$. In this case the resulting equations become quite simple, and one can readily concentrate to the underlying physics.

Substituting expressions (5.32) into Eqs. (4.35) results in the following linearized equations:

$$-i\omega \tilde{n}_e + in_0(\mathbf{k} \cdot \tilde{\mathbf{u}}_e) = 0,$$

$$-i\omega \tilde{n}_{\text{ion}} + in_0(\mathbf{k} \cdot \tilde{\mathbf{u}}_{\text{ion}}) = 0,$$

$$-i\omega m_e n_0 \tilde{\mathbf{u}}_e + en_0 \tilde{\mathbf{E}} = 0, \tag{5.33}$$

$$-i\omega m_{\text{ion}} n_0 \tilde{\mathbf{u}}_{\text{ion}} - en_0 \tilde{\mathbf{E}} = 0.$$

One can express the amplitudes of the density and velocity perturbations in terms of the electric field wave amplitude:

$$
\begin{aligned}
\tilde{n}_e &= -\frac{i}{\omega}\frac{en_0}{m_e}\left(\frac{\mathbf{k}}{\omega}\cdot\tilde{\mathbf{E}}\right), \\
\tilde{n}_{ion} &= \frac{i}{\omega}\frac{en_0}{m_{ion}}\left(\frac{\mathbf{k}}{\omega}\cdot\tilde{\mathbf{E}}\right), \\
\tilde{\mathbf{u}}_e &= -\frac{i}{\omega}\frac{e}{m_e}\tilde{\mathbf{E}}, \\
\tilde{\mathbf{u}}_{ion} &= \frac{i}{\omega}\frac{e}{m_{ion}}\tilde{\mathbf{E}}.
\end{aligned}
\tag{5.34}
$$

Next, we will use Maxwell's equations to define the electric field amplitude. Ampère's law is

$$
\nabla \times \mathbf{B} = \frac{1}{c^2}\frac{\partial \mathbf{E}}{\partial t} + \mu_0\,\mathbf{j}.
\tag{5.35}
$$

In our case the electric current density can be expressed to first-order accuracy as

$$
\mathbf{j} = e(n_{ion}\mathbf{u}_{ion} - n_e\mathbf{u}_e) \approx en_0(\tilde{\mathbf{u}}_{ion} - \tilde{\mathbf{u}}_e)\exp[i(\mathbf{k}\cdot\mathbf{r}-\omega t)]
$$
$$
= i\frac{e^2n_0}{\omega}\left(\frac{1}{m_{ion}}+\frac{1}{m_e}\right)\tilde{\mathbf{E}}\exp[i(\mathbf{k}\cdot\mathbf{r}-\omega t)].
\tag{5.36}
$$

This means that the current density also exhibits a plane wave–like behavior. The amplitude of the current density is

$$
\tilde{\mathbf{j}} = i\varepsilon_0\frac{\omega_p^2}{\omega}\tilde{\mathbf{E}},
\tag{5.37}
$$

where ω_p is the so-called *plasma frequency* defined as

$$
\omega_p^2 = \frac{e^2n_0}{\varepsilon_0}\left(\frac{1}{m_{ion}}+\frac{1}{m_e}\right).
\tag{5.38}
$$

With the help of expressions (5.32), (5.36), and (5.37), Ampère's law becomes

$$
\mathbf{k}\times\tilde{\mathbf{B}} = -\frac{\omega}{c^2}\left(1-\frac{\omega_p^2}{\omega^2}\right)\tilde{\mathbf{E}},
\tag{5.39}
$$

because we know that $\mu_0\varepsilon_0 = 1/c^2$. Finally, Faraday's law helps us to close the system. In the present approximation Faraday's law can be written in the following form:

$$
\mathbf{k}\times\tilde{\mathbf{E}} = \omega\tilde{\mathbf{B}}.
\tag{5.40}
$$

Substituting Eq. (5.40) into Eq. (5.39) yields the following equation for the electric field amplitude of the plane wave solution:

$$\mathbf{k} \times (\mathbf{k} \times \tilde{\mathbf{E}}) = -\frac{\omega^2}{c^2}\left(1 - \frac{\omega_p^2}{\omega^2}\right)\tilde{\mathbf{E}}. \tag{5.41}$$

The triple vector cross product can be written as $\mathbf{k} \times (\mathbf{k} \times \tilde{\mathbf{E}}) = (\mathbf{k} \cdot \tilde{\mathbf{E}})\mathbf{k} - k^2\tilde{\mathbf{E}}$, and therefore Eq. (5.41) can be written as

$$c^2(\mathbf{k} \cdot \tilde{\mathbf{E}})\mathbf{k} = \left[c^2 k^2 - \left(\omega^2 - \omega_p^2\right)\right]\tilde{\mathbf{E}}. \tag{5.42}$$

Equation (5.42) can also be written in matrix form:

$$\varepsilon \cdot \tilde{\mathbf{E}} = \mathbf{0}, \tag{5.43}$$

where the matrix ε is

$$\varepsilon = \begin{pmatrix} c^2 k^2 - \left(\omega^2 - \omega_p^2\right) & 0 & 0 \\ 0 & c^2 k^2 - \left(\omega^2 - \omega_p^2\right) & 0 \\ 0 & 0 & \omega^2 - \omega_p^2 \end{pmatrix}. \tag{5.44}$$

Here we have chosen the coordinate system in such a way that the wave propagation direction defines the direction of the z axis, $\mathbf{k} = (0, 0, k)$. Equation (5.43) has a nontrivial solution only if the determinant of the matrix ε vanishes. This requirement leads to the following equation:

$$\left[c^2 k^2 - \left(\omega^2 - \omega_p^2\right)\right]^2 \left(\omega^2 - \omega_p^2\right) = 0. \tag{5.45}$$

Only waves with wavenumber and frequency values satisfying Eq. (5.45) can exist in an unmagnetized stationary plasma. Equation (5.45) is called the *dispersion relation*, because it describes the way the plasma as a medium disperses plane waves.

5.6.1 Langmuir Waves

It is obvious that Eq. (5.45) has two different types of solutions. The first solution type is given by the dispersion relation

$$\omega^2 = \omega_p^2. \tag{5.46}$$

It is obvious from expression (5.44) that Eq. (5.46) corresponds to waves with $\tilde{\mathbf{E}} = (0, 0, \tilde{E})$, that is, these waves have their electric field vectors in the direction of $\mathbf{k}$. It immediately follows from Eq. (5.40) that for these waves $\tilde{\mathbf{B}} = \mathbf{0}$. Thus no magnetic field is associated with these waves, and they are *electrostatic* by nature.

It is also clear from Eq. (5.46) that these waves are dispersionless in cold plasmas: They have the same frequency (ω_p) for all wavenumbers. These waves are called *Langmuir waves*. It can also easily be seen that the phase and group velocities of

Langmuir waves are

$$
\begin{aligned}
v_{ph} &= \frac{\omega}{k} = \frac{\omega_p}{k}, \\
v_g &= \frac{\partial \omega}{\partial k} = 0.
\end{aligned}
\tag{5.47}
$$

Equation (5.47) indicates that Langmuir waves do not propagate (their group velocity is zero), but they represent local oscillations around charge equilibrium with the characteristic plasma frequency, ω_p.

5.6.2   Electromagnetic Waves

The second solution class of Eq. (5.45) leads to the following dispersion relation:

$$
\omega^2 = c^2 k^2 + \omega_p^2.
\tag{5.48}
$$

It can be seen with the help of expression (5.44) that waves described by Eq. (5.48) have electric field vectors perpendicular to the direction of propagation: $\tilde{\mathbf{E}} = (\tilde{E}_x, \tilde{E}_y, 0)$. For these waves Eq. (5.40) implies the existence of a nonzero magnetic field vector, and therefore these waves are *electromagnetic*. The electromagnetic waves can be polarized since Eq. (5.43) has two linearly independent solutions perpendicular to the magnetic field vector.

The phase and group velocities of these electromagnetic waves propagating in cold, unmagnetized plasmas are the following:

$$
\begin{aligned}
v_{ph} &= \frac{\omega}{k} = c\sqrt{1 + \frac{\omega_p^2}{c^2 k^2}} > c, \\
v_g &= \frac{\partial \omega}{\partial k} = c\frac{1}{\sqrt{1 + \frac{\omega_p^2}{c^2 k^2}}} < c.
\end{aligned}
\tag{5.49}
$$

It can be seen that the group velocity is smaller than the speed of the light and that

$$
v_{ph} v_g = c^2.
\tag{5.50}
$$

It is obvious that $\omega_p \to 0$ as the plasma density goes to zero. In this case the dispersion relation for these electromagnetic waves becomes $\omega = ck$, which is the dispersion relation for electromagnetic waves in vacuum.

It is also clear that electromagnetic plasma waves described by Eq. (5.48) can only exist above the plasma frequency, $\omega > \omega_p$. Waves below ω_p are only short-lived plasma oscillations which are rapidly damped in the medium.

5.7   MHD Waves

In the presence of a magnetic field the acoustic wave becomes quite a bit more complicated than it is in a pure hydrodynamic situation. For the sake of simplicity, let us consider the MHD equations with periodic oscillations around a background state characterized by a constant density, ρ_0, zero velocity vector, $\mathbf{u}_0 = \mathbf{0}$, and constant pressure, p_0. The background magnetic field vector is also assumed to be constant, and it points in the z direction, $\mathbf{B}_0 = (0, 0, B_0)$. All deviations from this equilibrium are assumed to be small. In this approximation the various physical quantities are expressed as follows:

$$\rho_m = \rho_0 + \tilde{\rho} \exp[i(\mathbf{k} \cdot \mathbf{r} - \omega t)],$$
$$\mathbf{u} = \tilde{\mathbf{u}} \exp[i(\mathbf{k} \cdot \mathbf{r} - \omega t)],$$
$$p = p_0 + \tilde{p} \exp[i(\mathbf{k} \cdot \mathbf{r} - \omega t)], \tag{5.51}$$
$$\mathbf{B} = \mathbf{B}_0 + \tilde{\mathbf{B}} \exp[i(\mathbf{k} \cdot \mathbf{r} - \omega t)],$$

where the quantities $\tilde{\rho}$, $\tilde{\mathbf{u}}$, $\tilde{p}$, and $\tilde{\mathbf{B}}$ are assumed to be first-order small constant wave amplitudes.

Expressions (5.51) can be substituted into the linearized MHD equations given by (5.2) and one can neglect the effects of gravity to obtain

$$-i\omega\tilde{\rho} + i\rho_0(\mathbf{k} \cdot \tilde{\mathbf{u}}) = 0,$$
$$-i\omega\rho_0\tilde{\mathbf{u}} + i\mathbf{k}\tilde{p} + i\frac{B_0}{\mu_0}\tilde{B}_z\mathbf{k} - i\frac{B_0}{\mu_0}k_z\tilde{\mathbf{B}} = \mathbf{0},$$
$$-i\omega\tilde{\mathbf{B}} + i(\mathbf{k} \cdot \tilde{\mathbf{u}})\mathbf{B}_0 - i(\mathbf{B}_0 \cdot \mathbf{k})\tilde{\mathbf{u}} = \mathbf{0}, \tag{5.52}$$
$$-\frac{3}{2}i\omega\tilde{p} + \frac{5}{2}ip_0(\mathbf{k} \cdot \tilde{\mathbf{u}}) = 0,$$

because $\mathbf{k} \cdot \tilde{\mathbf{B}} = 0$ (there are no magnetic monopoles).

One can combine the continuity and energy equations, and Eqs. (5.52) can be rewritten in terms of vector components:

$$-\omega\tilde{\rho} + \rho_0 k_x \tilde{u}_x + \rho_0 k_y \tilde{u}_y + \rho_0 k_z \tilde{u}_z = 0,$$
$$k_x a_s^2 \tilde{\rho} - \omega\rho_0\tilde{u}_x - \frac{B_0}{\mu_0}k_z\tilde{B}_x + \frac{B_0}{\mu_0}k_x\tilde{B}_z = 0,$$
$$k_y a_s^2 \tilde{\rho} - \omega\rho_0\tilde{u}_y - \frac{B_0}{\mu_0}k_z\tilde{B}_y + \frac{B_0}{\mu_0}k_y\tilde{B}_z = 0,$$
$$k_z a_s^2 \tilde{\rho} - \omega\rho_0\tilde{u}_z = 0,$$
$$-\omega\tilde{B}_x - B_0 k_z \tilde{u}_x = 0, \tag{5.53}$$
$$-\omega\tilde{B}_y - B_0 k_z \tilde{u}_y = 0,$$
$$-\omega\tilde{B}_z + B_0 k_x \tilde{u}_x + B_0 k_y \tilde{u}_y = 0,$$
$$a_s^2 \tilde{\rho} - \tilde{p} = 0,$$

where $a_s^2 = 5p_0/3\rho_0$ is the square of the sound speed.

Equations (5.53) can also be written in matrix form:

$$\mathcal{A} \cdot \mathcal{W} = 0, \tag{5.54}$$

where the matrix $\mathcal{A}$ and the vector $\mathcal{W}$ are given by

$$
\mathcal{A} = \begin{pmatrix}
-\omega & \rho_0 k_x & \rho_0 k_y & \rho_0 k_z & 0 & 0 & 0 & 0 \\
a_s^2 k_x & -\omega\rho_0 & 0 & 0 & -k_z \frac{\rho_0 V_A^2}{B_0} & 0 & k_x \frac{\rho_0 V_A^2}{B_0} & 0 \\
a_s^2 k_y & 0 & -\omega\rho_0 & 0 & 0 & -k_z \frac{\rho_0 V_A^2}{B_0} & k_y \frac{\rho_0 V_A^2}{B_0} & 0 \\
a_s^2 k_z & 0 & 0 & -\omega\rho_0 & 0 & 0 & 0 & 0 \\
0 & -B_0 k_z & 0 & 0 & -\omega & 0 & 0 & 0 \\
0 & 0 & -B_0 k_z & 0 & 0 & -\omega & 0 & 0 \\
0 & B_0 k_x & B_0 k_y & 0 & 0 & 0 & -\omega & 0 \\
a_s^2 & 0 & 0 & 0 & 0 & 0 & 0 & -1
\end{pmatrix}
\tag{5.55}
$$

and

$$
\mathcal{W} = \begin{pmatrix}
\tilde{\rho} \\
\tilde{u}_x \\
\tilde{u}_y \\
\tilde{u}_z \\
\tilde{B}_x \\
\tilde{B}_y \\
\tilde{B}_z \\
\tilde{p}
\end{pmatrix},
\tag{5.56}
$$

where V_A is the Alfvén speed.

Equation (5.54) has nontrivial solutions only if the determinant of the characteristic matrix, $\mathcal{A}$, vanishes. This requirement leads to the following dispersion relation:

$$
\omega\left(\omega^2 - k^2 V_A^2 \cos^2 \vartheta\right)\left[\omega^4 - \omega^2 k^2 \left(a_s^2 + V_A^2\right) + a_s^2 V_A^2 k^4 \cos^2 \vartheta\right],
\tag{5.57}
$$

where ϑ is the angle between the background magnetic field and the wave vector $\mathbf{k}$.

The dispersion relation for MHD waves (Eq. 5.57) describes several different waves. The first wave can be obtained from the first term in parentheses in Eq. (5.57):

$$
\omega^2 = V_A^2 k_z^2.
\tag{5.58}
$$

This wave is an Alfvén wave propagating along the background magnetic field.

The expression in the brackets has two solutions for ω^2:

$$
\frac{\omega^2}{k^2} = \frac{1}{2}\left[\left(a_s^2 + V_A^2\right) \pm \sqrt{\left(a_s^2 + V_A^2\right)^2 - 4 a_s^2 V_A^2 \cos^2 \vartheta}\right].
\tag{5.59}
$$

Here the $\pm$ signs correspond to the *fast* and *slow magnetosonic waves*.

The phase velocities of the Alfvén and magnetosonic waves are the following:

$$
v_{\mathrm{ph}_A}^2 = V_A^2 \cos^2 \vartheta,
$$

$$
v_{\mathrm{ph}_f}^2 = \frac{1}{2} \left[(a_s^2 + V_A^2) + \sqrt{(a_s^2 + V_A^2)^2 - 4a_s^2 V_A^2 \cos^2 \vartheta} \right],
$$

$$
v_{\mathrm{ph}_s}^2 = \frac{1}{2} \left[(a_s^2 + V_A^2) - \sqrt{(a_s^2 + V_A^2)^2 - 4a_s^2 V_A^2 \cos^2 \vartheta} \right].
$$

(5.60)

It can be shown that $v_{\mathrm{ph}_s}^2 \le v_{\mathrm{ph}_A}^2 \le v_{\mathrm{ph}_f}^2$. The expressions for the fast and slow magnetosonic speeds can be written in a more symmetric form:

$$
v_{\mathrm{ph}_s} = \frac{1}{2} \left(\sqrt{a_s^2 + V_A^2 + 2a_s V_A \cos \vartheta} - \sqrt{a_s^2 + V_A^2 - 2a_s V_A \cos \vartheta} \right),
$$

$$
v_{\mathrm{ph}_f} = \frac{1}{2} \left(\sqrt{a_s^2 + V_A^2 + 2a_s V_A \cos \vartheta} + \sqrt{a_s^2 + V_A^2 - 2a_s V_A \cos \vartheta} \right).
$$

(5.61)

5.8   Plasma Waves in a Cold Magnetized Plasma

Earlier we examined plasma waves in a cold unmagnetized plasma. These waves are relatively simple, because the medium is homogeneous, that is, there is no "preferred" direction in the background plasma. In this case the only "direction" in the plasma is associated with the direction of propagation of the wave. The situation becomes much more complicated in the case when the background plasma is magnetized ($\mathbf{B}_0 \ne \mathbf{0}$).

Let us consider again a mixture of stationary cold electron and ion gases, neglect the effects of collisions and external forces, and assume that there is no background electric field ($\mathbf{E}_0 = \mathbf{0}$). In this case the variation of physical quantities is assumed to take the following linearized form:

$$
n_e = n_0 + \tilde{n}_e \exp[i(\mathbf{k} \cdot \mathbf{r} - \omega t)],
$$

$$
n_{\mathrm{ion}} = n_0 + \tilde{n}_{\mathrm{ion}} \exp[i(\mathbf{k} \cdot \mathbf{r} - \omega t)],
$$

$$
\mathbf{u}_e = \tilde{\mathbf{u}}_e \exp[i(\mathbf{k} \cdot \mathbf{r} - \omega t)],
$$

$$
\mathbf{u}_{\mathrm{ion}} = \tilde{\mathbf{u}}_{\mathrm{ion}} \exp[i(\mathbf{k} \cdot \mathbf{r} - \omega t)],
$$

$$
\mathbf{B} = \mathbf{B}_0 + \tilde{\mathbf{B}} \exp[i(\mathbf{k} \cdot \mathbf{r} - \omega t)],
$$

$$
\mathbf{E} = \tilde{\mathbf{E}} \exp[i(\mathbf{k} \cdot \mathbf{r} - \omega t)].
$$

(5.62)

Here the background magnetic field, $\mathbf{B}_0$, is assumed to be constant (independent of both location and time).

Substituting these expressions into the two fluid continuity and momentum equations (4.35) yields the following:

$$-i\omega\tilde{n}_e + in_0(\mathbf{k} \cdot \tilde{\mathbf{u}}_e) = 0,$$

$$-i\omega\tilde{n}_{ion} + in_0(\mathbf{k} \cdot \tilde{\mathbf{u}}_{ion}) = 0,$$

$$-i\omega m_e n_0\tilde{\mathbf{u}}_e + en_0(\tilde{\mathbf{E}} + \tilde{\mathbf{u}}_e \times \mathbf{B}_0) = \mathbf{0},$$

$$-i\omega m_{ion}n_0\tilde{\mathbf{u}}_{ion} - en_0(\tilde{\mathbf{E}} + \tilde{\mathbf{u}}_{ion} \times \mathbf{B}_0) = 0.$$

$$(5.63)$$

The ion and electron momentum equations can be used to express the electric field perturbation in terms of the velocity perturbation:

$$\frac{e}{m_e}\tilde{\mathbf{E}} = i\omega\tilde{\mathbf{u}}_e - \Omega_e\tilde{\mathbf{u}}_e \times \mathbf{b},$$

$$\frac{e}{m_{ion}}\tilde{\mathbf{E}} = -i\omega\tilde{\mathbf{u}}_{ion} - \Omega_{ion}\tilde{\mathbf{u}}_{ion} \times \mathbf{b},$$

$$(5.64)$$

where $\mathbf{b}$ is again the unit vector along the magnetic field, and Ω_e and Ω_{ion} are the electron and ion gyrofrequencies. The coefficient matrix can be inverted to express the velocity perturbation in terms of the electric field:

$$\tilde{\mathbf{u}}_e = \frac{e}{m_e}\frac{\Omega_e}{\Omega_e^2 - \omega^2}\left[i\frac{\omega}{\Omega_e}\tilde{\mathbf{E}} - i\frac{\Omega_e}{\omega}\tilde{E}_{\parallel}\mathbf{b} + \tilde{\mathbf{E}} \times \mathbf{b}\right],$$

$$\tilde{\mathbf{u}}_{ion} = \frac{e}{m_{ion}}\frac{\Omega_{ion}}{\Omega_{ion}^2 - \omega^2}\left[-i\frac{\omega}{\Omega_{ion}}\tilde{\mathbf{E}} + i\frac{\Omega_{ion}}{\omega}\tilde{E}_{\parallel}\mathbf{b} + \tilde{\mathbf{E}} \times \mathbf{b}\right],$$

$$(5.65)$$

where $\tilde{E}_{\parallel} = \tilde{\mathbf{E}} \cdot \mathbf{b}$ is the parallel (magnetic field aligned) component of the electric field. One can define the z axis along the background magnetic field vector and express Eq. (5.65) in tensor form:

$$\tilde{\mathbf{u}}_e = \frac{e}{m_e}\begin{bmatrix} i\frac{\omega}{\Omega_e^2 - \omega^2} & \frac{\Omega_e}{\Omega_e^2 - \omega^2} & 0 \\ -\frac{\Omega_e}{\Omega_e^2 - \omega^2} & i\frac{\omega}{\Omega_e^2 - \omega^2} & 0 \\ 0 & 0 & -i\frac{1}{\omega} \end{bmatrix}\tilde{\mathbf{E}},$$

$$\tilde{\mathbf{u}}_{ion} = \frac{e}{m_{ion}}\begin{bmatrix} -i\frac{\omega}{\Omega_{ion}^2 - \omega^2} & \frac{\Omega_{ion}}{\Omega_{ion}^2 - \omega^2} & 0 \\ -\frac{\Omega_{ion}}{\Omega_{ion}^2 - \omega^2} & -i\frac{\omega}{\Omega_{ion}^2 - \omega^2} & 0 \\ 0 & 0 & i\frac{1}{\omega} \end{bmatrix}\tilde{\mathbf{E}}.$$

$$(5.66)$$

Equations (5.66) express the electron and ion velocities in terms of the perturbation electric field. With the help of these equations, one can express the amplitude of the electric current density with first-order accuracy:

$$\tilde{\mathbf{j}} = en_0(\tilde{\mathbf{u}}_{\text{ion}} - \tilde{\mathbf{u}}_{\text{e}})$$

$$= \varepsilon_0 \begin{bmatrix} -i\omega\left(\dfrac{\omega_{\text{pi}}^2}{\Omega_{\text{ion}}^2 - \omega^2} + \dfrac{\omega_{\text{pe}}^2}{\Omega_{\text{e}}^2 - \omega^2}\right) & \dfrac{\omega_{\text{pi}}^2 \Omega_{\text{ion}}}{\Omega_{\text{ion}}^2 - \omega^2} - \dfrac{\omega_{\text{pe}}^2 \Omega_{\text{e}}}{\Omega_{\text{e}}^2 - \omega^2} & 0 \\[3mm] -\dfrac{\omega_{\text{pi}}^2 \Omega_{\text{ion}}}{\Omega_{\text{ion}}^2 - \omega^2} + \dfrac{\omega_{\text{pe}}^2 \Omega_{\text{e}}}{\Omega_{\text{e}}^2 - \omega^2} & -i\omega\left(\dfrac{\omega_{\text{pi}}^2}{\Omega_{\text{ion}}^2 - \omega^2} + \dfrac{\omega_{\text{pe}}^2}{\Omega_{\text{e}}^2 - \omega^2}\right) & 0 \\[3mm] 0 & 0 & i\dfrac{\omega_{\text{p}}^2}{\omega} \end{bmatrix} \tilde{\mathbf{E}},$$

$$(5.67)$$

where $\omega_{\text{pe}}^2 = n_0 e^2/\varepsilon_0 m_{\text{e}}$ and $\omega_{\text{pi}}^2 = n_0 e^2/\varepsilon_0 m_{\text{ion}}$ are the electron and ion plasma frequencies, while $\omega_{\text{p}}^2 = \omega_{\text{pi}}^2 + \omega_{\text{pe}}^2$ is the plasma frequency.

So far we have expressed the amplitude of the current density in terms of the electric field magnitude and the direction of the background magnetic field. Next, we will use this relation in Faraday's law and Ampère's law and to obtain a wave equation for the electric field amplitude, $\tilde{\mathbf{E}}$.

In a first-order approximation Ampère's law becomes

$$i\mathbf{k} \times \tilde{\mathbf{B}} = -\frac{i\omega}{c^2}\tilde{\mathbf{E}} + \mu_0 \tilde{\mathbf{j}}$$

$$= -i\frac{\omega}{c^2} \begin{bmatrix} 1 + \dfrac{\omega_{\text{pi}}^2}{\Omega_{\text{ion}}^2 - \omega^2} + \dfrac{\omega_{\text{pe}}^2}{\Omega_{\text{e}}^2 - \omega^2} & \dfrac{i}{\omega}\left(\dfrac{\omega_{\text{pi}}^2 \Omega_{\text{ion}}}{\Omega_{\text{ion}}^2 - \omega^2} - \dfrac{\omega_{\text{pe}}^2 \Omega_{\text{e}}}{\Omega_{\text{e}}^2 - \omega^2}\right) & 0 \\[3mm] -\dfrac{i}{\omega}\left(\dfrac{\omega_{\text{pi}}^2 \Omega_{\text{ion}}}{\Omega_{\text{ion}}^2 - \omega^2} - \dfrac{\omega_{\text{pe}}^2 \Omega_{\text{e}}}{\Omega_{\text{e}}^2 - \omega^2}\right) & 1 + \dfrac{\omega_{\text{pi}}^2}{\Omega_{\text{ion}}^2 - \omega^2} + \dfrac{\omega_{\text{pe}}^2}{\Omega_{\text{e}}^2 - \omega^2} & 0 \\[3mm] 0 & 0 & 1 - \dfrac{\omega_{\text{p}}^2}{\omega^2} \end{bmatrix} \tilde{\mathbf{E}}.$$

$$(5.68)$$

Finally, we use Faraday's law to relate the magnetic and electric field amplitudes:

$$i\mathbf{k} \times \tilde{\mathbf{E}} = i\omega\tilde{\mathbf{B}}. \qquad (5.69)$$

Taking the curl of Eq. (5.69) yields the following:

$$\mathbf{k} \times (\mathbf{k} \times \tilde{\mathbf{E}}) = \omega\mathbf{k} \times \tilde{\mathbf{B}} \qquad (5.70)$$

Using the vector identity $\mathbf{k} \times (\mathbf{k} \times \tilde{\mathbf{E}}) = \mathbf{k}(\mathbf{k} \cdot \tilde{\mathbf{E}}) - k^2\tilde{\mathbf{E}}$ (see Appendix B), we obtain

$$\mathbf{k}(\mathbf{k} \cdot \tilde{\mathbf{E}}) - k^2\tilde{\mathbf{E}}$$

$$= -\frac{\omega^2}{c^2} \begin{bmatrix} 1 + \dfrac{\omega_{\text{pi}}^2}{\Omega_{\text{ion}}^2 - \omega^2} + \dfrac{\omega_{\text{pe}}^2}{\Omega_{\text{e}}^2 - \omega^2} & \dfrac{i}{\omega}\left(\dfrac{\omega_{\text{pi}}^2 \Omega_{\text{ion}}}{\Omega_{\text{ion}}^2 - \omega^2} - \dfrac{\omega_{\text{pe}}^2 \Omega_{\text{e}}}{\Omega_{\text{e}}^2 - \omega^2}\right) & 0 \\[3mm] -\dfrac{i}{\omega}\left(\dfrac{\omega_{\text{pi}}^2 \Omega_{\text{ion}}}{\Omega_{\text{ion}}^2 - \omega^2} - \dfrac{\omega_{\text{pe}}^2 \Omega_{\text{e}}}{\Omega_{\text{e}}^2 - \omega^2}\right) & 1 + \dfrac{\omega_{\text{pi}}^2}{\Omega_{\text{ion}}^2 - \omega^2} + \dfrac{\omega_{\text{pe}}^2}{\Omega_{\text{e}}^2 - \omega^2} & 0 \\[3mm] 0 & 0 & 1 - \dfrac{\omega_{\text{p}}^2}{\omega^2} \end{bmatrix} \tilde{\mathbf{E}}.$$

$$(5.71)$$

Finally, this equation can be rewritten to obtain

$$
\begin{bmatrix}
k^2 - k_x^2 - \frac{\omega^2}{c^2}\varepsilon_1 & -k_xk_y + i\frac{\omega^2}{c^2}\varepsilon_2 & -k_xk_z \\
-k_xk_y - i\frac{\omega^2}{c^2}\varepsilon_2 & k^2 - k_y^2 - \frac{\omega^2}{c^2}\varepsilon_1 & -k_yk_z \\
-k_xk_z & -k_yk_z & k^2 - k_z^2 - \frac{\omega^2}{c^2}\varepsilon_3
\end{bmatrix}
\tilde{\mathbf{E}} = \mathbf{0},
\tag{5.72}
$$

where

$$
\varepsilon_1 = 1 + \frac{\omega_{\mathrm{pi}}^2}{\Omega_{\mathrm{ion}}^2 - \omega^2} + \frac{\omega_{\mathrm{pe}}^2}{\Omega_{\mathrm{e}}^2 - \omega^2},
\tag{5.73}
$$

$$
\varepsilon_2 = \frac{\Omega_{\mathrm{e}}}{\omega}\frac{\omega_{\mathrm{pe}}^2}{\Omega_{\mathrm{e}}^2 - \omega^2} - \frac{\Omega_{\mathrm{ion}}}{\omega}\frac{\omega_{\mathrm{pi}}^2}{\Omega_{\mathrm{ion}}^2 - \omega^2},
\tag{5.74}
$$

and

$$
\varepsilon_3 = 1 - \frac{\omega_p^2}{\omega^2}.
\tag{5.75}
$$

A nontrivial solution of Eq. (5.72) exists only if the determinant of the matrix vanishes. This condition gives us the following dispersion relation:

$$
\cos^2\vartheta = \frac{\left[\frac{\omega^2}{c^2k^2}\varepsilon_{\mathrm{L}}\varepsilon_{\mathrm{R}} - \frac{1}{2}(\varepsilon_{\mathrm{L}} + \varepsilon_{\mathrm{R}})\right]\left[\frac{\omega^2}{c^2k^2}\varepsilon_3 - 1\right]}{\left[\frac{\omega^2}{c^2k^2}\frac{1}{2}(\varepsilon_{\mathrm{L}} + \varepsilon_{\mathrm{R}})\varepsilon_3 - \varepsilon_3 - \frac{\omega^2}{c^2k^2}\varepsilon_{\mathrm{L}}\varepsilon_{\mathrm{R}} + \frac{1}{2}(\varepsilon_{\mathrm{L}} + \varepsilon_{\mathrm{R}})\right]},
\tag{5.76}
$$

where ϑ is the angle between the $\mathbf{k}$ and the $\mathbf{B}_0$ vectors and ε_{L} and ε_{R} are defined the following way:

$$
\varepsilon_{\mathrm{L}} = \varepsilon_1 - \varepsilon_2,
\tag{5.77}
$$

$$
\varepsilon_{\mathrm{R}} = \varepsilon_1 + \varepsilon_2.
\tag{5.78}
$$

Equation (5.76) is the dispersion relation for plasma waves propagating in cold, collisionless, homogeneous plasmas in the presence of a uniform background magnetic field. This dispersion relation is also called the *Appleton–Hartle* relation. Given the wavenumber, k, and the direction of propagation, ϑ, the Appleton–Hartle relation can be solved for the frequency of the wave, ω. Note that in general the Appleton–Hartle relation is quite complicated, and it allows for a wide variety of plasma waves. In this book we limit ourselves to the examination of several relatively simple waves that propagate along the ambient magnetic field lines ($\vartheta = 0$). General examination of the potential solutions of Eq. (5.76) is left to plasma physics textbooks.

5.9   Parallel Plasma Waves

Let us restrict ourselves to plasma waves that propagate along the magnetic field lines ($\Theta = 0$ or $k_z = \pm k$). In this case the wave equation (5.72) simplifies to

$$
\begin{bmatrix}
\frac{c^2 k^2}{\omega^2} - \varepsilon_1 & i\varepsilon_2 & 0 \\
-i\varepsilon_2 & \frac{c^2 k^2}{\omega^2} - \varepsilon_1 & 0 \\
0 & 0 & -\varepsilon_3
\end{bmatrix}
\tilde{\mathbf{E}} = \mathbf{0}.
\tag{5.79}
$$

The coefficient matrix of Eq. (5.79) has three orthogonal eigenvectors and three corresponding eigenvalues. These quantities are the following:

$$
\lambda_1 = \frac{c^2 k^2}{\omega^2} - \varepsilon_{\mathrm{L}}, \qquad
\tilde{\mathbf{E}}_{\mathrm{L}} = \tilde{E}_{\mathrm{L}}
\begin{bmatrix} 1 \\ -i \\ 0 \end{bmatrix}.
\tag{5.80}
$$

$$
\lambda_2 = \frac{c^2 k^2}{\omega^2} - \varepsilon_{\mathrm{R}}, \qquad
\tilde{\mathbf{E}}_{\mathrm{R}} = \tilde{E}_{\mathrm{R}}
\begin{bmatrix} 1 \\ i \\ 0 \end{bmatrix},
\tag{5.81}
$$

$$
\lambda_3 = -\varepsilon_3 \qquad
\tilde{\mathbf{E}}_3 =
\begin{bmatrix} 0 \\ 0 \\ \tilde{E}_3 \end{bmatrix}.
\tag{5.82}
$$

With the help of these quantities, Eq. (5.79) simplifies to the following three equations:

$$
\left(\frac{c^2 k^2}{\omega^2} - \varepsilon_{\mathrm{L}} \right) \tilde{E}_{\mathrm{L}} = 0,
\tag{5.83}
$$

$$
\left(\frac{c^2 k^2}{\omega^2} - \varepsilon_{\mathrm{R}} \right) \tilde{E}_{\mathrm{R}} = 0,
\tag{5.84}
$$

$$
\varepsilon_3 \tilde{E}_3 = 0.
\tag{5.85}
$$

These equations characterize three independent solutions for the electric field amplitude of parallel propagating plasma waves. The conditions for nontrivial solutions are

$$
\frac{c^2 k^2}{\omega^2} - \varepsilon_{\mathrm{L}} = 0,
\tag{5.86}
$$

$$
\frac{c^2 k^2}{\omega^2} - \varepsilon_{\mathrm{R}} = 0,
\tag{5.87}
$$

$$
\varepsilon_3 = 0.
\tag{5.88}
$$

It can be easily seen that Eq. (5.88) corresponds to an electrostatic solution. In this case $\tilde{\mathbf{B}} = \mathbf{0}$ (see Eq. 5.70) and the dispersion relation becomes

$$
\omega^2 = \omega_{\mathrm{p}}^2,
\tag{5.89}
$$

which describes the well-known Langmuir oscillation.

The solutions of Eqs. (5.86) and (5.87) are electromagnetic waves (the amplitude of magnetic perturbation is not zero). The dispersion relations for these waves become

$$k_{\text{L}} = \pm \frac{\omega}{c} \sqrt{1 - \frac{\omega_{\text{pe}}^2}{\omega(\omega + \Omega_{\text{e}})} - \frac{\omega_{\text{pi}}^2}{\omega(\omega - \Omega_i)}}$$

$$= \pm \frac{\omega}{c} \sqrt{1 - \frac{\omega_{\text{p}}^2}{(\omega - \Omega_{\text{ion}})(\omega + \Omega_{\text{e}})}} \tag{5.90}$$

and

$$k_{\text{R}} = \pm \frac{\omega}{c} \sqrt{1 - \frac{\omega_{\text{pe}}^2}{\omega(\omega - \Omega_{\text{e}})} - \frac{\omega_{\text{pi}}^2}{\omega(\omega + \Omega_i)}}$$

$$= \pm \frac{\omega}{c} \sqrt{1 - \frac{\omega_{\text{p}}^2}{(\omega + \Omega_{\text{ion}})(\omega - \Omega_{\text{e}})}}. \tag{5.91}$$

5.9.1   Circularly Polarized Waves

It can be seen that both of these electromagnetic waves have their electric field vectors rotate in the plane perpendicular to the background magnetic field. For this reason both $\tilde{\mathbf{E}}_{\text{R}}$ and $\tilde{\mathbf{E}}_{\text{L}}$ describe *circularly polarized* electomagnetic waves. Waves with complex amplitude $\tilde{\mathbf{E}}_{\text{R}}$ rotate in the right-hand sense with respect to the ambient magnetic field (counterclockwise), whereas waves with complex amplitude $\tilde{\mathbf{E}}_{\text{L}}$ rotate in the left-hand sense (clockwise). This is illustrated in Figure 5.2. It can be seen that right circularly polarized waves rotate in the direction of the electron cyclotron motion, whereas left circularly polarized waves rotate in the direction of the ion gyration.

Next, we explore some interesting properties of the right- and left-hand circularly polarized parallel electromagnetic waves.

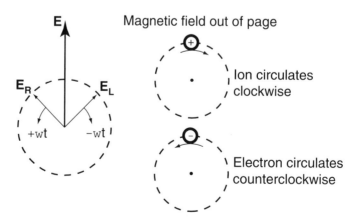

Figure 5.2 Rotation of circularly polarized plasma waves.

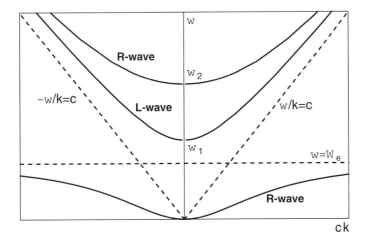

Figure 5.3 Dispersion diagram of parallel propagating electromagnetic waves.

Inspection of Eqs. (5.90) and (5.91) reveals several interesting properties of parallel propagating circularly polarized electromagnetic waves. It is obvious that these can exist only if the expressions under the square root are positive, otherwise the waves are quickly damped. This means that right-hand polarized waves can exist below the electron gyrofrequency, whereas left-hand polarized waves can exist below the ion gyrofrequency.

It should also be pointed out that k_R and k_L become singular at $\omega = \Omega_e$ and $\omega = \Omega_{ion}$, respectively. At these frequency values the electrons (ions) can absorb significant amounts of energy from the waves. Near these *cyclotron resonances* our linear approximation loses validity and much more complicated methods need to be applied. These methods go well beyond the scope of this book.

At high frequencies the dispersion relations (5.90) and (5.91) once again give real values for the wavenumber. For

$$\omega \geq \omega_1 = \frac{1}{2}\Omega_e \left[\sqrt{4\frac{\omega_p^2}{\Omega_e^2} + 1} - 1 \right] \tag{5.92}$$

the left-hand polarized waves reappear, whereas for

$$\omega \geq \omega_2 = \frac{1}{2}\Omega_e \left[\sqrt{4\frac{\omega_p^2}{\Omega_e^2} + 1} + 1 \right] \tag{5.93}$$

the right-hand waves can also exist (here we made use of the fact that $\Omega_e \gg \Omega_{ion}$).

We conclude that there are certain frequency domains where no parallel propagating electromagnetic waves can exist (between Ω_e and ω_1) or only right-hand (between Ω_{ion} and Ω_e) or left-hand polarized waves (ω_1 and ω_2) can propagate. This is illustrated in Figure 5.3, which shows the dispersion diagram of parallel propagating electromagnetic waves. This feature of parallel electromagnetic waves forms the basis of several valuable plasma diagnostic techniques widely used in space physics.

5.9.2   Whistler Waves

We finish our discussion of parallel electromagnetic waves by examining a wave that plays a very important role in ionospheric and magnetospheric physics. Let us restrict ourselves to waves with frequencies above the ion gyrofrequency but well below the electron gyrofrequency ($\Omega_{ion} < \omega \ll \Omega_e$). According to our previous discussion, these are right circularly polarized electromagnetic waves which propagate along the background magnetic field. For these waves the dispersion relations (5.90) and (5.91) simplify to the following:

$$k_R = \pm \frac{\omega}{c} \sqrt{1 - \frac{\omega_{pe}^2}{\omega(\omega - \Omega_e)} - \frac{\omega_{pi}^2}{\omega(\omega + \Omega_{ion})}} \approx \pm \frac{\omega_{pe}}{c\sqrt{\Omega_e}}\sqrt{\omega}, \qquad (5.94)$$

because the plasma frequency is much larger than the ion gyrofrequency (or ω in this case). The group and phase velocities of these waves are

$$
\begin{aligned}
v_{ph} &= \frac{\omega}{k_R} = \pm c \frac{\sqrt{\Omega_e}}{\omega_{pe}}\sqrt{\omega}, \\
v_g &= \frac{d\omega}{dk_R} = \pm 2c \frac{\sqrt{\Omega_e}}{\omega_{pe}}\sqrt{\omega} = 2v_{ph}.
\end{aligned}
\qquad (5.95)
$$

If one generates a pulse containing many frequencies, the higher frequency components of the pulse propagate faster along the magnetic field line than the lower frequency ones. A broadband receiver placed far enough from the origin of the pulse will detect the high frequencies first and the lower frequencies later.

This phenomenon was first observed by field radio operators in World War I using radios operating around 10 kHz. Occasionally they heard short duration signals which started at high frequencies and finished at low frequencies. The explanation of these observations is that lightning generates a pulse of electromagnetic waves containing many frequencies. Some of the waves propagate along the geomagnetic field lines to the opposite hemisphere of the earth, where it is observed as whistling noise, starting at high frequencies and gradually moving to lower frequencies. These waves are called *whistlers*. The generation and propagation of a whistler is schematically depicted in Figure 5.4.

5.9.3   Low-Frequency Waves

The parallel propagating circularly polarized electromagnetic wave dispersion relation can be greatly simplified at low frequencies ($\omega \ll \Omega_{ion}$). In this case we obtain

$$k_R = k_L = \pm \frac{\omega}{c}\sqrt{1 + \frac{\omega_p^2}{\Omega_{ion}\Omega_e}} = \pm \frac{\omega}{c}\sqrt{1 + \frac{c^2}{V_A^2}} \approx \pm \frac{\omega}{V_A} \qquad (5.96)$$

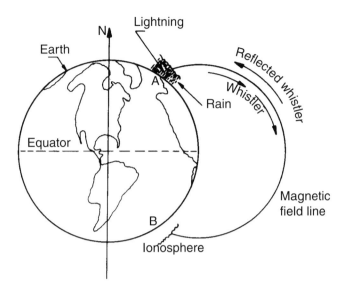

Figure 5.4 Generation of whistler waves in the atmosphere.

since we limited ourselves to nonrelativistic situations ($V_A \ll c$). It can easily be seen that for these waves

$$v_{ph} = v_g = \pm \frac{c}{\sqrt{1 + \frac{c^2}{V_A^2}}} = \pm \frac{V_A}{\sqrt{1 + \frac{V_A^2}{c^2}}} \approx V_A. \tag{5.97}$$

This result indicates that low-frequency parallel propagating electromagnetic waves are linearly polarized (the sum of two counterrotating electric field vectors with equal amplitudes is constant since $k_R = k_L$), and the magnitude of their propagation velocity is the Alfvén speed.

5.10   Problems

Problem 5.1 Derive the phase and group velocities for internal gravity waves.

Problem 5.2 Calculate the wave amplitudes, $\tilde{\rho}$ and $\tilde{\mathbf{u}}$, for gravity waves

1. for horizontal waves ($k_z = 0$) and
2. for vertical waves ($k_x = k_y = 0$).

Problem 5.3 Consider a cold uniform plasma with no magnetic field. The plasma is moving with a constant velocity u_0 in the z direction.

1. Derive the dispersion relation of Langmuir waves.
2. Derive the group and phase velocities of these waves.

Problem 5.4 In the ionosphere a spacecraft detects a broadband low-frequency electromagnetic wave packet propagating along the magnetic field line. The time delay between the 10 kHz and 40 kHz waves is 0.00355 s. If the magnetic field strength is 10^4 nT and the plasma density is uniformly 10^6 m^{-3}, how far was the lightning from the spacecraft?

Chapter 6

Shocks and Discontinuities

It has been observed under certain conditions that most mediums in the space environment can experience abrupt changes of macroscopic parameters. In a broad
sense shocks and discontinuities are defined as transition layers where the state of the
fluid changes from one that is near an equilibrium state to a different one. Examples
involve detonation waves in the atmosphere, shocks, and transition layers in the magnetosphere, the interplanetary medium, and in the Sun. In all these cases the transition
layer is very narrow compared to the characteristic scale of the problem.

In this chapter we consider the fundamental theory of shocks and discontinuities in
neutral gases and quasineutral plasmas.

6.1 Normal Shock Waves in Perfect Gases

In the perfect gas approximation shock waves are discontinuity surfaces separating two
distinct gas states. In higher order approximations (such as the Navier–Stokes equations) the shock wave comprises a region where physical quantities change smoothly
but rapidly. In this case the shock has a finite thickness, generally of the order of the
mean free path.

Because the shock wave is a more or less instantaneous compression of the gas, it
cannot be a reversible process. The energy for compressing the gas flowing through
the shock wave is derived from the kinetic energy of the bulk flow upstream of the
shock wave.

The simplest case for studying shock waves is a normal plane shock wave, when the
gas flows parallel to the x axis and all physical quantities depend only on the x coordinate. In this case the normal vector of the shock surface is parallel to the flow direction.

First, we consider the hydrodynamic description of shock waves in perfect neutral gases ($q = 0$). For the sake of mathematical simplicity, let us consider a steady-state one-dimensional flow in the x direction with constant flow area and neglect the effects of external forces. Such a flow occurs for instance in a very long tube. Also, we assume that the shock wave is perpendicular to the flow direction. The direction of shock waves is usually characterized by the shock normal vector. Shock waves with normal vectors parallel to the flow direction are usually referred to as normal shock waves.

It is assumed that far upstream and far downstream of the shock the gas is very close to thermodynamic equilibrium and the velocity distibution function can be approximated by Maxwellians. This means that in these two regions the macroscopic conservation equations can be obtained using the Euler equations (4.33). In the present one-dimensional, steady-state case these equations simplify to the following:

$$
\begin{aligned}
&\frac{d(nu)}{dx} = 0, \\
&mnu\frac{du}{dx} + \frac{dp}{dx} = 0, \\
&\frac{3}{2}u\frac{dp}{dx} + \frac{5}{2}p\frac{du}{dx} = 0,
\end{aligned}
\tag{6.1}
$$

where m is the particle mass and $n(x)$, $u(x)$, and $p(x)$ represent the particle number density, flow velocity, and pressure, respectively.

The one-dimensional, steady-state continuity and momentum equations can immediately be integrated to yield

$$
\begin{aligned}
&n_1 u_1 = n_2 u_2, \\
&mn_1 u_1^2 + p_1 = mn_2 u_2^2 + p_2,
\end{aligned}
\tag{6.2}
$$

where the subscripts 1 and 2 refer to far upstream and far downstream conditions, respectively. The energy equation can be rewritten into a total derivative using the momentum equation:

$$
\frac{3}{2}u\frac{dp}{dx} + \frac{5}{2}p\frac{du}{dx} = \frac{5}{2}\frac{d(up)}{dx} - u\frac{dp}{dx} = \frac{d}{dx}\left(\frac{5}{2}up + \frac{1}{2}mnu^3\right).
\tag{6.3}
$$

This form of the energy equation can easily be integrated to yield the following conservation relation:

$$
\frac{\gamma}{\gamma - 1}u_1 p_1 + \frac{1}{2}mn_1 u_1^3 = \frac{\gamma}{\gamma - 1}u_2 p_2 + \frac{1}{2}mn_2 u_2^3,
\tag{6.4}
$$

where γ is the specific heat ratio (in our case naturally $\gamma = 5/3$). Next, we want to rewrite the conservation equations (6.2) and (6.4) in terms of a single characteristic parameter, the upstream *Mach number*, M_1. The Mach number is the ratio of the flow speed to the local sound speed, and it can be expressed as

$$M^2 = \frac{mnu^2}{\gamma p}. \tag{6.5}$$

With the help of this definition, the integral form of the momentum and energy equations (given by 6.2 and 6.4) can be rewritten as

$$
p_1\left(1 + \gamma M_1^2\right) = p_2\left(1 + \gamma M_2^2\right),
$$
$$
u_1 p_1 \left(1 + \frac{\gamma - 1}{2} M_1^2\right) = u_2 p_2 \left(1 + \frac{\gamma - 1}{2} M_2^2\right). \tag{6.6}
$$

Also, with the help of the continuity and energy equations one can immediately express the ratio of the upstream and downstream temperatures:

$$\frac{T_1}{T_2} = \frac{n_1 u_1}{n_2 u_2} \frac{T_1}{T_2} = \frac{p_1 u_1}{p_2 u_2} = \frac{2 + (\gamma - 1)M_2^2}{2 + (\gamma - 1)M_1^2}. \tag{6.7}$$

It follows from the definition of the Mach number (Eq. 6.5) that the gas velocity, u, is proportional to the Mach number times the square root of the temperature, $u \propto M T^{1/2}$. With the help of this relation and of Eq. (6.7), one obtains the following expression for the velocity ratio:

$$\frac{u_1}{u_2} = \frac{M_1}{M_2} \sqrt{\frac{T_1}{T_2}} = \frac{M_1}{M_2} \sqrt{\frac{2 + (\gamma - 1)M_2^2}{2 + (\gamma - 1)M_1^2}}. \tag{6.8}$$

The same velocity ratio can also be expressed independently with the help of the continuity and momentum equations (6.1 and 6.6) and the temperature ratio (6.7):

$$\frac{u_1}{u_2} = \frac{1 + \gamma M_1^2}{1 + \gamma M_2^2} \frac{2 + (\gamma - 1)M_2^2}{2 + (\gamma - 1)M_1^2}. \tag{6.9}$$

Equations (6.8) and (6.9) can be combined to obtain an algebraic equation connecting the upstream and downstream Mach numbers:

$$\frac{1 + \gamma M_1^2}{1 + \gamma M_2^2} = \frac{M_1}{M_2} \sqrt{\frac{2 + (\gamma - 1)M_1^2}{2 + (\gamma - 1)M_2^2}}. \tag{6.10}$$

This equation has two solutions. The first solution describes no change in the gas parameters, $M_2 = M_1$. This trivial solution, of course, is not a shock. The second solution is nontrivial and relates the upstream and downstream Mach numbers for a normal shock wave:

$$M_2 = \sqrt{\frac{2 + (\gamma - 1)M_1^2}{2\gamma M_1^2 - (\gamma - 1)}}. \tag{6.11}$$

It is important to note that this solution has a singularity at $M_1^2 = (\gamma - 1)/2\gamma$. Also, the limiting value of M_2^2 as M_1 goes to infinity turns out to be the same value,

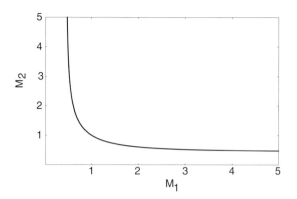

Figure 6.1 The dependence of the downstream Mach number (M_2) on the upstream Mach number (M_1) for one-dimensional normal shocks.

$(\gamma - 1)/2\gamma$. The $M_2(M_1)$ function is shown in Figure 6.1. Inspection of Figure 6.1 reveals that for supersonic upstream flows ($M_1 > 1$) the downstream Mach number is subsonic. This is a physically meaningful solution describing a normal shock wave. For subsonic upstream Mach numbers, however, the solution becomes unphysical exhibiting supersonic downstream Mach numbers. This result essentially means that normal shocks can only be formed in supersonic flows, where the $M_1 > 1$ condition is fulfilled.

The normal shock wave is a sharp (infinitely thin) discontinuity in the gas parameters. Equation (6.11) gives a unique relation between the upstream and downstream Mach numbers. With the help of this equation one can express the change of macroscopic physical quantities through a normal shock wave:

$$\frac{T_2}{T_1} = \frac{\left(2 + (\gamma - 1)M_1^2\right)\left(2\gamma M_1^2 - (\gamma - 1)\right)}{(\gamma + 1)^2 M_1^2},$$

$$\frac{n_2}{n_1} = \frac{u_1}{u_2} = \frac{(\gamma + 1)M_1^2}{2 + (\gamma - 1)M_1^2}, \tag{6.12}$$

$$\frac{p_2}{p_1} = \frac{2\gamma M_1^2 - (\gamma - 1)}{\gamma + 1}.$$

Equations (6.12) are the well-known *Rankine–Hugoniot relations* for normal shock waves in perfect gases.

6.2   MHD Shocks and Discontinuities

In order to consider MHD discontinuities let us start from the conservative form of the ideal MHD equations (4.90) and neglect the effects of external forces (such as gravity):

$$\frac{\partial \rho_m}{\partial t} + \nabla \cdot (\rho_m \mathbf{u}) = 0,$$

$$\frac{\partial(\rho_m \mathbf{u})}{\partial t} + \nabla \cdot \left(\rho_m \mathbf{uu} + p\mathbf{I} + \frac{B^2}{2\mu_0}\mathbf{I} - \frac{\mathbf{BB}}{\mu_0} \right) = \mathbf{0},$$

$$\frac{\partial}{\partial t}\left(\frac{1}{2}\rho_m u^2 + \frac{1}{\gamma-1}p + \frac{B^2}{2\mu_0} \right)$$

$$+ \nabla \cdot \left(\frac{1}{2}\rho_m u^2 \mathbf{u} + \frac{\gamma}{\gamma-1}p\mathbf{u} + \frac{(\mathbf{B}\cdot\mathbf{B})\mathbf{u} - \mathbf{B}(\mathbf{B}\cdot\mathbf{u})}{\mu_0} \right) = 0. \qquad (6.13)$$

These equations are supplemented by the induction equation and the equation expressing the absence of magnetic monopoles:

$$\frac{\partial \mathbf{B}}{\partial t} = \nabla \times (\mathbf{u} \times \mathbf{B}),$$

$$\nabla \cdot \mathbf{B} = 0 \qquad\qquad\qquad\qquad\qquad\qquad\qquad\qquad\qquad\qquad (6.14)$$

Let us consider steady-state conditions and assume that there is an infinitesimally thin discontinuity in the plasma. For the sake of simplicity, let us assume that the discontinuity is planar (i.e., the radius of curvature of the discontinuity surface is very large compared to all other characteristic scales). One can also assume for the sake of mathematical simplicity that all quantities vary only in the direction perpendicular to the discontinuity surface. In this case the continuity, momentum, and energy equations simply state that the total fluxes of mass, momentum, and energy are conserved across the discontinuity:

$$[\rho_m u_n] = 0,$$

$$\left[\rho_m u_n \mathbf{u_t} - \frac{B_n \mathbf{B_t}}{\mu_0} \right] = 0,$$

$$\left[\rho_m u_n^2 + p + \frac{B_t^2 - B_n^2}{2\mu_0} \right] = 0, \qquad\qquad\qquad\qquad (6.15)$$

$$\left[\frac{1}{2}\rho_m \left(u_n^2 + u_t^2 \right) u_n + \frac{\gamma}{\gamma-1}p u_n + \frac{B_t^2}{\mu_0}u_n - \frac{B_n}{\mu_0}(\mathbf{B_t}\cdot\mathbf{u_t}) \right] = 0,$$

where subscripts n and t refer to the normal and tangential components of vectors (with respect to the discontinuity surface). The square brackets denote jumps across the discontinuity:

$$[A] = A_2 - A_1, \qquad\qquad\qquad\qquad\qquad\qquad\qquad\qquad (6.16)$$

where the subscripts 1 and 2 refer to upstream and downstream conditions. Equations (6.15) are called the *Rankine–Hugoniot jump conditions* for ideal MHD. The jump conditions from Eqs. (6.14) are the following:

$$[u_n\mathbf{B_t} - B_n\mathbf{u_t}] = 0,$$

$$[B_n] = 0. \qquad\qquad\qquad\qquad\qquad\qquad\qquad\qquad\qquad (6.17)$$

Equations (6.15) and (6.17) describe several different types of MHD discontinuities.

6.2.1   Contact and Tangential Discontinuities

Let us first consider the case in which no particle transport occurs across the thin discontinuity. In this case evidently $u_{1n} = u_{2n} = 0$. The jump conditions simplify to the following:

$$[B_n \mathbf{B}_t] = \mathbf{0},$$

$$\left[p + \frac{B_t^2 - B_n^2}{2\mu_0} \right] = 0,$$

$$[B_n(\mathbf{B}_t \cdot \mathbf{u}_t)] = 0, \qquad\qquad (6.18)$$

$$[B_n \mathbf{u}_t] = \mathbf{0},$$

$$[B_n] = 0.$$

The last equation tells us that the normal component of the magnetic field does not change across the discontinuity. This can be achieved in two distinct ways.

The first (and simplest) case is when $B_{n1} = B_{n2} = 0$. In this case the only remaining jump condition is

$$\left[p + \frac{B_t^2}{2\mu_0} \right] = 0. \qquad\qquad (6.19)$$

This *tangential discontinuity* occurs when the tangential component of the magnetic field, as well as the density and pressure, is discontinuous across the thin transition layer. However, Eq. (6.19) tells us that the sum of the kinetic and magnetic pressures must remain constant across a tangential discontinuity.

The second case is when $B_{n1} = B_{n2} \neq 0$. In this case Eqs. (6.18) simplify to

$$[\mathbf{B}_t] = \mathbf{0},$$

$$[p] = 0, \qquad\qquad (6.20)$$

$$[\mathbf{u}_t] = \mathbf{0}.$$

In this case there is no mass flux through the discontinuity, but the velocity and magnetic field vectors, as well as the thermal pressure, are continuous across the discontinuity. Such discontinuities are called *contact surfaces* or *contact discontinuities*.

6.2.2   MHD Shocks

Discontinuities when there is transport across the transition layer are called shocks. In the case of ideal MHD there are three distinct types of shocks.

The first type of shock is when both the normal velocity and the density are continuous (and nonzero) across the shock: $u_{1n} = u_{2n} = u_n \neq 0$ and $\rho_{m1} = \rho_{m2} = \rho_m$. This type of shock is *noncompressive*. In this case the jump conditions become the following:

$$\rho_m u_n [\mathbf{u}_t] - \frac{B_n}{\mu_0} [\mathbf{B}_t] = \mathbf{0}, \tag{6.21}$$

$$\left[p + \frac{B_t^2}{2\mu_0} \right] = 0, \tag{6.22}$$

$$\frac{1}{2} \rho_m u_n \left[u_t^2 \right] + \frac{\gamma}{\gamma - 1} u_n [p] + u_n \left[\frac{B_t^2}{\mu_0} \right] - \frac{B_n}{\mu_0} [(\mathbf{B}_t \cdot \mathbf{u}_t)] = 0, \tag{6.23}$$

$$u_n [\mathbf{B}_t] - B_n [\mathbf{u}_t] = \mathbf{0}. \tag{6.24}$$

Equations (6.21) and (6.24) can immediately be combined to obtain

$$\left(u_n^2 - \frac{B_n^2}{\rho_m \mu_0} \right) [\mathbf{B}_t] = \mathbf{0}. \tag{6.25}$$

If the magnetic field changes across the shock, $[\mathbf{B}_t] \neq \mathbf{0}$, and therefore the *plasma velocity across these shocks* can only be

$$u_n = \pm \frac{B_n}{\sqrt{\rho_m \mu_0}} = \pm V_{An}. \tag{6.26}$$

In other words, the *plasma speed* in this case is the normal component of the Alfvén speed. In the next step we multiply Eq. (6.21) by $\mathbf{u}_{t1} + \mathbf{u}_{t1}$ and Eq. (6.24) by $(\mathbf{B}_{t1} + \mathbf{B}_{t1})/\mu_0$ and add the resulting equations together to get

$$\frac{1}{2} \rho_m u_n \left[u_t^2 \right] + u_n \left[\frac{B_t^2}{2\mu_0} \right] - \frac{B_n}{\mu_0} [(\mathbf{B}_t \cdot \mathbf{u}_t)] = 0. \tag{6.27}$$

Substituting this result into Eq. (6.23) yields the following relation:

$$\left[\frac{\gamma}{\gamma - 1} p + \frac{B_t^2}{2\mu_0} \right] = 0. \tag{6.28}$$

However, the only way to satisfy Eqs. (6.22) and (6.28) simultaneously is if

$$[p] = 0 \quad \text{and} \quad \left[B_t^2 \right] = 0. \tag{6.29}$$

In summary, this *shock propagates at the Alfvén speed* (with respect of the plasma), and the pressure and the magnitude of the magnetic field are conserved across the shock. However, the tangential component of the magnetic field can "rotate" across the shock. These shocks are called *Alfvén shocks* or *intermediate shocks*. Since all thermodynamic quantities, such as density and pressure, are continuous across the shock, they are also called *rotational discontinuities*.

The other types of shocks are characterized by jumps in the normal velocity component and the density across the shock. These are called *compressive shocks*. However, the mass flux, Φ_m, is conserved ($\Phi_m = \rho_{m1} u_{1n} = \rho_{m2} u_{2n}$). In this case the jump conditions become

$$[\mathbf{u}_t] - \frac{B_n}{\Phi_m \mu_0}[\mathbf{B}_t] = \mathbf{0}, \tag{6.30}$$

$$\Phi_m[u_n] + [p] + \left[\frac{B_t^2}{2\mu_0}\right] = 0, \tag{6.31}$$

$$\frac{1}{2}\Phi_m\left[u_n^2\right] + \frac{1}{2}\Phi_m\left[u_t^2\right] + \frac{\gamma}{\gamma-1}[pu_n] + \left[\frac{B_t^2}{\mu_0}u_n\right] - \frac{B_n}{\mu_0}[(\mathbf{B}_t \cdot \mathbf{u}_t)] = 0, \tag{6.32}$$

$$[u_n\mathbf{B}_t] - B_n[\mathbf{u}_t] = \mathbf{0}. \tag{6.33}$$

Equations (6.30) and (6.33) can be combined to obtain

$$[u_n\mathbf{B}_t] - \frac{B_n^2}{\Phi_m \mu_0}[\mathbf{B}_t] = \mathbf{0}. \tag{6.34}$$

This relation indicates that the direction of $\mathbf{B}_t$ does not change across the shock, but its magnitude may change. Equation (6.34) is also called the *coplanarity relation*, which expresses that the magnetic field vector remains in the same plane normal to the shock, $\mathbf{e}_n \cdot (\mathbf{B}_1 \times \mathbf{B}_2)$. This coplanarity enables us to rewrite Eq. (6.34) in a scalar form:

$$[u_n B_t] - \frac{B_n^2}{\Phi_m \mu_0}[B_t] = 0. \tag{6.35}$$

The coplanarity of $\mathbf{B}_t$ also implies the coplanarity of $\mathbf{u}_t$ (see Eq. 6.30), and therefore Eq. (6.30) can also be rewritten in a scalar form:

$$[u_t] - \frac{B_n}{\Phi_m \mu_0}[B_t] = 0. \tag{6.36}$$

In the next step we multiply Eq. (6.30) first by $\mathbf{u}_{t1} + \mathbf{u}_{t2}$ and then by $B_n(\mathbf{B}_{t1} + \mathbf{B}_{t2})/\mu_0\Phi_m$, and then we subtract the resulting equations from each other. This yields the following equation:

$$\frac{1}{2}\Phi_m\left[u_t^2\right] + \frac{B_n^2}{2\mu_0^2\Phi_m}\left[B_t^2\right] - \frac{B_n}{\mu_0}[(\mathbf{B}_t \cdot \mathbf{u}_t)] = 0. \tag{6.37}$$

Substituting this relation into Eq. (6.32) yields the following:

$$\frac{1}{2}\Phi_m\left[u_n^2\right] + \frac{\gamma}{\gamma-1}[pu_n] + \left[\frac{B_t^2}{\mu_0}u_n\right] - \frac{B_n^2}{2\mu_0^2\Phi_m}\left[B_t^2\right] = 0. \tag{6.38}$$

Equations (6.31), (6.35), (6.36), and (6.38) represent a complete set of equations for the jump conditions. It can be shown that there are two compressive solutions to these equations. The condition that a nontrivial solution must exist in both of these cases constrains the normal velocity of the plasma flow to the following values:

$$a_{\text{slow}}^2 = \frac{1}{2}\left[\left(a_s^2 + V_A^2 \right) - \sqrt{\left(a_s^2 + V_A^2 \right)^2 - 4a_s^2 V_{A_n}^2} \right] \tag{6.39}$$

and

$$a_{\text{fast}}^2 = \frac{1}{2}\left[\left(a_s^2 + V_A^2 \right) + \sqrt{\left(a_s^2 + V_A^2 \right)^2 - 4a_s^2 V_{A_n}^2} \right], \tag{6.40}$$

where $a_s^2 = \gamma p / \rho_m$ is the acoustic speed, while V_A and V_{A_n} are the Alfvén speed and the normal component of the Alfvén speed. The speeds a_{slow} and a_{fast} are the *slow magnetosonic* and *fast magnetosonic speeds*, respectively. Compressive shocks can have plasma flowing across the shock.

The fast and slow magnetosonic speeds can again be expressed in a more symmetric way:

$$a_{\text{slow}} = \frac{1}{2}\left| \sqrt{a_s^2 + V_A^2 + 2a_s V_{A_n}} - \sqrt{a_s^2 + V_A^2 - 2a_s V_{A_n}} \right|, \tag{6.41}$$

$$a_{\text{fast}} = \frac{1}{2}\left(\sqrt{a_s^2 + V_A^2 + 2a_s V_{A_n}} + \sqrt{a_s^2 + V_A^2 - 2a_s V_{A_n}} \right). \tag{6.42}$$

Note that in Eq. (6.41) the absolute value must be taken since V_{A_n} can be negative.

It can be shown that *slow shocks* (compressive shocks where the plasma flows through the discontinuity with slow magnetosonic speed) are characterized by the decrease of the tangential magnetic field across the shock, whereas *fast shocks* are characterized by an increase of the tangential magnetic field.

Finally, we observe that the following relation always holds between the various characteristic speeds:

$$a_{\text{slow}} \leq V_A \leq a_{\text{fast}}. \tag{6.43}$$

6.3   Problems

Problem 6.1 Consider a monatomic perfect gas in an infinitely long tube with constant cross section. There is a steady-state flow through the tube with a shock wave somewhere in the tube. Far upstream from the shock the distribution function is Maxwellian and supersonic ($M_1 > 1$) with density n_1 and temperature T_1. Far downstream the distribution function is Maxwellian and subsonic.

1. The entropy density per unit mass can be expressed in terms of the phase-space distribution function F:

$$s = -\frac{k}{mn} \iiint_{\infty} d^3v \, F \ln(F).$$

Show that

$$\Delta S = \frac{s_2 - s_1}{C_v} = \frac{5}{3}\ln\left(\frac{T_2}{T_1}\right) - \frac{2}{3}\ln\left(\frac{p_2}{p_1}\right),$$

where C_v is the specific heat at constant volume.

2. Show that in the upstream region the entropy is smaller than in the downstream region.

Problem 6.2 A discontinuity is detected in the solar wind and interplanetary magnetic field. The radial velocity of the solar wind remains constant, but the density jumps from $5\ \text{cm}^{-3}$ to $10\ \text{cm}^{-3}$. The proton temperature jumps from $5\ \text{eV}$ before the discontinuity to $13.8\ \text{eV}$ afterward, whereas the electron temperature remains unchanged at $15\ \text{eV}$. The magnetic field vector of $\mathbf{B} = (0, -8, 6)$ nT before the discontinuity rotates and drops in strength to $\mathbf{B} = (0, 3, 4)$ nT. What type of discontinuity might this be, and why?

Problem 6.3 Consider the following electric and magnetic fields in the rest frame:

$$\mathbf{E} = E_0 \mathbf{e}_y,$$

$$\mathbf{B} = B_0\left(\frac{z}{L}\mathbf{e}_x + \delta\mathbf{e}_z\right).$$

Find the velocity of the moving frame where the electric field vanishes.

Chapter 7

Transport of Superthermal Particles

In this chapter we will briefly consider some of the basic theoretical tools used in describing the transport of superthermal particles. By superthermal particles we mean a very small fraction of the total particle population with energies far exceeding the average thermal energy. These superthermal particles contribute negligibly to the particle density and bulk velocity (due to their very small number compared to the total number of particles), but in some cases they may represent a significant contribution to the pressure and heat flow.

We will consider the basic transport equations describing two kinds of superthermal particles: energetic solar particles and photoelectrons. Since our goal is to provide an introduction to the theoretical tools of space physics, we will constrain our derivations to the most fundamental processes. More sophisticated treatments can be found in the literature.

7.1 Transport of Energetic Particles

As in most cases, we start from the Boltzmann equation describing the evolution of the particle distribution function. The main difference this time is that because superthermal particles can be relativistic, we need to derive a transport equation that is valid for relativistic particles as well. To achieve this we use the form of the Boltzmann equation given by Eq. (2.36), where the variables of the distribution function are time, location, and full (inertial) velocity. In this section we replace velocity with total momentum, $\mathbf{P}$. Therefore the distribution function becomes $F(t, \mathbf{r}, \mathbf{P})$ (where $\mathbf{P}$ is measured in the inertial frame).

To apply Eq. (2.36) to energetic particles and cosmic rays in the heliosphere, we make several assumptions:

- The bulk flow velocity, $\mathbf{u}$, is not relativistic, and it is not affected by the energetic particles. Also, the flow velocity is assumed to be independent of time on timescales comparable to the transport time. In other words, $\mathbf{u}(\mathbf{r})$ is an externally imposed function.

- Collisions with other particles are typically negligible for energetic particles. Most changes in the distribution function are caused by small-angle scattering on magnetic irregularities. It is well established that in the plasma frame these "collisions" change the direction of the particles relatively efficiently, but they are fairly inefficient in changing the particle energy. In light of this observation the collision term is approximated by a special form of the Fokker–Planck term (in the frame of reference of the background plasma):

$$\frac{\delta F}{\delta t} = \frac{F_0 - F}{\tau_c} + \frac{1}{p^2}\frac{\partial}{\partial p}\left(p^2 D_p \frac{\partial F}{\partial p}\right). \tag{7.1}$$

Here p is the magnitude of the particle momentum, $\mathbf{p}$, measured in the plasma frame, D_p is the diffusion coefficient of momentum diffusion (which depends on the magnitude of the momentum), τ_c is the characteristic time of angular isotropization of the momentum distribution, and $F_0(t, \mathbf{r}, p)$ is the momentum-space solid angle–averaged distribution function:

$$F_0(t, \mathbf{r}, p) = \frac{1}{4\pi} \oint_{4\pi} F(t, \mathbf{r}, \mathbf{p})\, d\tilde{\Omega}_p. \tag{7.2}$$

(Here the solid angle element is denoted by $d\tilde{\Omega}_p$.) The relation between $\mathbf{p}$ and $\mathbf{P}$ will be discussed below.

- The external acceleration acting on the particles is due to effects of an externally imposed magnetic field, $\mathbf{B}$. It should be noted that $\mathbf{B}$ is the "average" magnetic field, which excludes turbulence or small modifications generated by the superthermal particles themselves. The electric conductivity of the background plasma is assumed to be infinitely high and therefore $\mathbf{E} + \mathbf{u} \times \mathbf{B} = \mathbf{0}$ (for details see Section 4.4 about the ideal MHD equations). Gravity is neglected in these calculations.

With the help of these assumptions the Boltzmann equation for energetic particles can be written in the following form:

$$\frac{\partial F}{\partial t} + \left(\frac{\mathbf{P}}{m^*} \cdot \nabla\right)F + (\mathbf{f}_L \cdot \nabla_P)F = \frac{\delta F}{\delta t}, \tag{7.3}$$

where $\mathbf{f}_L$ is the Lorentz force acting on the particle and m^* is the relativistic mass ($m^* = \gamma m$, where γ is the Lorentz factor and m is the rest mass).

Because the scattering process conserves energy in the fluid frame, it is useful to transform our transport equation into the fluid frame. Because the bulk flow velocity, $\mathbf{u}$, is not relativistic, the transformation of momentum is a simple Galilean transformation:

$$\mathbf{P} = m^*\mathbf{u} + \mathbf{p} = \gamma m\mathbf{u} + \mathbf{p}. \tag{7.4}$$

Transformation (7.4) is accurate to first order in u/c. With the help of the random momentum, $\mathbf{p}$, one can express the Boltzmann equation in the fluid frame:

$$\left(1 + \frac{\mathbf{u} \cdot \mathbf{p}}{mc^2}\right) \frac{\partial F}{\partial t} + (\mathbf{u} \cdot \nabla) F + \left(\frac{\mathbf{p}}{m} \cdot \nabla\right) F$$
$$- m[(\mathbf{u} \cdot \nabla)\mathbf{u}] \cdot \nabla_p F - [(\mathbf{p} \cdot \nabla)\mathbf{u}] \cdot \nabla_p F$$
$$+ \frac{q}{m} (\mathbf{p} \times \mathbf{B}) \cdot \nabla_p F = \frac{F_0 - F}{\tau_c} + \frac{1}{p^2} \frac{\partial}{\partial p} \left(p^2 D_p \frac{\partial F}{\partial p}\right), \tag{7.5}$$

where q is the electric charge of the energetic particle.

It is usually assumed that due to the relatively rapid isotropization the angular distribution of the energetic particles is close to isotropic. In this case the distribution function can be written as the sum of a momentum-space solid angle–averaged distribution, $F_0(t, \mathbf{r}, p)$, and a small momentum-space direction-dependent function, $F_1(t, \mathbf{r}, \mathbf{p}) = 3m\mathbf{p} \cdot \mathbf{S}(t, \mathbf{r}, p)/p^2$:

$$F(t, \mathbf{r}, \mathbf{p}) = F_0(t, \mathbf{r}, p) + F_1(t, \mathbf{r}, \mathbf{p}) = F_0(t, \mathbf{r}, p) + 3m \frac{\mathbf{p} \cdot \mathbf{S}(t, \mathbf{r}, p)}{p^2}, \tag{7.6}$$

where $\mathbf{S}$ is the flux function. It follows from the definition that the velocity-space solid angle average of F_1 must vanish:

$$\langle F_1(t, \mathbf{r}, \mathbf{p})\rangle_{\tilde{\Omega}_p} = \frac{1}{4\pi} \oint_{4\pi} 3m \frac{\mathbf{p} \cdot \mathbf{S}(t, \mathbf{r}, p)}{p^2} d\tilde{\Omega}_p = 0. \tag{7.7}$$

Substituting Eq. (7.6) into the Boltzmann equation yields the following:

$$\left(1 + \frac{\mathbf{u} \cdot \mathbf{p}}{mc^2}\right) \frac{\partial F_0}{\partial t} + (\mathbf{u} \cdot \nabla) F_0 + \left(\frac{\mathbf{p}}{m} \cdot \nabla\right) F_0$$
$$- m[(\mathbf{u} \cdot \nabla)\mathbf{u}] \cdot \nabla_p F_0 - [(\mathbf{p} \cdot \nabla)\mathbf{u}] \cdot \nabla_p F_0$$
$$+ \frac{q}{m} (\mathbf{p} \times \mathbf{B}) \cdot \nabla_p F_0 + \left(1 + \frac{\mathbf{u} \cdot \mathbf{p}}{mc^2}\right) \frac{\partial F_1}{\partial t} + (\mathbf{u} \cdot \nabla) F_1 + \left(\frac{\mathbf{p}}{m} \cdot \nabla\right) F_1$$
$$- m[(\mathbf{u} \cdot \nabla)\mathbf{u}] \cdot \nabla_p F_1 - [(\mathbf{p} \cdot \nabla)\mathbf{u}] \cdot \nabla_p F_1 + \frac{q}{m} (\mathbf{p} \times \mathbf{B}) \cdot \nabla_p F_1$$
$$= -\frac{F_1}{\tau_c} + \frac{1}{p^2} \frac{\partial}{\partial p} \left(p^2 D_p \frac{\partial F_0}{\partial p}\right) + \frac{1}{p^2} \frac{\partial}{\partial p} \left(p^2 D_p \frac{\partial F_1}{\partial p}\right). \tag{7.8}$$

Next, we take the momentum-space angular average of Eq. (7.8) by applying the following integral operator:

$$\langle\rangle_{\tilde{\Omega}_p} = \frac{1}{4\pi} \oint_{4\pi} d\tilde{\Omega}_p. \tag{7.9}$$

This leads to the following result:

$$
\frac{\partial F_0}{\partial t} + \frac{\mathbf{u}}{c^2} \cdot \frac{\partial \mathbf{S}}{\partial t} + (\mathbf{u} \cdot \nabla) F_0 + (\nabla \cdot \mathbf{S}) - \frac{p}{3} (\nabla \cdot \mathbf{u}) \frac{\partial F_0}{\partial p}
$$

$$
- m^2 [(\mathbf{u} \cdot \nabla) \mathbf{u}] \cdot \frac{1}{p^2} \frac{\partial}{\partial p} (p \mathbf{S}) = \frac{1}{p^2} \frac{\partial}{\partial p} \left(p^2 D_p \frac{\partial F_0}{\partial p} \right). \tag{7.10}
$$

In Eq. (7.10) the time derivative and the momentum derivative of the flux function can be neglected with respect of similar terms involving F_0, leaving

$$
\frac{\partial F_0}{\partial t} + (\mathbf{u} \cdot \nabla) F_0 - \frac{p}{3} (\nabla \cdot \mathbf{u}) \frac{\partial F_0}{\partial p} + (\nabla \cdot \mathbf{S}) = \frac{1}{p^2} \frac{\partial}{\partial p} \left(p^2 D_p \frac{\partial F_0}{\partial p} \right). \tag{7.11}
$$

The individual terms in Eq. (7.11) describe the time derivative and convection of the momentum-space solid angle–averaged distribution function, adiabatic deceleration (or acceleration) due to the changing bulk velocity, contribution from the directional particle flow, and the effect of velocity diffusion.

Next, we multiply Eq. (7.8) by $\mathbf{p}/m$ and take the random momentum solid angle average. This operation leads to the following result:

$$
\mathbf{u} \frac{p^2}{3m^2 c^2} \frac{\partial F_0}{\partial t} + \frac{p^2}{3m^2} \nabla F_0 - [(\mathbf{u} \cdot \nabla) \mathbf{u}] \frac{p}{3} \frac{\partial F_0}{\partial p} + \frac{\partial \mathbf{S}}{\partial t}
$$

$$
+ (\mathbf{u} \cdot \nabla) \mathbf{S} + (\mathbf{S} \cdot \nabla) \mathbf{u} - \frac{p}{3} (\nabla \cdot \mathbf{u}) \frac{\partial \mathbf{S}}{\partial p} - \frac{1}{5} \Lambda \cdot \frac{1}{p^2} \frac{\partial}{\partial p} (p^3 \mathbf{S})
$$

$$
- \frac{q}{m} (\mathbf{S} \times \mathbf{B}) = -\frac{\mathbf{S}}{\tau_c} + \frac{1}{p^2} \frac{\partial}{\partial p} \left(p^2 D_p \frac{\partial \mathbf{S}}{\partial p} \right), \tag{7.12}
$$

where the flow velocity gradient tensor is defined as

$$
\Lambda_{ij} = \frac{\partial u_i}{\partial x_j} + \frac{\partial u_j}{\partial x_i} - \frac{2}{3} \delta_{ij} (\nabla \cdot \mathbf{u}). \tag{7.13}
$$

Next, we will greatly simplify Eq. (7.12). First of all, we recognize that the quantity eB/m is the particle's gyrofrequency. We also recall that the distribution function is assumed to be nearly isotropic, and therefore the flux function is much smaller than $\mathbf{u}F_0$. In space plasmas the gyroperiod and the angular isotropization time-scale are quite short, and so one can neglect all terms containing $\mathbf{S}$, except when they are multiplied by eB/m or $1/\tau_c$. One can also neglect the time derivatives and the adiabatic term containing the velocity derivative of F_0. Equation (7.12) then simplifies to

$$
\mathbf{S} = -\frac{p^2}{3m^2} \tau_c \nabla F_0 + \left(\mathbf{S} \times \frac{\tau_c q \mathbf{B}}{m} \right). \tag{7.14}
$$

Equation (7.14) can be rewritten to

$$
A \cdot \mathbf{S} = -\frac{p^2}{3m^2} \tau_c \nabla F_0, \tag{7.15}
$$

where

$$A_{ij} = \delta_{ij} - \frac{\tau_c q}{m} \epsilon_{ijk} B_k. \tag{7.16}$$

The matrix A can be readily inverted, and we obtain

$$\mathbf{S} = -\frac{\tau_c p^2}{3m^2} \frac{1}{1 + \left(\frac{\tau_c q B}{m}\right)^2} \left[\nabla F_0 + \left(\frac{\tau_c q}{m}\right)^2 \mathbf{B}(\mathbf{B} \cdot \nabla) F_0 + \frac{\tau_c q}{m} (\nabla F_0 \times \mathbf{B}) \right]. \tag{7.17}$$

It can be shown that, for energetic particles, gyration is much faster than isotropization; therefore, $(\tau_c q B)/m \gg 1$. In this approximation Eq. (7.17) becomes

$$\mathbf{S} = -\frac{p^2}{3m^2} \frac{1}{\tau_c} \left(\frac{m}{q B}\right)^2 \left[\nabla F_0 + \left(\frac{\tau_c q}{m}\right)^2 \mathbf{B}(\mathbf{B} \cdot \nabla) F_0 + \frac{\tau_c q}{m} (\nabla F_0 \times \mathbf{B}) \right]. \tag{7.18}$$

This equation can now be substituted into Eq. (7.11), and one can obtain a transport equation for F_0:

$$\frac{\partial F_0}{\partial t} + (\mathbf{u} \cdot \nabla) F_0 - \frac{p}{3} (\nabla \cdot \mathbf{u}) \frac{\partial F_0}{\partial p} - \frac{p^2}{3m} \frac{1}{q} \nabla \cdot \left(\nabla F_0 \times \frac{\mathbf{B}}{B^2} \right)$$
$$= \nabla \cdot (\kappa \cdot \nabla F_0) + \frac{1}{p^2} \frac{\partial}{\partial p} \left(p^2 D_p \frac{\partial F_0}{\partial p} \right), \tag{7.19}$$

where the tensor κ is defined as

$$\kappa = \frac{p^2}{3m^2} \tau_c \left[\frac{1}{\tau_c^2 \Omega_c^2} \mathbf{I} + \mathbf{b} \mathbf{b} \right]. \tag{7.20}$$

Here $\Omega_c = |q| B/m$ is the particle gyrofrequency and $\mathbf{b} = \mathbf{B}/B$ is the unit vector along the magnetic field line. The last term on the left-hand side of Eq. (7.19) can be manipulated to obtain the following equation:

$$\frac{\partial F_0}{\partial t} + (\mathbf{u} \cdot \nabla) F_0 - \frac{p}{3} (\nabla \cdot \mathbf{u}) \frac{\partial F_0}{\partial p} + (\mathbf{V}_d \cdot \nabla) F_0$$
$$= \nabla \cdot (\kappa \cdot \nabla F_0) + \frac{1}{p^2} \frac{\partial}{\partial p} \left(p^2 D_p \frac{\partial F_0}{\partial p} \right), \tag{7.21}$$

where we introduced the guiding-center drift velocity vector of energetic particles:

$$\mathbf{V}_d = \frac{p^2}{3m} \frac{1}{q} \left(\nabla \times \frac{\mathbf{B}}{B^2} \right). \tag{7.22}$$

Inspection of Eq. (7.21) reveals two new terms in the transport equation for the velocity-space solid angle–averaged distribution function. The last term on the left-hand side describes cosmic ray drift: This term is responsible for some very interesting cosmic ray behavior. Note that because the drift velocity contains q, the direction of the cosmic ray drift depends on the sign of the particle charge. The first term on the right-hand side describes spatial diffusion. (Note that diffusion along the magnetic field line is much more efficient than perpendicular diffusion.) This is a very important feature of energetic particle propagation. It is important to note that spatial diffusion is a consequence of a combination of angular scattering and anisotropic streaming. This form of the energetic particle transport equation was first put forward by Parker[1] (the original version of the equation did not contain the velocity diffusion term).

In a more sophisticated treatment the scattering process cannot be described with a simple relaxation such as τ_c. The accurate connection between the proper parallel and perpendicular diffusion coefficients and the properties of the stochastic magnetic fields[2] are still not fully understood.

7.2   Guiding Center Transport

It was shown earlier in this book that in the presence of a magnetic field the motion of charged particles can be decomposed into two parts: gyration around the guiding center and the motion of the guiding center itself. In the case when the particle gyroradius is small compared to the characteristic length of magnetic field variation, the gyromotion becomes quite unimportant, and the transport of the particles can be characterized by the transport of their guiding centers. This approximation is particularly important for photoelectron transport in the ionospheres of magnetized planets (such as Earth). As we shall see later in Section 10.8, photoelectron transport plays a crucial role in the energetics of the terrestrial ionosphere. Here we will briefly outline the derivation of the kinetic transport equation that can be used to describe the transport of photoelectrons. In this derivation we follow the work of Khazanov and his coworkers[3], and we will extensively refer back to quantities and concepts defined in Chapter 1 concerning single particle motion in electric and magnetic fields.

Let us start from the Boltzmann equation as expressed in the inertial phase space (Eq. 2.36):

$$\frac{\partial F(t, \mathbf{r}, \mathbf{v})}{\partial t} + (\mathbf{v} \cdot \nabla) F(t, \mathbf{r}, \mathbf{v}) + (\mathbf{a} \cdot \nabla_v) F(t, \mathbf{r}, \mathbf{v}) = \frac{\delta F(t, \mathbf{r}, \mathbf{v})}{\delta t}. \qquad (7.23)$$

Here the function $F(t, \mathbf{r}, \mathbf{v})$ describes the phase-space distribution of superthermal particles. However, we are not really interested in F as a function of actual particle location and velocity, but we want to know the distribution of particles as a function of the location and the velocity of their guiding center. The transformation between phase-space coordinates and guiding center coordinates requires knowledge of the

[1] Parker, E. N., "The passage of energetic particles through interplanetary space," *Planet. Space Sci.*, **13**, 9, 1965.
[2] Jokipii, J. R., "Propagation of cosmic rays in the solar wind," *Rev. Geophys. Space Phys.*, **9**, 27, 1971.
[3] Khazanov, G. V., Neubert, T., and Gefan, G. D., "Kinetic theory of ionosphere–plasmasphere transport of suprathermal electrons," *IEEE Trans. Plasma Sci.*, **22**, 187, 1994.

magnetic field vector **B**. The location of the guiding center can be written as (cf. Eq. 1.30):

$$\mathbf{r}_g = \mathbf{r} - \frac{m}{qB^2}(\mathbf{v} \times \mathbf{B}). \tag{7.24}$$

Next, we express the distribution function, F, in terms of the guiding center parameters, $s, \mathbf{r}_\perp, \Phi_g, v_\parallel, v_\perp$, and ψ_g, where s is distance along the magnetic field line containing the guiding center at some initial time (we will call this field line the reference field line), $r_\perp$ and Φ_g are the perpendicular distance and azimuth angle characterizing the location of the guiding center with respect to the reference field line, while $v_\parallel, v_\perp$, and ψ_g are the cylindrical velocity components of the guiding center with respect to the reference magnetic field line. It should be noted that all these quantities depend on the phase-space coordinates, $\mathbf{r}$ and $\mathbf{v}$. With the help of these quantities the Boltzmann equation can be written as follows:

$$\frac{\partial F}{\partial t} + \dot{s}\frac{\partial F}{\partial s} + \dot{r}_\perp\frac{\partial F}{\partial r_\perp} + \dot{\Phi}_g\frac{\partial F}{\partial \Phi_g} + \dot{v}_\parallel\frac{\partial F}{\partial v_\parallel} + \dot{v}_\perp\frac{\partial F}{\partial v_\perp} + \dot{\psi}_g\frac{\partial F}{\partial \psi_g} = \frac{\delta F}{\delta t}. \tag{7.25}$$

At this point we make an important simplifying assumption. It is usually assumed that the distribution of particles is independent of the gyration phase angles (both location and velocity azimuth angles, Φ_g and ψ_g). Thus Eq. (7.25) simplifies to

$$\frac{\partial F}{\partial t} + \dot{s}\frac{\partial F}{\partial s} + \dot{r}_\perp\frac{\partial F}{\partial r_\perp} + \dot{v}_\parallel\frac{\partial F}{\partial v_\parallel} + \dot{v}_\perp\frac{\partial F}{\partial v_\perp} = \frac{\delta F}{\delta t}. \tag{7.26}$$

In the next step we average this equation over a gyration period (i.e., we integrate over the azimuth angle and divide the result by 2π). This operation results in the following:

$$\frac{\partial F}{\partial t} + \langle \dot{s} \rangle_{\Phi_g}\frac{\partial F}{\partial s} + \langle \dot{r}_\perp \rangle_{\Phi_g}\frac{\partial F}{\partial r_\perp} + \langle \dot{v}_\parallel \rangle_{\Phi_g}\frac{\partial F}{\partial v_\parallel} + \langle \dot{v}_\perp \rangle_{\Phi_g}\frac{\partial F}{\partial v_\perp} = \frac{\delta F}{\delta t}. \tag{7.27}$$

The averaged quantities, $\langle \dot{s} \rangle_{\Phi_g}, \langle \dot{r}_\perp \rangle_{\Phi_g}, \langle \dot{v}_\parallel \rangle_{\Phi_g}$, and $\langle \dot{v}_\perp \rangle_{\Phi_g}$ were discussed extensively in Chapter 1, in connection with guiding center drifts. These results can be directly applied here. For the sake of simplicity, we neglect any externally imposed gravitational fields; this can be readily incorporated when needed.

It is quite obvious that with very good accuracy the average guiding center velocity along the reference magnetic field line is just the parallel velocity of the particle (because the guiding center drift velocities are much smaller than the velocity of superthermal particles):

$$\langle \dot{s} \rangle_{\Phi_g} = v_\parallel. \tag{7.28}$$

In the perpendicular direction only drift motions can transport the guiding center, because most of the particle's perpendicular velocity is associated with gyration around the magnetic field lines. The average guiding center drift perpendicular to the reference field line, $\mathbf{V}_{\mathrm{drift}\perp}$, is the sum of all drifts discussed in Chapter 1. This means that

$$\langle \dot{r}_\perp \rangle_{\Phi_g} = V_{\mathrm{drift}\perp}. \tag{7.29}$$

In most space plasmas this drift is small, and therefore the term describing the perpendicular transport of the guiding center is usually neglected. The gyration-averaged acceleration of the particles can be obtained from the magnetic mirror force (Eq. 1.44) plus the acceleration caused by externally imposed parallel electric fields:

$$\langle \dot{v}_\parallel \rangle_{\Phi_g} = \frac{q}{m} E_\parallel - \frac{1}{2} \frac{v_\perp^2}{B} \frac{dB}{ds}. \tag{7.30}$$

Finally, $\langle \dot{v}_\perp \rangle_{\Phi_g}$ can be calculated by noting that the total particle energy is conserved in a magnetic field, and therefore

$$v_\perp \frac{dv_\perp}{dt} = -v_\parallel \frac{dv_\parallel}{dt} \tag{7.31}$$

(the effects of the perpendicular electric field components are neglected for the sake of simplicity). Substituting Eq. (7.30) into Eq. (7.31) yields the last desired equation for $\langle \dot{v}_\perp \rangle_{\Phi_g}$:

$$\langle \dot{v}_\perp \rangle_{\Phi_g} = \frac{1}{2} \frac{v_\parallel v_\perp}{B} \frac{dB}{ds}. \tag{7.32}$$

Now one can substitute relations (7.30) and (7.32) into Eq. (7.27) to obtain the following transport equation:

$$\frac{\partial F}{\partial t} + v_\parallel \frac{\partial F}{\partial s} + \left(\frac{q}{m} E_\parallel - \frac{1}{2} \frac{v_\perp^2}{B} \frac{dB}{ds} \right) \frac{\partial F}{\partial v_\parallel} + \frac{1}{2} \frac{v_\parallel v_\perp}{B} \frac{dB}{ds} \frac{\partial F}{\partial v_\perp} = \frac{\delta F}{\delta t}. \tag{7.33}$$

The parallel and perpendicular velocity components frequently are expressed in terms of the magnitude of velocity (speed), v, and the cosine of the pitch angle, $\mu = \cos \Theta$. Because $v_\parallel = \mu v$ and $v_\perp = v \sqrt{1 - \mu^2}$, we have

$$v = \sqrt{v_\parallel^2 + v_\perp^2} \tag{7.34}$$

and

$$\mu = \cos \left[\arctan \left(\frac{v_\perp}{v_\parallel} \right) \right]. \tag{7.35}$$

Using these relations, one obtains

$$
\begin{aligned}
\frac{\partial F}{\partial v_\parallel} &= \frac{\partial v}{\partial v_\parallel} \frac{\partial F}{\partial v} + \frac{\partial \mu}{\partial v_\parallel} \frac{\partial F}{\partial \mu} = \mu \frac{\partial F}{\partial v} + \frac{1 - \mu^2}{v} \frac{\partial F}{\partial \mu}, \\
\frac{\partial F}{\partial v_\perp} &= \frac{\partial v}{\partial v_\perp} \frac{\partial F}{\partial v} + \frac{\partial \mu}{\partial v_\perp} \frac{\partial F}{\partial \mu} = \sqrt{1 - \mu^2} \frac{\partial F}{\partial v} - \frac{\mu \sqrt{1 - \mu^2}}{v} \frac{\partial F}{\partial \mu}.
\end{aligned} \tag{7.36}
$$

Substituting these relations into Eq. (7.33) yields the following transport equation for superthermal particles:

$$\frac{\partial F}{\partial t} + \mu v \frac{\partial F}{\partial s} + \frac{q}{m} E_{\parallel} \mu \frac{\partial F}{\partial v} + \left(\frac{q}{m} E_{\parallel} - \frac{v^2}{2B} \frac{dB}{ds} \right) \frac{1 - \mu^2}{v} \frac{\partial F}{\partial \mu} = \frac{\delta F}{\delta t}. \quad (7.37)$$

In many practical applications we consider the flux of superthermal particles and not the distribution function. The flux is defined as

$$\Phi = \frac{2\varepsilon F}{m^2}, \quad (7.38)$$

where ε is the kinetic energy: $\varepsilon = mv^2/2$. Assuming that the particle velocity changes only slowly along the field line, one obtains the following equation for the flux in terms of the kinetic energy:

$$\sqrt{\frac{m}{2\varepsilon}} \frac{\partial \Phi}{\partial t} + \mu \frac{\partial \Phi}{\partial s} + q E_{\parallel} \mu \varepsilon \frac{\partial}{\partial \varepsilon} \left(\frac{\Phi}{\varepsilon} \right) +$$
$$\left(q \frac{E_{\parallel}}{\varepsilon} - \frac{1}{B} \frac{dB}{ds} \right) \frac{1 - \mu^2}{2} \frac{\partial \Phi}{\partial \mu} = \sqrt{\frac{m}{2\varepsilon}} \frac{\delta \Phi}{\delta t}. \quad (7.39)$$

This is the standard form of the transport equation for the guiding center motion of superthermal particles.

7.3   Problems

Problem 7.1 Near the orbit of Earth the magnitude of the interplanetary magnetic field is about 5 nT. Assume that the magnetic field is in the z direction, and consider energetic particles with a mean free path of 1 AU. Evaluate the diffusion tensor, κ, for 100 keV, 1 MeV, 10 MeV, 100 MeV, and 1 GeV protons.

Problem 7.2 Evaluate the guiding center drift velocity vector of energetic particles, V_d, in a radial magnetic field ($\mathbf{B} = B_0 \mathbf{e}_r$).

Part II

The Upper Atmosphere

Chapter 8

The Terrestrial Upper Atmosphere

Sidney Chapman introduced the nomenclature used to describe the various regions of the upper atmosphere. The classification is primarily based on the variation of temperature with altitude. In this system the regions are called "spheres" and the boundaries between the regions are called "pauses."

The *troposphere* (in Greek it means "turning sphere") is the lowest atmospheric region. It begins at the surface (which provides the major heat source for the atmosphere) and extends to about 10–12 km. This region is mainly characterized by a negative temperature gradient (≈ -10 K/km). The troposphere is bounded by the *tropopause*, which separates the troposphere from the *stratosphere* (Greek word for "layered sphere"). The temperature at the tropopause is about 200 K. Originally the stratosphere was thought to be isothermal, but in fact, in this region the temperature increases about 2 K/km due to the absorption of solar UV radiation by stratospheric ozone. Stratospheric ozone is particularly important because it absorbs UV radiation harmful to life.

The maximum temperature (≈ 270 K) is reached at the *stratopause*, which is located at around 50 km altitude. Above the stratosphere lies the *mesosphere* (middle atmosphere), where the temperature again decreases with altitude. The temperature reaches its minimum (≈ 180 K) at the *mesopause*, located at an altitude of about 85 km. The mesopause is the coldest region in the terrestrial atmosphere. The mesosphere is a region where complex chemical and aeronomical processes (such as airglow) take place. Because it extends from about 50 to 85 km altitudes, it is quite difficult to carry out *in situ* observations in the mesosphere (sounding rockets and remote observations are the primary sources of information about this important region).

The three lowest regions of the terrestrial atmosphere have similar hydrodynamic characteristics, and therefore they together constitute the *homosphere*, within which the mean molecular mass does not change significantly (the mean molecular mass in

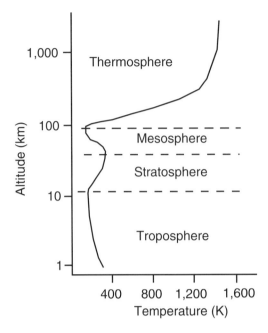

Figure 8.1 Typical profile of neutral atmospheric temperature with the various atmospheric layers.

the homosphere is about 29 amu (atomic mass units)). In this region the rate of mixing of the atmospheric gases is fast enough to produce a more or less uniform relative composition for the major atmospheric constituents (N_2, O_2, Ar, CO_2).

Above the mesopause the solar UV radiation is fairly efficiently absorbed, whereas the reradiation processes are rather inefficient. The interplay between these processes results in a dramatic increase of temperature. This region is called the *thermosphere*. The thermospheric temperature eventually becomes nearly constant at values that are usually well over 1,000 K (the actual temperature varies with solar activity). This is the "hottest" part of the terrestrial atmosphere. The main regions of the neutral atmosphere are schematically shown in Figure 8.1.

In the thermosphere the mixing of constituents is inhibited by the positive temperature gradient. Therefore the various atmospheric constituents may separate due to gravity, leading to compositional variations with altitude. This region is also called the *heterosphere*. Above about 600 km altitude individual atoms can escape from the Earth's gravitational field without collisions. This region is called the *exosphere* or *geocorona*.

In what follows we will apply some of the theoretical tools discussed in Part I and explore the physical and chemical processes controlling the main features of the terrestrial upper atmosphere.

8.1   Hydrostatic Equilibrium

One of the most important features of the atmosphere is the vertical profile of density and pressure. The altitude variation of these quantities is described by the assumption of hydrostatic equilibrium, which can be obtained from the gas transport equations. The main assumptions of hydrostatic equilibrium are the following:

- The atmosphere is well mixed and can be treated as a single fluid with a mean molecular mass of $\bar{m}$.
- The atmosphere is collisional, and the distribution function of molecular velocities can locally be approximated by Maxwellians. This assumption is often referred to as assuming local thermodynamic equilibrium (LTE). This means that the transport of each atmospheric constituent can be described by a set of Euler equations coupled to the other constituents via collisions (see Eq. 4.35).
- The atmosphere is in a steady-state, static equilibrium. This means that the velocities of the constituents can be neglected.
- Heat transfer between species and heat conduction effects are neglected.
- Local mass production and loss terms compensate each other (i.e., there is no net particle production).

Using these approximations the continuity and energy equations are automatically satisfied, while the momentum equation yields the following:

$$\nabla p = \bar{m} n \mathbf{g} = \rho_m \mathbf{g}. \tag{8.1}$$

Let us introduce a coordinate system with the altitude axis, z, pointing upward in the vertical direction. In this case $\mathbf{g} = (0, 0, -g)$, and Eq. (8.1) can be written as

$$\frac{dp}{dz} = -\bar{m} n g \tag{8.2}$$

or

$$\frac{dp}{dz} = -\frac{\bar{m} g}{kT} p = -\frac{1}{H} p, \tag{8.3}$$

where we introduced the *scale height*, $H = kT / \bar{m} g$. Observations indicate that H is a slowly varying function of location and in Eq. (8.3) it can be considered locally constant. In this case Eq. (8.3) can be easily integrated to obtain the well-known *barometric altitude formula*:

$$p_{\text{hs}}(z) = p_0 \exp\left(-\frac{z - z_0}{H}\right), \tag{8.4}$$

where p_0 is the pressure at the reference altitude z_0 (it may slowly vary with location). Because it was assumed that H varies only slowly with location, and because the scale height is a linear function of the gas temperature, one can conclude that the temperature is also a slowly varying function of location. In this approximation one obtains a similar barometric altitude formula for the gas density:

$$\rho_{\text{hs}} = \rho_{m0} \exp\left(-\frac{z - z_0}{H}\right). \tag{8.5}$$

It should be noted that the scale height for a given species, $H_s = kT_s/m_s g$, varies linearly with the temperature and inversely with the molecular mass of the species. Thus heavier molecules have smaller scale heights, and consequently, their density decreases more rapidly with altitude. This effect becomes important only in the heterosphere, where molecular mixing can no longer compensate for this gravitational separation of lighter and heavier species.

8.2   Stability of the Atmosphere

The equation of hydrostatic equilibrium given by Eq. (8.1) can be satisfied with a wide range of temperature altitude profiles, $T(z)$. However, we are interested in temperature profiles that lead to a stable atmosphere. By stability we mean that a small parcel of air displaced adiabatically from its original location by a small altitude of Δz experiences forces that return it to its original location.

We assume that the atmosphere is characterized by altitude profiles of pressure, density, and temperature, $p(z)$, $\rho_m(z)$, and $T(z)$. These quantities are related by some gas law. Therefore one can express the density in terms of pressure and temperature: $\rho_m = \rho_m(p, T)$.

When a small parcel of air is displaced from altitude z_0 to $z' = z_0 + \Delta z$, its pressure adjusts to the new altitude instantaneously, while its temperature lags behind: $\rho'_m = \rho_m(p', T_0 + \delta T)$ (here $p' = p(z')$, $T_0 = T(z_0)$, and δT is the adiabatic temperature change due to change of pressure from $p_0 = p(z_0)$ to p'). If $\Delta z > 0$ and ρ'_m is smaller than the surrounding density of $\rho_m(p', T')$, the buoyancy force will drive the parcel upward even more instead of returning it to its original location. In this case the atmosphere is *hydrostatically unstable*.

In light of the above argument the condition of hydrostatic stability is the following:

$$\rho_m(p', T_0 + \delta T) - \rho_m(p', T') > 0. \tag{8.6}$$

This expression can be expanded into a Taylor series around T_0 to yield the following condition for hydrostatic stability:

$$\left(\frac{\partial \rho_m(p', T_0)}{\partial T} \right)_p \left[\delta T - \frac{dT_0}{dz} \Delta z \right] > 0. \tag{8.7}$$

Next, we will express the adiabatic temperature change, δT, in terms of the pressure gradient. To do so one has to recall that in adiabatic processes the quantity p/ρ_m^γ remains constant (see Eq. 4.32). Assuming that the perfect gas law can be applied during the adiabatic compression or expansion of the gas, we conclude that the quantity $T^\gamma/p^{(\gamma-1)}$ is constant for the displaced parcel of air. This means that

$$\frac{T_0^\gamma}{p_0^{(\gamma-1)}} = \frac{(T_0 + \delta T)^\gamma}{p'^{(\gamma-1)}}. \tag{8.8}$$

The right-hand side can be manipulated to obtain with first-order accuracy

$$
\frac{(T_0 + \delta T)^\gamma}{p'^{(\gamma-1)}} = \frac{T_0^\gamma}{p_0^{(\gamma-1)}} \frac{\left(1 + \frac{\delta T}{T_0}\right)^\gamma}{\left(1 + \frac{1}{p_0}\frac{dp_0}{dz}\Delta z\right)^{(\gamma-1)}}
$$
$$
= \frac{T_0^\gamma}{p_0^{(\gamma-1)}} \left[1 + \gamma \frac{\delta T}{T_0} - (\gamma - 1)\frac{1}{p_0}\frac{dp_0}{dz}\Delta z\right]. \tag{8.9}
$$

Substituting expression (8.9) into Eq. (8.8) yields the following expression for δT:

$$
\delta T = \frac{\gamma - 1}{\gamma}\frac{T_0}{p_0}\frac{dp_0}{dz}\Delta z. \tag{8.10}
$$

Next, we substitute expression (8.10) into the stability condition given by Eq. (8.7):

$$
\left(\frac{\partial \rho_m(p', T_0)}{\partial T}\right)_p \left[\frac{\gamma - 1}{\gamma}\frac{T_0}{p_0}\frac{dp_0}{dz} - \frac{dT_0}{dz}\right]\Delta z > 0. \tag{8.11}
$$

Using the ideal gas law, it is easy to show that

$$
\left(\frac{\partial \rho_m(p', T_0)}{\partial T}\right)_p \propto \frac{\partial}{\partial T}\left(\frac{p'}{T}\right)_{p=p', T=T_0} = -\frac{p'}{T_0^2} < 0. \tag{8.12}
$$

Using relation (8.12) and assuming $\Delta z > 0$, Eq. (8.11) can be written as

$$
\frac{dT_0}{dz} > \frac{\gamma - 1}{\gamma}\frac{T_0}{p_0}\frac{dp_0}{dz}. \tag{8.13}
$$

Finally, one can use the condition for hydrostatic equilibrium (Eq. 8.2) to obtain the following condition for hydrostatic stability:

$$
\frac{dT}{dz} > -\frac{\gamma - 1}{\gamma}\frac{\bar{m}}{k}g = -\frac{g}{C_p}, \tag{8.14}
$$

where C_p is the specific heat at constant pressure. The quantity $\Gamma_{ad} = g/C_p$ is called the *adiabatic lapse rate*, and its numerical value is $\Gamma_{ad} = 9.8$ K/km.

   This result means that a small parcel of displaced air will be forced to move toward its original altitude if the temperature gradient is larger than the adiabatic lapse rate. However, the parcel may overshoot its original location and oscillate around it. The frequency of this oscillation is the Brunt–Väisälä frequency (see Eq. 5.24). For temperature gradients below the adiabatic lapse rate, the atmosphere becomes vertically unstable and large-scale vertical motions are generated.

8.3   Winds and Waves

There are wind systems in the upper atmosphere, as well as periodic motions, such as waves. The winds and waves play a very important role in the energy transport in the upper atmosphere. Here we will try to explain the main features of these phenomena, without going into too many details.

The upper atmosphere is treated as a single neutral fluid in local thermodynamic equilibrium. We apply a modified version of the Euler equations, where the effects due to planetary rotation are also incorporated.

8.3.1 Acceleration Due to Planetary Rotation

Let us consider a coordinate system that rotates with a constant angular velocity $\mathbf{\Omega}_r$. Taking the time derivative of an arbitrary vector $\mathbf{A}$ in the rotating system requires an adjustment for the rotation:

$$\left(\frac{d\mathbf{A}}{dt}\right)_{\text{inr}} = \left(\frac{d\mathbf{A}}{dt}\right)_{\text{rot}} + \mathbf{\Omega}_r \times \mathbf{A}, \tag{8.15}$$

where the subscripts "inr" and "rot" refer to derivatives in the inertial and rotating frames, respectively.

In our particular case we obtain

$$\left(\frac{d\mathbf{r}}{dt}\right)_{\text{inr}} = \left(\frac{d\mathbf{r}}{dt}\right)_{\text{rot}} + \mathbf{\Omega}_r \times \mathbf{r} \tag{8.16}$$

or in another form

$$\mathbf{v}_{\text{inr}} = \mathbf{v}_{\text{rot}} + \mathbf{\Omega}_r \times \mathbf{r}. \tag{8.17}$$

Now one can express the inertial velocity in terms of the velocity in the rotating system:

$$\begin{aligned}
\left(\frac{d\mathbf{v}_{\text{inr}}}{dt}\right)_{\text{inr}} &= \left(\frac{d\mathbf{v}_{\text{inr}}}{dt}\right)_{\text{rot}} + \mathbf{\Omega}_r \times \mathbf{v}_{\text{inr}} \\
&= \left(\frac{d(\mathbf{v}_{\text{rot}} + \mathbf{\Omega}_r \times \mathbf{r})}{dt}\right)_{\text{rot}} + \mathbf{\Omega}_r \times (\mathbf{v}_{\text{rot}} + \mathbf{\Omega}_r \times \mathbf{r}) \\
&= \left(\frac{d\mathbf{v}_{\text{rot}}}{dt}\right)_{\text{rot}} + 2\mathbf{\Omega}_r \times \mathbf{v}_{\text{rot}} + \mathbf{\Omega}_r \times (\mathbf{\Omega}_r \times \mathbf{r}).
\end{aligned} \tag{8.18}$$

Here the first term describes the acceleration in the rotating planetary system, the second term is the Coriolis acceleration, while the third term describes the centrifugal acceleration.

Using these results, one obtains the Euler equations for a neutral species in the rotating coordinate system (this is the coordinate system rotating with the Earth):

$$\begin{aligned}
&\frac{\partial m_s n_s}{\partial t} + \nabla \cdot (m_s n_s \mathbf{u}_s) = m_s \frac{\delta n_s}{\delta t}, \\[1ex]
&m_s n_s \frac{\partial \mathbf{u}_s}{\partial t} + m_s n_s (\mathbf{u}_s \cdot \nabla)\mathbf{u}_s + \nabla p_s - m_s n_s \mathbf{g} \\[1ex]
&+ 2m_s n_s \mathbf{\Omega}_r \times \mathbf{u}_s + m_s n_s \mathbf{\Omega}_r \times (\mathbf{\Omega}_r \times \mathbf{r}) = m_s n_s \mathbf{a}_s + m_s n_s \frac{\delta \mathbf{u}_s}{\delta t}, \\[1ex]
&\frac{1}{\gamma - 1}\frac{\partial p_s}{\partial t} + \frac{1}{\gamma - 1}(\mathbf{u}_s \cdot \nabla)p_s + \frac{\gamma}{\gamma - 1}p_s(\nabla \cdot \mathbf{u}_s) - \nabla \cdot (\kappa_s \nabla T_s) = \frac{1}{\gamma - 1}\frac{\delta p_s}{\delta t}.
\end{aligned} \tag{8.19}$$

Here we used the specific heat ratio of the gas, γ (earlier it was assumed that the gas did not have internal degrees of freedom). We also added an externally imposed acceleration, $\mathbf{a}_s$ (this acceleration will help us to describe the effects of electromagnetic forces in the case of charged particles).

It is costomary to incorporate the centrifugal acceleration into the gravitational acceleration and define an "effective" gravitational acceleration:

$$\mathbf{g}_{\text{eff}} = \mathbf{g} - \mathbf{\Omega}_r \times (\mathbf{\Omega}_r \times \mathbf{r}). \tag{8.20}$$

With this the Euler equations become

$$\frac{\partial m_s n_s}{\partial t} + \nabla \cdot (m_s n_s \, \mathbf{u}_s) = m_s \frac{\delta n_s}{\delta t},$$

$$m_s n_s \frac{\partial \mathbf{u}_s}{\partial t} + m_s n_s \, (\mathbf{u}_s \cdot \nabla)\mathbf{u}_s + \nabla p_s - m_s n_s \mathbf{g}_{\text{eff}} + 2 m_s n_s \mathbf{\Omega}_r \times \mathbf{u}_s$$

$$= m_s n_s \mathbf{a}_s + m_s n_s \frac{\delta \mathbf{u}_s}{\delta t}, \tag{8.21}$$

$$\frac{1}{\gamma - 1} \frac{\partial p_s}{\partial t} + \frac{1}{\gamma - 1} (\mathbf{u}_s \cdot \nabla) p_s + \frac{\gamma}{\gamma - 1} p_s (\nabla \cdot \mathbf{u}_s)$$

$$- \nabla \cdot (\kappa_s \nabla T_s) = \frac{1}{\gamma - 1} \frac{\delta p_s}{\delta t}.$$

8.3.2   Linearized Equations

Winds, waves, and tides represent relatively small perturbations over the hydrostatic equilibrium solution. We first linearize Eqs. (8.21) by assuming that the density, velocity, and pressure can be expressed as first-order small perturbations on top of the hydrostatic solution. In the linearized equations all second-order terms are neglected.

For the sake of simplicity, we consider the upper atmosphere as a single-species gas; the results can be readily generalized to multispecies cases. Mathematically speaking, it is assumed that

$$\rho_m = \rho_{\text{hs}} + \rho_1,$$
$$\mathbf{u} = \mathbf{u}_1, \tag{8.22}$$
$$p = p_{\text{hs}} + p_1,$$

where the subscript "hs" refers to hydrostatic equilibrium given by Eqs. (8.4) and (8.5) (note that $\mathbf{u}_{\text{hs}} = \mathbf{0}$). Substituting these expressions into the Euler equations, neglecting the collision terms, and linearizing yield the following result:

$$\frac{\partial \rho_1}{\partial t} + \nabla \cdot (\rho_{\text{hs}} \mathbf{u}_1) = 0,$$

$$\rho_{\text{hs}} \frac{\partial \mathbf{u}_1}{\partial t} + \nabla_{\text{h}} p_{\text{hs}} + \nabla p_1 - \rho_1 \mathbf{g} + 2 \rho_{\text{hs}} \mathbf{\Omega}_r \times \mathbf{u}_1 = \rho_{\text{hs}} \mathbf{a}, \tag{8.23}$$

$$\frac{1}{\gamma - 1} \frac{\partial p_1}{\partial t} + \frac{1}{\gamma - 1} (\mathbf{u}_1 \cdot \nabla) p_{\text{hs}} + \frac{\gamma}{\gamma - 1} p_{\text{hs}} (\nabla \cdot \mathbf{u}_1) = 0.$$

Here $\nabla_h p_{hs}$ is the small horizontal gradient in the background pressure (small compared to the hydrostatic pressure gradient in the vertical direction) due to the nonhydrostatic variation of pressure (the exponential variation of the hydrostatic pressure with altitude is already taken into account in this equation). Also, it was assumed that the external acceleration $\mathbf{a}$ is a first-order small quantity, and we neglected the small effects of the centrifugal acceleration.

Now we are ready to consider some fundamental perturbations in the upper atmosphere.

8.3.3   Geostrophic and Thermal Winds

The *geostrophic approximation* considers the atmospheric motion to be steady state ($\partial \mathbf{u}_1 / \partial t = \mathbf{0}$). For the sake of mathematical simplicity, we make several additional simplifications: We neglect all external accelerations other than gravity ($\mathbf{a} = \mathbf{0}$) and neglect the small contributions arising from the curvature of the Earth. These simplifications help us to understand the fundamental physics while keeping the mathematics as simple as possible. We also assume exact hydrostatic balance in the vertical (z) direction (this means that we consider only horizontal winds) and neglect the vertical component of the Coriolis force (this is naturally not a good approximation near the equator). With these assumptions the momentum equation becomes the following:

$$\nabla_h p_{hs} + \nabla p_1 - \rho_1 \mathbf{g} + 2\rho_{hs} \mathbf{\Omega}_r \times \mathbf{u}_1 = \mathbf{0}, \tag{8.24}$$

where ∇_h is the gradient operator in the horizontal direction:

$$\nabla_h = \mathbf{e}_x \frac{\partial}{\partial x} + \mathbf{e}_y \frac{\partial}{\partial y}. \tag{8.25}$$

In a first approximation the pressure and density perturbations balance each other ($\nabla p_1 - \rho_1 \mathbf{g} \approx \mathbf{0}$), and therefore Eq. (8.25) simplifies to

$$\nabla_h p_{hs} + 2\rho_{hs} \mathbf{\Omega}_r \times \mathbf{u}_1 = \mathbf{0}. \tag{8.26}$$

The components of the unit vector in the vertical direction can be readily expressed in terms of geographic latitude, and the scalar product of the vertical unit vector and the angular velocity vector is simply $\mathbf{\Omega}_r \cdot \mathbf{e}_z = \Omega_r \sin\theta$ (θ being the geographic latitude). Because in the present approximation the wind is horizontal, one can solve Eq. (8.26) for the horizontal components of the wind velocity:

$$
\begin{aligned}
u_{1x} &= -\frac{1}{2\Omega_r \sin\theta} \frac{1}{\rho_{hs}} \frac{\partial p_{hs}}{\partial y}, \\
u_{1y} &= \frac{1}{2\Omega_r \sin\theta} \frac{1}{\rho_{hs}} \frac{\partial p_{hs}}{\partial x}.
\end{aligned}
\tag{8.27}
$$

This approximation works well for large-scale (also called *synoptic-scale*) flows away from the equator. Equation (8.27) can also be written in vector form:

$$\mathbf{u}_1 = \frac{1}{2\rho_{hs}\Omega_r \sin\theta} \mathbf{e}_z \times \nabla_h p_{hs}. \tag{8.28}$$

This wind does not flow from high pressure to low pressure regions along the pressure gradient. Instead, the wind is horizontal, but it is perpendicular to the horizontal pressure gradient: In effect it follows the lines of constant pressure (*isobars*). This wind is called *geosptrophic wind*.

Next, we find the vertical shear of the geostrophic wind (the horizontal velocity is still a function of altitude). This is given by

$$\frac{\partial \mathbf{u}_1}{\partial z} = \frac{\mathbf{e}_z}{2\Omega_r \sin\theta} \times \frac{\partial}{\partial z}\left(\frac{\nabla_h p_{hs}}{\rho_{hs}}\right). \tag{8.29}$$

We will evaluate Eq. (8.29) using the ideal gas law and assuming that the temperature gradient is horizontal. First, the derivative on the right-hand side can be manipulated to obtain the following expression:

$$\frac{\partial}{\partial z}\left(\frac{\nabla_h p_{hs}}{\rho_{hs}}\right) = \frac{kT}{m}\frac{\partial}{\partial z}\left(\frac{\nabla_h p_{hs}}{p_{hs}}\right) = \frac{kT}{m}\nabla_h\frac{\partial \ln(p_{hs})}{\partial z}$$

$$= -\frac{kT}{m}\nabla_h\left(\frac{1}{H}\right) = g\frac{\nabla_h T}{T}. \tag{8.30}$$

Substituting this result into Eq. (8.29) yields the vertical shear of the geostrophic wind:

$$\frac{\partial \mathbf{u}_1}{\partial z} = \frac{g}{2\Omega_r \sin\theta}\mathbf{e}_z \times \frac{\nabla_h T}{T}. \tag{8.31}$$

This velocity shear is called the *thermal wind*. The name is a little misleading, since the thermal wind is not a wind but rather the vertical gradient of the horizontal geostrophic wind. Thermal winds are mainly important in the homosphere.

8.3.4   Acoustic–Gravity Waves

Under some conditions, perturbations in the upper atmosphere will result in oscillatory motion. This means that first-order perturbed quantities in the linearized Euler equations can be written as periodic functions. However, the fact that the background hydrostatic equilibrium solution varies exponentially with altitude requires the use of altitude-corrected periodic perturbations:

$$\rho_m = \hat{\rho}_0 \exp\left(-\frac{z}{H}\right) + \hat{\rho}\exp\left[-\frac{z}{2H} + i(\mathbf{k}\cdot\mathbf{r} - \omega t)\right],$$

$$\mathbf{u} = \hat{\mathbf{u}}\exp\left[\frac{z}{2H} + i(\mathbf{k}\cdot\mathbf{r} - \omega t)\right], \tag{8.32}$$

$$p = \hat{p}_0 \exp\left(-\frac{z}{H}\right) + \hat{p}\exp\left[-\frac{z}{2H} + i(\mathbf{k}\cdot\mathbf{r} - \omega t)\right],$$

where $\hat{\rho}_0$ and $\hat{p}_0$ are the hydrostatic density and pressure values at $z = 0$, $H = \hat{p}_0/\hat{\rho}_0 g$ is the hydrostatic scale height, ω is the frequency, and $\mathbf{k}$ is the wavenumber. It is again assumed that $\hat{\rho}_0 \gg \hat{\rho}$ and $\hat{p}_0 \gg \hat{p}$. Substituting these expressions into the linearized

Euler equations (8.23) and neglecting the Coriolis acceleration, we get

$$-i\omega\hat{\rho} - \frac{\hat{\rho}_0}{2H}\hat{u}_z + i\hat{\rho}_0(\mathbf{k}\cdot\hat{\mathbf{u}}) = 0,$$

$$-i\omega\hat{\rho}_0\hat{\mathbf{u}} + i\mathbf{k}\hat{p} - \left(\frac{\hat{p}}{2H} - \hat{\rho}g\right)\mathbf{e}_z = \mathbf{0}, \tag{8.33}$$

$$-i\omega\hat{p} + (\gamma - 2)\frac{\hat{p}_0}{2H}\hat{u}_z + i\gamma\hat{p}_0(\mathbf{k}\cdot\hat{\mathbf{u}}) = 0.$$

Here we assumed that $\mathbf{g} = (0, 0, -g)$, and $\mathbf{e}_z$ is the unit vector in the z direction. The continuity equation can be combined with the energy equation to eliminate the $\mathbf{k}\cdot\hat{\mathbf{u}}$ term, giving

$$-i\omega\hat{\rho} - \frac{\hat{\rho}_0}{2H}\hat{u}_z + i\hat{\rho}_0(\mathbf{k}\cdot\hat{\mathbf{u}}) = 0,$$

$$-i\omega\hat{\rho}_0\hat{\mathbf{u}} + i\mathbf{k}\hat{p} - \left(\frac{\hat{p}}{2H} - \hat{\rho}g\right)\mathbf{e}_z = \mathbf{0}, \tag{8.34}$$

$$-i\omega\hat{p} + (\gamma - 1)\frac{\hat{p}_0}{H}\hat{u}_z + i\omega\frac{\gamma\hat{p}_0}{\hat{\rho}_0}\hat{\rho} = 0.$$

Equation (8.34) can also be written in matrix form:

$$\mathcal{A}\cdot\mathcal{W} = 0, \tag{8.35}$$

where the matrix $\mathcal{A}$ and the vector $\mathcal{W}$ are given by

$$\mathcal{A} = \begin{pmatrix} -i\omega & ik_x\hat{\rho}_0 & ik_y\hat{\rho}_0 & ik_z\hat{\rho}_0 - \frac{\hat{\rho}_0}{2H} & 0 \\ 0 & -i\omega\hat{\rho}_0 & 0 & 0 & ik_x \\ 0 & 0 & -i\omega\hat{\rho}_0 & 0 & ik_y \\ g & 0 & 0 & -i\omega\hat{\rho}_0 & ik_z - \frac{1}{2H} \\ i\omega\frac{\gamma\hat{p}_0}{\hat{\rho}_0} & 0 & 0 & (\gamma - 1)\frac{\hat{p}_0}{H} & -i\omega \end{pmatrix} \tag{8.36}$$

and

$$\mathcal{W} = \begin{pmatrix} \hat{\rho} \\ \hat{u}_x \\ \hat{u}_y \\ \hat{u}_z \\ \hat{p} \end{pmatrix}. \tag{8.37}$$

The matrix $\mathcal{A}$ can be expressed in terms of the sound speed, $a_s^2 = \gamma\hat{p}_0/\hat{\rho}_0$, and the *Brunt–Väisälä frequency*, $N^2 = (\gamma - 1)g^2/a_s^2$:

$$\mathcal{A} = \begin{pmatrix} -i\omega & ik_x\hat{\rho}_0 & ik_y\hat{\rho}_0 & ik_z\hat{\rho}_0 - \frac{\gamma\hat{\rho}_0 N}{2a_s\sqrt{\gamma-1}} & 0 \\ 0 & -i\omega\hat{\rho}_0 & 0 & 0 & ik_x \\ 0 & 0 & -i\omega\hat{\rho}_0 & 0 & ik_y \\ \frac{a_s N}{\sqrt{\gamma-1}} & 0 & 0 & -i\omega\hat{\rho}_0 & ik_z - \frac{\gamma N}{2a_s\sqrt{\gamma-1}} \\ i\omega a_s^2 & 0 & 0 & \sqrt{\gamma-1}\hat{\rho}_0 N a_s & -i\omega \end{pmatrix}.$$
$$\tag{8.38}$$

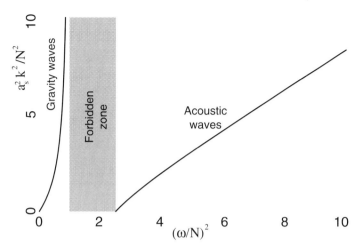

Figure 8.2 Plot of dispersion relation (8.42) for perpendicular propagation ($\vartheta = \pi/2$ and $\gamma = 5/3$).

Equation (8.35) has nontrivial (not identically zero) solutions only if the determinant of the characteristic matrix $\mathcal{A}$ vanishes. This requirement leads to the following condition:

$$\omega^4 - \frac{1}{4}\frac{\gamma^2}{\gamma-1}N^2\omega^2 - a_s^2 k^2\omega^2 + N^2 a_s^2\left(k_x^2 + k_y^2\right) = 0. \tag{8.39}$$

Equation (8.39) expresses the condition for a wave characterized by the frequency ω and wavenumber $\mathbf{k}$ to exist and propagate in the medium. If a given pair of ω and $\mathbf{k}$ does not satisfy Eq. (8.39), that particular wave cannot propagate in the gas characterized by ρ_{hs} and p_{hs}. Equation (8.39) is called the *dispersion relation* of these waves.

One can introduce the angle of wave propagation with respect to the vertical direction

$$\sin^2\vartheta = \frac{k_x^2 + k_y^2}{k^2}. \tag{8.40}$$

With these quantities the dispersion relation becomes

$$\omega^2\left(\omega^2 - \frac{\gamma^2}{4(\gamma-1)}N^2\right) - a_s^2 k^2(\omega^2 - N^2\sin^2\vartheta) = 0. \tag{8.41}$$

One can solve for $a_s^2 k^2$ to obtain

$$a_s^2 k^2 = \frac{\omega^2\left(\omega^2 - \frac{\gamma^2}{4(\gamma-1)}N^2\right)}{(\omega^2 - N^2\sin^2\vartheta)}. \tag{8.42}$$

The dispersion relation (8.42) has several interesting features (see Figure 8.2). First of all, we know that for existing waves k^2 must be positive (or zero); therefore no wave can exist in regions where expression (8.42) is negative. Moreover, inspection of Eq. (8.42) reveals that it is singular at $\omega = N\sin\vartheta$ and is negative for ω values between

$\omega_1 = N \sin \vartheta$ and $\omega_2 = \sqrt{\gamma^2/4(\gamma - 1)}N$. This means that no waves (described by this dispersion relation) can exist between ω_1 and ω_2.

Let us examine the type of waves below and above the forbidden region. First, notice that the location of the singularity depends on the angle of propagation. For strictly vertical propagating waves $\sin \vartheta = 0$, and therefore there is no singularity and only the upper branch exists. For oblique or perpendicular waves the two branches of the dispersion relation describe two distinct kinds of waves. The physical nature of the upper branch can be easily understood by examining frequencies well above the Brunt–Väisälä frequency, $\omega \gg N$:

$$\lim_{\omega \gg N} a_s^2 k^2 = \omega^2. \tag{8.43}$$

These are clearly the well-known acoustic waves since both the phase velocity (ω/k) and the group velocity ($\partial \omega/\partial k$) are the acoustic speed, a_s.

The more interesting (and less known) waves are related to the branch below the Brunt–Väisälä frequency. Let us examine this mode in the $\omega \ll N$ (and $\vartheta = \pi/2$) limit. In this case we get

$$\lim_{\omega \ll N} a_s^2 k^2 = \frac{\gamma^2}{4(\gamma - 1)}\omega^2. \tag{8.44}$$

These low-frequency waves are called gravity waves. It can be seen that the phase and group velocities of gravity waves are identical, and their value is $\omega/k = \partial \omega/\partial k = \pm\sqrt{4(\gamma - 1)/\gamma^2}a_s$. Thus gravity waves are low-frequency oblique waves, propagating with velocities below the local sound speed.

In the terrestrial upper atmosphere the Brunt–Väisälä frequency is of the order of 0.01 Hz, and typical gravity wave periods are minutes to tens of minutes.

8.4   Diffusion

8.4.1   Molecular Diffusion

Above 100 km the atmospheric density decreases relatively slowly, indicating a dramatic increase of the scale height. In a situation when the hydrostatic approximation is more or less valid such a scale height increase can be the result of either increasing temperature or decreasing mean molecular mass ($H = kT/mg$). Such a decrease in the mean molecular mass might be the result of diffusive separation of atmospheric constituents. In this section we discuss the basic assumptions and limitations of the diffusion approximation and will apply the results to minor atmospheric constituents.

We start from the multispecies Euler equation (4.35) for neutral particles. For the sake of simplicity, we neglect all reactive collisions (including ionization, charge exchange, recombination, etc.), taking only elastic collisions into account. In this approximation the momentum equation for the individual species is the following:

$$m_s n_s \frac{\partial \mathbf{u}_s}{\partial t} + m_s n_s (\mathbf{u}_s \cdot \nabla)\mathbf{u}_s + \nabla p_s - m_s n_s \mathbf{g} = \sum_t m_s n_s \bar{\nu}_{st}(\mathbf{u}_t - \mathbf{u}_s). \tag{8.45}$$

We also use the single-fluid momentum equation (4.56) for a neutral gas mixture:

$$mn\frac{\partial \mathbf{u}}{\partial t} + mn(\mathbf{u} \cdot \nabla)\mathbf{u} + \nabla p - mn\mathbf{g} = \mathbf{0}, \tag{8.46}$$

where m is the mean molecular mass, n is the total gas concentration, and $\mathbf{u}$ is the bulk flow velocity of the gas mixture. We multiply Eq. (8.46) by $m_s n_s / mn$ and subtract it from the single-species momentum equation. This yields the following:

$$m_s n_s \left[\frac{\partial \mathbf{u}_s}{\partial t} - \frac{\partial \mathbf{u}}{\partial t}\right] + m_s n_s [(\mathbf{u}_s \cdot \nabla)\mathbf{u}_s - (\mathbf{u} \cdot \nabla)\mathbf{u}] + \nabla p_s - \frac{m_s n_s}{mn}\nabla p$$

$$= \sum_t m_s n_s \bar{\nu}_{st}(\mathbf{u}_t - \mathbf{u}_s). \tag{8.47}$$

Next, we make the assumption that the atmosphere is near steady state and that the bulk velocities are small (second-order terms in the flow velocity can be neglected). In this approximation the terms in the square brackets can be neglected and one obtains the following equation:

$$\nabla p_s - \frac{m_s n_s}{mn}\nabla p = \sum_t m_s n_s \bar{\nu}_{st}(\mathbf{u}_t - \mathbf{u}_s). \tag{8.48}$$

We are considering a situation when molecules of a minor species diffuse through the background of major species molecules. In this case "t" refers to the abundant molecules of a major species, and consequently one can make the approximation that $\mathbf{u}_t \approx \mathbf{u}$. Thus we get

$$\mathbf{w}_{sD} = -D_s \left[\frac{\nabla n_s}{n_s} + \frac{\nabla T}{T} - \frac{m_s}{m}\frac{\nabla p}{p}\right], \tag{8.49}$$

where $\mathbf{w}_{sD} = \mathbf{u}_s - \mathbf{u}$ is the *diffusion velocity* of species s, and we introduced the *diffusion coefficient*,

$$D_s = \frac{1}{\sum_t \bar{\nu}_{st}} \frac{kT}{m_s}. \tag{8.50}$$

Here we also assumed that the mixture of neutral gases can still be characterized by a single temperature T.

More sophisticated approximations of the collision term lead to a somewhat different form of the diffusion Equation (8.49) that also includes the effects of thermal diffusion (cf. *Banks & Kockart 1973*):

$$\mathbf{w}_{sD} = -D_s \left[\frac{\nabla n_s}{n_s} + (1 + \alpha_{\mathrm{T}})\frac{\nabla T}{T} - \frac{m_s}{m}\frac{\nabla p}{p}\right]. \tag{8.51}$$

Here α_{T} is the *thermal diffusion coefficient*. This correction factor is only important for the lightest molecules (H and He). For helium one can use $\alpha_T = -0.40$, for hydrogen we use $\alpha_T = -0.25$, whereas $\alpha_{\mathrm{T}} = 0$ is used for heavier molecules.

In a nearly hydrostatic stratified atmosphere (in which the atmospheric parameters only vary with altitude) Eq. (8.51) can be written as

$$w_{sD} = -D_s \left[\frac{1}{n_s} \frac{dn_s}{dz} + (1 + \alpha_T) \frac{1}{T} \frac{dT}{dz} + \frac{m_s g}{kT} \right]. \tag{8.52}$$

The values of diffusion coefficients can be calculated using various approximations. In the simplest case one can use classical kinetic theory to approximate the molecular diffusion coefficient in the following form (cf. *Banks & Kockart 1973*):

$$D_s = 1.52 \times 10^{18} \sqrt{\frac{1}{M_s} + \frac{1}{M}} \frac{\sqrt{T}}{n}, \tag{8.53}$$

where D_s is given in $cm^2\,s^{-1}$, M and M_s are the mean molecular mass of the gas and the mass of species s (measured in amu), while T and n are measured in units of K and cm^{-3}.

It is very important to note that the discussed approximation is valid only for minor species. The transport of major atmospheric constituents must be described by more sophisticated methods.

8.4.2   Eddy Diffusion

Molecular diffusion is not the only diffusion process that can be important in the upper atmosphere. Another physical mechanism that can lead to vertical transport is *eddy diffusion*. Molecular diffusion separates the various species according to their mass. Turbulence, however, tends to mix the atmospheric constituents and it works against separation. Turbulent mixing dominates in the homosphere, whereas diffusive separation occurs in the thermosphere. There is, however, a transition region near 100 km where both processes are of the same order of magnitude, and consequently, both processes must be taken into account.

The vertical eddy diffusion velocity for an atmospheric constituent s can be defined as

$$w_{sE} = -K_s \frac{n}{n_s} \frac{d}{dz} \left(\frac{n_s}{n} \right) = -K_s \left[\frac{1}{n_s} \frac{dn_s}{dz} - \frac{1}{n} \frac{dn}{dz} \right], \tag{8.54}$$

where K_s is the *eddy diffusion coefficient*. We also know that in a nearly hydrostatic atmosphere the pressure satisfies the following relation (see Eq. 8.3):

$$\frac{dp}{p} = \frac{dn}{n} + \frac{dT}{T} = -\frac{dz}{H}. \tag{8.55}$$

Substituting Eq. (8.55) into (8.54) yields

$$w_{sE} = -K_s \left[\frac{1}{n_s} \frac{dn_s}{dz} + \frac{1}{T} \frac{dT}{dz} + \frac{mg}{kT} \right]. \tag{8.56}$$

It is obvious that a perfectly mixed gas (when $H_s = H$) leads to $w_E = 0$. Using

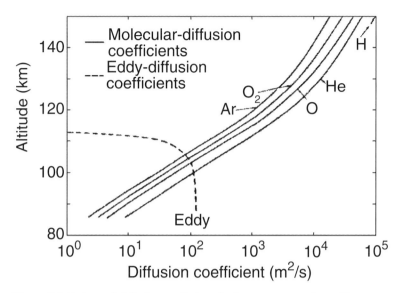

Figure 8.3 Estimated diffusion coefficients in the upper atmosphere. (From *Richmond.*[1])

Eqs. (8.56) and (8.52), one can express the total vertical drift velocity of molecular species s:

$$w_s = w_{sD} + w_{sE} = -D_s \left[\frac{1}{n_s} \frac{dn_s}{dz} + (1 + \alpha_T) \frac{1}{T} \frac{dT}{dz} + \frac{m_s g}{kT} \right]$$

$$- K_s \left[\frac{1}{n_s} \frac{dn_s}{dz} + \frac{1}{T} \frac{dT}{dz} + \frac{mg}{kT} \right]. \tag{8.57}$$

It is very difficult to estimate eddy diffusion coefficients. Figure 8.3 shows an estimated eddy diffusion profile together with molecular diffusion profiles for some selected species. Inspection of the figure reveals that above ≈ 100 km altitude molecular diffusion dominates.

8.4.3   Diffusive Equilibrium

A gas constituent is in diffusive equilibrium when the total vertical diffusion velocity, w_s, vanishes. In this case Eq. (8.57) becomes the following:

$$\frac{1}{n_s} \frac{dn_s}{dz} = -\left(1 + \frac{\alpha_T}{1 + \Lambda_s} \right) \frac{1}{T} \frac{dT}{dz} - \frac{m_s/m + \Lambda_s}{1 + \Lambda_s} \frac{mg}{kT}, \tag{8.58}$$

where we introduced the ratio of the diffusion coefficients, $\Lambda_s = K_s/D_s$. The solution of this equation is

$$n_{sDE} = n_{s0} \frac{T_0}{T} \exp \left[- \int_{z_0}^{z} \frac{m_s/m + \Lambda_s}{1 + \Lambda_s} \frac{mg}{kT} dz' - \int_{T_0}^{T} \frac{\alpha_T}{1 + \Lambda_s} \frac{dT'}{T'} \right], \tag{8.59}$$

where n_{s0} and T_0 are the concentration and temperature at the reference altitude z_0.

[1] Richmond, A. D., "Thermospheric dynamics and electrodynamics," in *Solar Terrestrial Physics*, ed. R. L. Carovilano and J. M. Forbes, p. 523, D. Reidel, Dordrecht, The Netherlands, 1983.

Equation (8.59) is the general solution of diffusive equilibrium when both eddy diffusion and molecular diffusion are important. In the homosphere eddy diffusion dominates and consequently $\Lambda_s \to \infty$. In this limit we obtain the perfect mixing altitude profile:

$$n_{s\,\text{mix}} = n_{s0}\frac{T_0}{T}\exp\left[-\int_{z_0}^{z}\frac{mg}{kT}dz'\right] \approx n_{s0}\frac{T_0}{T}\exp\left(-\frac{z-z_0}{H}\right), \tag{8.60}$$

since the temperature is a slowly varying function of altitude. This solution is the familiar hydrostatic equilibrium solution.

In the thermosphere molecular diffusion dominates and $\Lambda_s \to 0$. In this case the diffusive equilibrium solution simplifies to the following expression:

$$n_{s\,\text{DE}} = n_{s0}\left(\frac{T_0}{T}\right)^{1+\alpha_T}\exp\left[-\int_{z_0}^{z}\frac{dz'}{H_s}\right] \approx n_{s0}\left(\frac{T_0}{T}\right)^{1+\alpha_T}\exp\left[-\frac{z-z_0}{H_s}\right]. \tag{8.61}$$

This result tells us that the altitude profile of minor species in diffusive equilibrium is characterized by exponential decrease, but the scale height is different for the different species. As a matter of fact, the scale height, H_s, is proportional to $1/m_s$, and therefore light particles (such as hydrogen) have a much larger scale height than heavier molecules. The result is that the concentration of heavy particles decreases very rapidly with altitude, while the light particles (particularly hydrogen and helium) become more and more important.

8.4.4   Maximum Diffusion Velocities

One can estimate the maximum diffusion velocities (and consequently fluxes) for minor atmospheric constituents. We start by introducing the concentration gradient ratio for species s:

$$\frac{1}{n_s}\frac{dn_s}{dz} = X_s(z)\frac{1}{n}\frac{dn}{dz}. \tag{8.62}$$

Note that X_s is a function of altitude z. One can use Eq. (8.55) to express the gradient of the total concentration. With this we obtain

$$\frac{1}{n_s}\frac{dn_s}{dz} = -X_s(z)\left[\frac{1}{H} + \frac{1}{T}\frac{dT}{dz}\right]. \tag{8.63}$$

It is obvious that for light minor species X_s varies from $X_s = 1$ (perfectly mixed atmosphere) to a value corresponding to diffusive equilibrium (see Eq. 8.58):

$$(X_s)_{\text{DE}} = 1 + \frac{\dfrac{\alpha_T}{1+\Lambda_s}\dfrac{1}{T}\dfrac{dT}{dz} - \dfrac{1-m_s/m}{1+\Lambda_s}\dfrac{1}{H}}{\dfrac{1}{H} + \dfrac{1}{T}\dfrac{dT}{dz}}. \tag{8.64}$$

Substituting Eq. (8.63) into our expression for the diffusion velocity (Eq. 8.57) yields the following expression:

$$w_s = \frac{D_s}{H}\left[X_s - \frac{H}{H_s} - (1 + \alpha_T - X_s)\frac{H}{T}\frac{dT}{dz}\right] + \frac{K_s}{H}(X_s - 1)\left[1 + \frac{H}{T}\frac{dT}{dz}\right].$$

$$(8.65)$$

We see that w_s is a linear function of X_s, and we know that w_s vanishes for $X_s = (X_s)_{DE}$ (because diffusive equilibrium is defined by zero velocity). This means that the maximum value of the diffusion velocity is obtained for perfect mixing, $X_s = 1$:

$$(w_s)_{\max} = \frac{D_s}{H}\left[1 - \frac{m_s}{m} - \alpha_T\frac{H}{T}\frac{dT}{dz}\right].$$

$$(8.66)$$

In the case when the thermal diffusion can be neglected, we obtain the following very simple expression for the maximum diffusion velocity:

$$(w_s)_{\max} = \frac{D_s}{H}\left(1 - \frac{m_s}{m}\right).$$

$$(8.67)$$

Hence light particles ($m_s < m$) diffusive upward, whereas heavy constituents ($m_s > m$) diffuse downward. At around 100 km altitude the maximum diffusion velocities of H and He atoms are typically a few centimeters per second. Once the density of the minor species is known, one can immediately obtain the maximum diffusive flux of the given species. This limiting flux is often used in upper atmospheric model calculations. For instance, the density of hydrogen at around 100 km is about 10^8 particles per cubic centimeter; consequently the limiting upward flux is a few times 10^8 cm^{-2} s^{-1}. Eventually this hydrogen escapes from the atmosphere; and observed values agree well with this value of the loss flux.

8.5   Thermal Structure

Next, we examine the thermal structure of the upper atmosphere above ≈ 100 km, where the atmospheric temperature increases from about 200 K to over 1,000 K. Let us start by examining the single-fluid energy equation (Eq. 4.57). We assume that the presence of an external heating rate, Q, which includes chemical, radiative, turbulent, Joule, and viscous heating. In this case the single-fluid energy equation can be written as

$$\frac{1}{\gamma - 1}\frac{\partial p}{\partial t} + \frac{1}{\gamma - 1}\nabla \cdot (p\mathbf{u}) + p(\nabla \cdot \mathbf{u}) + (\nabla \cdot \mathbf{h}) = Q.$$

$$(8.68)$$

Next, we consider steady-state conditions and assume that the thermosphere is horizontally stratified (we only take into account the altitude dependence and neglect horizontal variations). In this approximation Eq. (8.68) becomes

$$\frac{1}{\gamma - 1}\frac{d}{dz}(pu_z) + p\frac{du_z}{dz} + \frac{dh_z}{dz} = Q.$$

$$(8.69)$$

In this case the single-fluid continuity equation is very simple:

$$\frac{d}{dz}(\rho_m u_z) = 0. \tag{8.70}$$

After some manipulation these two equations yield the following:

$$\frac{dh_z}{dz} = Q - u_z\left[\rho_m \frac{d}{dz}\left(\frac{\gamma}{\gamma-1}\frac{k}{m}T\right) - \frac{dp}{dz}\right]. \tag{8.71}$$

In a nearly hydrostatic atmosphere (where the vertical velocity is much smaller than the local sound speed) $dp/dz = -\rho g$ (see Eq. 8.2). This leads to the following equation:

$$\frac{dh_z}{dz} = Q - u_z\rho_m\left[\frac{d}{dz}\left(\frac{\gamma}{\gamma-1}\frac{k}{m}T\right) + g\right]. \tag{8.72}$$

In the atmosphere turbulence also transports energy. When considering the total heat transport, h_z, one must take into account not only the well-known Fourier heat flux, $h_{zF} = -\kappa dT/dz$, but also the eddy heat transfer, h_{zE}. The eddy heat transfer can be approximated in the following form:[2]

$$h_{zE} = -K_E\rho_m\left[\frac{d}{dz}\left(\frac{\gamma}{\gamma-1}\frac{k}{m}T\right) + g\right], \tag{8.73}$$

where K_E is the mean eddy diffusion coefficient. With this approximation the energy balance equation (8.72) becomes the following:

$$-\frac{d}{dz}\left\{\kappa\frac{dT}{dz} + K_E\rho_m\left[\frac{d}{dz}\left(\frac{\gamma}{\gamma-1}\frac{k}{m}T\right) + g\right]\right\}$$

$$= Q - u_z\rho_m\left[\frac{d}{dz}\left(\frac{\gamma}{\gamma-1}\frac{k}{m}T\right) + g\right]. \tag{8.74}$$

Equation (8.74) helps us to explain the thermal structure of the thermosphere.

In the upper thermosphere ρ_m becomes very small and the eddy diffusion term is negligible. At these altitudes the heat sources are also negligible, and so here Eq. (8.74) simply becomes

$$\frac{d}{dz}\left(\kappa\frac{dT}{dz}\right) \approx 0. \tag{8.75}$$

Thus the conductive heat flux is nearly constant in the upper thermosphere. Because there are no significant sources or sinks in the upper thermosphere, the heat flux itself must also be nearly zero:

$$\kappa\frac{dT}{dz} \approx 0. \tag{8.76}$$

[2] Richmond, A. D., "Thermospheric dynamics and electrodynamics," in *Solar Terrestrial Physics*, ed. R. L. Carovilano and J. M. Forbes, p. 523, D. Reidel, Dordrecht, The Netherlands, 1983.

This leads to the conclusion that the upper thermosphere is nearly isothermal, which is in excellent agreement with observations.

Lower in the thermosphere the vertical motion is quite insignificant, so the convective term can be neglected in Eq. (8.74). In this approximation Eq. (8.74) can be integrated, and we obtain the following expression:

$$\kappa \frac{dT}{dz} + K_E \rho_m \left[\frac{d}{dz} \left(\frac{\gamma}{\gamma - 1} \frac{k}{m} T \right) + g \right] = \int_z^\infty Q(z') \, dz'. \qquad (8.77)$$

The net heating rate, Q, is positive at all but the lowest portion of the thermosphere due to the significant heat inputs and to the lack of significant radiative cooling. Thus the integral on the right-hand side of Eq. (8.77) is positive at most altitudes. Also, one can neglect the effects of eddy diffusion above 150 km or so. At these altitudes, one obtains

$$\kappa \frac{dT}{dz} = \int_z^\infty Q(z') \, dz' > 0. \qquad (8.78)$$

This means that the temperature in the thermosphere steadily (and rapidly) increases above approximately 150 km or so. The temperature gradient remains positive down to the mesopause, but in the lower thermosphere eddy heat conduction is also important. At low altitudes (90–110 km) the positive temperature gradient is mainly the result of local heat balance (positive net local heating) and downward heat transport is not very important in this region.

8.6 The Exosphere

The exosphere (sometimes also called geocorona) comprises the uppermost region of the atmosphere where collisions play only a negligible role in determining the vertical profiles of neutral gas parameters. We define the exosphere as the atmospheric region where the mean free path exceeds the scale height. The *exobase* (the base of the exosphere) is defined as the critical level where the mean free path and the scale height are equal.

If we assume the exosphere to be isothermal (this is the upper thermosphere where the temperature is nearly constant), then the concentration decreases exponentially with a constant scale height H. Using Eq. (2.14), one can relate the density at the exobase, n_c, to the scale height:

$$\lambda = \frac{1}{\sqrt{2} n_c \sigma} = H, \qquad (8.79)$$

where σ is the average cross section of molecular collisions. This equation can be solved for n_c. For typical exospheric temperatures ($\approx 1{,}500$ K) and densities, the Earth's exobase turns out to be near $z_c \approx 500$ km.

A particle of mass m and vertical velocity v_z will escape from the gravitational attraction of Earth if its vertical kinetic energy is larger than the gravitational potential energy at the exobase. This means that for escaping particles the $v_z > v_e$ condition must be satisfied, where v_e is given by

$$\frac{1}{2}mv_e^2 = mg_e \frac{R_e^2}{R_e + z_c}.$$ (8.80)

Here $g_e = 9.81$ m/s^2 is the gravitational acceleration at the surface, $R_e = 6,378$ km is Earth's radius, and z_c is the altitude of the exobase. Solving Eq. (8.80) yields $v_e = 10.4$ km/s. (We note that the escape velocity at the surface of the Earth is 11.2 km/s.) Exospheric particles are usually classified according to their orbits. The various categories are:

- Ballistic particles coming from the exobase along elliptic trajectories. These particles will eventually fall back to the atmosphere, since their initial vertical velocity was less than v_e.

- Trapped particles in bound orbits. These particles simply orbit the Earth and do not come down to the critical level. Such particles can only be created by collisions. In detailed calculations these particles must be taken into account, which means that collisional effects must be included.

- Escaping particles coming from the critical level along hyperbolic orbits with $v_z > v_e$.

- Interplanetary neutral particles crossing the exobase and absorbed by the atmosphere. They are typically insignificant.

- Interplanetary particles not crossing the exobase, just "passing through" the geocorona. They are typically insignificant.

The flux of escaping particles can be estimated by assuming that neutral particles are in thermodynamic equilibrium at the exobase. In this case the distribution of molecular velocities at the exobase is Maxwellian and one can calculate the upward flux of escaping particles (particles with $v_z > v_e$):

$$\Phi_e = \int_0^{2\pi} d\phi \int_0^{\pi/2} \sin\vartheta \, d\vartheta \int_{v_e}^{\infty} dv v^3 \cos\vartheta \, n_c \left(\frac{m}{2\pi kT}\right)^{3/2} \exp\left(-\frac{mv^2}{2kT}\right).$$ (8.81)

Note that we integrate over upward moving particles ($\vartheta > 0$). This simulates the spherical geometry of the terrestrial atmosphere, because it assumes that any upward moving particle with speed above v_e leaves the atmosphere. In a plane parallel atmosphere the upward vertical velocity must exceed v_e.

The integral in expression (8.81) can be evaluated to obtain

$$\Phi_e = \frac{1}{2}n_c \sqrt{\frac{2kT}{\pi m}} \left(1 + \frac{mv_e^2}{2kT}\right) \exp\left(-\frac{mv_e^2}{2kT}\right).$$ (8.82)

Equation (8.82) gives us the *thermal escape flux* from an isothermal exosphere. This process was first discussed by J. H. Jeans, and it is also called *Jeans escape*. This formula is only accurate within about 20 to 30%, because of all the simplifying assumptions we made (such as neglecting collisions above the exobase and assuming a Maxwellian distribution).

One can calculate the thermal escape fluxes of primary atmospheric constituents. It turns out that the Jeans escape flux of atomic hydrogen is about $10^8 \text{cm}^{-2} \text{s}^{-1}$ and that of atomic helium is approximately $10^5 \text{cm}^{-2} \text{s}^{-1}$, whereas it is totally negligible for heavier particles. Observations show much higher escape fluxes for all neutral species: This indicates that nonthermal escape processes are dominant in the upper atmosphere. For instance, the upward flux of hydrogen due to eddy diffusion is a factor of 3 to 5 higher than the Jeans escape flux.

8.7   Some Concepts of Atmospheric Chemistry

Next, we briefly discuss some fundamental chemical concepts relevant to the aeronomy of the upper atmosphere, and then we briefly summarize the composition of the stratosphere and mesosphere (concentrating on the minor species, since the major constituents are well mixed).

8.7.1   Thermodynamics

The reactivity of atmospheric constituents is determined by the *enthalpy* and the *entropy* of the chemical processes involving these substances. The enthalpy (or *heat content*) of a molecule represents the energy necessary to make or break the chemical bonds holding together the atoms in the molecule. By convention, zero enthalpy is assigned to the lowest energy form of the chemical compounds. For instance, at room temperature about 120 kcal/mole is required to break the bond of an O_2 molecule and produce two oxygen atoms in their ground state. Therefore zero enthalpy is assigned to the O_2 molecule, and each oxygen atom has an enthalpy of about 60 kcal/mole.

Let us examine the following generic reaction:

$$A + B \rightarrow C + D. \tag{8.83}$$

The enthalpy change associated with this reaction is given by the difference between the enthalpies of the products and of the reactants:

$$\Delta H_R^0 = \Delta H_f^0(C) + \Delta H_f^0(D) - \Delta H_f^0(A) - \Delta H_f^0(B), \tag{8.84}$$

where the superscript 0 refers to the enthalpy at a standard temperature, which is by convention 298.15 K (tabulations of ΔH in the literature generally refer to this temperature[3]). In general, ΔH is a function of temperature, but the variation is relatively small over the range of atmospheric temperatures. If ΔH_R^0 is negative, heat is

---

[3] Tabulations in the literature generally use units of kcal/mole. Note that $1\,\text{eV} = 1.6022 \times 10^{-19}\,\text{J} = 11,604\,\text{K}$. This amount of energy per molecule corresponds to $23.05\,\text{kcal/mole}$ (see the Table A.2 in Appendix A).

released by the reaction, and it is called *exothermic*. However, if ΔH_R^0 is positive, external energy must be supplied for the reaction to proceed, and it is called *endothermic*. It is obvious that exothermic reactions can take place spontaneously in the atmosphere, whereas endothermic processes need additional energy (such as the absorption of a photon).

Entropy change (which in a general sense is the increase of randomness associated with a particular process) also influences the spontaneity of chemical reactions, as a reaction favors particular products only if the change in the *Gibbs free energy* is negative, $\Delta G = \Delta H - T\Delta S < 0$ (where S is entropy and T is temperature). The change of free energy associated with a particlar reaction [such as the one given by Eq. (8.84)] is defined similarly to ΔH_R^0:

$$\Delta G_R^0 = \Delta G_f^0(C) + \Delta G_f^0(D) - \Delta G_f^0(A) - \Delta G_f^0(B). \tag{8.85}$$

If ΔG_R^0 is negative, then the reaction can take place spontaneously at the reference temperature. Table 8.1 presents values of ΔH_f^0 and ΔG_f^0 for a number of atmospheric constituents.

The quantities ΔH_R^0 and ΔG_R^0 are particularly useful in determining the spontaneity of possible chemical reactions and whether a reaction is exothermic or endothermic. As an example let us consider the ozone producing reaction

$$O + O_2 + M \rightarrow O_3 + M, \tag{8.86}$$

Table 8.1. *Thermodynamic data for some atmospheric constituents.*

Species	ΔH_f^0 kcal/mole	ΔG_f^0 kcal/mole	Species	ΔH_f^0 kcal/mole	ΔG_f^0 kcal/mole
O	59.553	55.389	NH	82.0	80.6
O_2	0	0	NH_2	45.5	47.8
O_3	34.1	39.0	NH_3	−10.98	−3.93
H	52.103	48.588	HNO	23.8	26.859
H_2	0	0	HNO_2	−19.0	−11.0
OH	9.31	8.18	HNO_3	−32.28	−17.87
HO_2	3.5	4.4	CO	−26.416	−32.780
H_2O	−57.796	−54.634	CO_2	−94.051	−94.254
H_2O_2	−32.58	−25.24	CH_3	34.8	35.3
N	112.979	108.883	CH_4	−17.88	−12.13
N_2	0	0	HCO	−25.95	−24.51
NO	21.57	20.69	H_2CO	−25.95	−24.51
NO_2	7.93	12.26	CH_3O	3.9	6.4
NO_3	17.0	27.7	CH_3OOH	−30.8	−17.4
N_2O	19.61	24.90	CH_3NO_2	−17.86	−1.65
N_2O_4	2.19	23.38	CH_3ONO	−16.5	−1.5
N_2O_5	2.7	27.5	CH_3NO_3	−29.8	−9.4

Source: From *Brasseur & Solomon (1984).*

where M can be any arbitrary atom or molecule. For this reaction

$$\Delta H_R^0 = \Delta H_f^0(O_3) - \Delta H_f^0(O) - \Delta H_f^0(O_2) = -25.4 \text{ kcal/mole},$$
$$\Delta G_R^0 = \Delta G_f^0(O_3) - \Delta G_f^0(O) - \Delta G_f^0(O_2) = -16.4 \text{ kcal/mole}. \tag{8.87}$$

This reaction is exothermic and can proceed spontaneously. As a matter of fact, reaction (8.86) is one of the fundamental reactions in the upper atmosphere.

Another interesting example is the photolysis of molecular oxygen to form two oxygen atoms in their ground state (the electronic states of atoms and molecules will be discussed in detail in Section 9.2):

$$O_2 + h\nu \rightarrow 2O, \tag{8.88}$$

where h is the Planck constant and ν is the frequency of the absorbed photon. Now the enthalpy change is

$$\Delta H_R^0 = \Delta H_f^0(O) - \Delta H_f^0(O_2) - h\nu = 119.10 \text{ kcal/mole} - h\nu. \tag{8.89}$$

For this reaction to take place, ΔH_R^0 must be negative (the photon must have enough energy to break the chemical bonds); therefore the photon energy, $h\nu$, must be larger than 5.167 eV. This means that photons with wavelength $\lambda \leq 240$ nm can initiate reaction (8.89).

8.7.2   Chemical Kinetics

Chemical kinetics addresses the question of how fast the potential chemical reactions occur. It is reaction kinetics that ultimately controls the rate of formation and destruction of atmospheric constituents. The density of various species is determined by a complicated interplay between rates of production and loss as well as by the rate of transport processes.

Reactions involving dissociation of a single particle, $A \rightarrow B + C$, are called *first-order reactions* (or *unimolecular reactions*). Reactions involving two reacting particles (such as $A + B \rightarrow C + D$) are called *second-order*, or *two-body*, or *bimolecular reactions*. Processes involving three reacting particles are referred to as *third-order*, or *three-body*, or *termolecular reactions*. The rate of a chemical reaction describes the rate of formation of the products (or the rate of destruction of the reactants).

Let us first discuss unimolecular reaction rates. In this case the reaction rate can be written in the following form:

$$-\frac{d[A]}{dt} = \frac{d[B]}{dt} = \frac{d[C]}{dt} = k_u[A]. \tag{8.90}$$

Here k_u is the *reaction rate constant* and square brackets denote the concentration of the given species. For unimolecular reactions k_u, has units of s^{-1}. Equation (8.90) can be readily integrated to obtain the time evolution of the concentration of particles A:

$$[A] = [A]_0 e^{-k_u t}, \tag{8.91}$$

where $[A]_0$ is the concentration of molecules A at $t = 0$. It is obvious that the density of species A exponentially decreases with time. The time constant of this decrease, $\tau_A = 1/k_u$, is called the *chemical lifetime* of species A.

For the bimolecular reaction $A + B \rightarrow C + D$, the reaction rate is given by

$$-\frac{d[A]}{dt} = -\frac{d[B]}{dt} = \frac{d[C]}{dt} = \frac{d[D]}{dt} = k_b[A][B], \qquad (8.92)$$

where the two-body reaction rate constant, k_b, has units of $cm^3 \, s^{-1}$. The lifetime for bimolecular reactions is

$$\tau_A = \frac{1}{k_b[B]}. \qquad (8.93)$$

Similarly, the reaction rate for the $A + B + C \rightarrow D + E$ process is

$$-\frac{d[A]}{dt} = -\frac{d[B]}{dt} = -\frac{d[C]}{dt} = \frac{d[D]}{dt} = \frac{d[E]}{dt} = k_t[A][B][C], \qquad (8.94)$$

where the three-body reaction rate constant, k_t, has units of $cm^6 \, s^{-1}$. The lifetime for termolecular reactions is

$$\tau_A = \frac{1}{k_t[B][C]}. \qquad (8.95)$$

In a complicated system, like the atmosphere, numerous chemical reactions involving a particular species can take place simultaneously. In this case one must take into consideration all the possible production and loss processes. As an example let us consider species A and assume that it has the following reactions:

$$
\begin{aligned}
A + B &\xrightarrow{k_1} \text{Products}, \\
A + C + M &\xrightarrow{k_2} \text{Products}, \\
A + F &\xrightarrow{k_3} \text{Products}, \\
G + H &\xrightarrow{k_4} A + \text{Products}.
\end{aligned}
\qquad (8.96)
$$

The reaction rate for molecules A is the following:

$$\frac{d[A]}{dt} = -(k_1[B] + k_2[C][M] + k_3[F])[A] + k_4[G][H]. \qquad (8.97)$$

Assuming that the concentrations of the other compounds are nearly constant, the chemical lifetime of molecule A is the following:

$$\tau_A = \frac{1}{k_1[B] + k_2[C][M] + k_3[F]}. \qquad (8.98)$$

In *chemical equilibrium* $d[A]/dt = 0$ and one can express the concentration of species A:

$$[A] = \frac{k_4[G][H]}{k_1[B] + k_2[C][M] + k_3[F]}. \qquad (8.99)$$

8.8   Atmospheric Composition and Chemistry

The upper atmosphere is composed of various major and minor atomic and molecular constituents. The major species are nitrogen and oxygen (in atomic and various molecular forms) at lower altitudes and the light atoms hydrogen and helium at high altitudes. Minor constituents typically have nearly negligible number densities, but some of them play roles well beyond their insignificant densities. The critical minor constituents include ozone, various nitrogen oxides, alkali metals, carbon dioxide, and, naturally, water. Table 8.2 summarizes the composition of the atmosphere at ground level (this composition remains unchanged throughout the homosphere, up to nearly 100 km).

Figure 8.4 shows the composition and density profile in the upper atmosphere. It shows how the densities of different species vary with altitude. Below the thermosphere the atmosphere is well mixed by eddy diffusion (see Section 8.4.2), so that the decrease of concentration with altitude is similar for all species. In the thermosphere atomic oxygen and hydrogen become dominant.

8.8.1   Stratosphere and Mesosphere

Even though the dominant constituents in the stratosphere and mesosphere are still N_2 and O_2, photochemical and collisional reactions produce a series of minor constituents that play a very important role in the atmospheric energetics (heat sources and heat transport). Also, stratospheric ozone is very important, because it efficiently interacts with solar UV radiation and absorbs potentially harmful (for life) photons. In this section we will consider sources of sinks of stratospheric and mesospheric ozone and some other important minor species.

Figure 8.5 shows the temperature, scale height, and concentration profile in the upper homosphere (from 15 to 120 km) for average dayside conditions. One can see that the scale height is relatively constant and the atmospheric concentration approximately decreases exponentially. These plots show the basic structure of the stratosphere and mesosphere. It was assumed that as a result of strong mixing the average molecular weight is about 29 throughout these regions.

Table 8.2. *Composition of the atmosphere at ground level.*

Molecule	Mass	Volume %	Concentration (cm^{-3})
Nitrogen (N_2)	28.02	78.1	2.1×10^{19}
Oxygen (O_2)	32.00	20.9	5.6×10^{18}
Argon (Ar)	39.96	0.9	2.5×10^{17}
Carbon dioxide (CO_2)	44.02	0.03	8.9×10^{15}
Neon (Ne)	20.17	0.002	4.9×10^{14}
Helium (He)	4.00	0.0005	1.4×10^{14}
Water (H_2O)	18.02	variable	variable

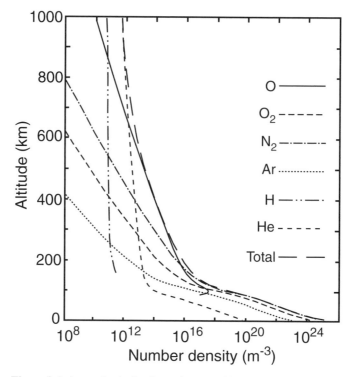

Figure 8.4 Atmospheric density and composition profile. (From *Richmond.*[4])

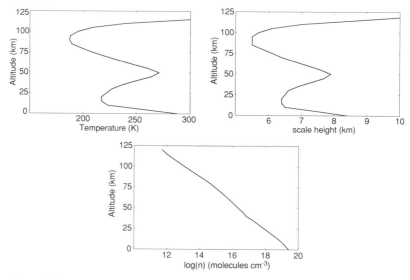

Figure 8.5 Temperature, scale height, and concentration profiles in the homosphere.

[4] Richmond, A. D., "Thermospheric dynamics and electrodynamics," in *Solar Terrestrial Physics*, ed. R. L. Carovilano and J. M. Forbes, p. 523, D. Reidel, Dordrecht, The Netherlands, 1983.

In order to understand the formation and destruction of ozone in the stratosphere and mesosphere, we start with the "Chapman reactions," (first considered by Sidney Chapman[5]) which involve only oxygen reactions.

Ozone production starts with the photodissociation of molecular oxygen by absorption in the Herzberg continuum (200–242 nm or 2,000–2,420 Å). Shorter wavelength UV radiation (which is also capable of causing O_2 photodissociation) is mainly absorbed at higher altitudes. Therefore in this region the dominant reaction is

$$O_2 + h\nu \rightarrow O(^3P) + O(^3P). \tag{8.100}$$

Here $O(^3P)$ refers to the ground state of the oxygen atom. Atomic (and molecular) electronic states will be discussed in detail in Section 9.2. Atomic oxygen then attaches to O_2 in a three-body collision with an unspecified atom or molecule, M, to form ozone:

$$O + O_2 + M \rightarrow O_3 + M. \tag{8.101}$$

It should be noted that this reaction takes place only in relatively dense gases, since it involves three-body collisions, which have low probability. The reaction rate of this three-body collision is $k_{12} = 6.0 \times 10^{-34}(T/300)^{-2.3}$ cm^6/s.

Both kinds of "odd oxygen" (i.e., O and O_3) are destroyed by the reaction

$$O + O_3 \rightarrow O_2 + O_2. \tag{8.102}$$

The reaction rate for this process is $k_{13} = 8.0 \times 10^{-12} \exp(-2,060/T)$ cm^3/s. Also, ozone is converted back to atomic and molecular oxygen by the following photochemical reaction:

$$O_3 + h\nu \rightarrow O + O_2. \tag{8.103}$$

The cross section of this reaction peaks in the strong Hartley continuum between 240 and 280 nm. This photodissociation reaction is the principal heat source in the stratosphere, so the minor species O_3 plays a critical role in stratospheric energetics.

One can calculate equilibrium altitude profiles of O, O_2, and O_3 using the four Chapman reactions. This can be done by neglecting transport altogether and assuming that in equilibrium, production and loss balance for the minor species (O and O_3). This means that at an altitude h, the following two equations must hold:

$$2J_2(h)[O_2] + J_3(h)[O_3] = k_{13}[O][O_3] + k_{12}[O][O_2][M],$$
$$J_3(h)[O_3] = -k_{13}[O][O_3] + k_{12}[O][O_2][M], \tag{8.104}$$

where square brackets denote concentrations (in units of molecule cm^{-3}) and J_2 and J_3 are the altitude dependent photoreaction rates for the O_2 and O_3 photodissociations (reactions 8.100 and 8.103). These rates naturally depend on the solar UV production.

---

[5] Chapman, S., "A theory of upper-atmosphere ozone," *Mem. Roy. Meteorol. Soc.*, **3**, 103, 1930.

In chemical equilibrium these equations lead to the following expressions:

$$[O] = \frac{J_2[O_2]}{k_{13}[O_3]},$$

$$[O_3] = \frac{k_{12}[O][O_2][M]}{k_{13}[O] + J_3}.$$

(8.105)

When obtaining these expressions, we recognized that $J_2[O_2] \gg J_3[O_3]$ due to the small concentration of ozone compared to the concentration of molecular oxygen. On the dayside $k_{13}[O] \ll J_3$ below about 60 km because of the low O abundance, and therefore one obtains

$$\frac{[O_3]_{day}}{[O_2]} = \sqrt{\frac{k_{12}J_2[M]}{k_{13}J_3}}$$

(8.106)

and

$$[O]_{day} = \sqrt{\frac{J_2 J_3}{k_{12}k_{13}[M]}}.$$

(8.107)

At nighttime the two photodissociation reactions are shut down and there is no new production of odd oxygen. However, reaction (8.101) converts most of the O to ozone. This is a fast reaction because of the large concentration of O_2. Nevertheless, because $[O] \ll [O_3]$ in the stratosphere, the resulting increase in the ozone density is relatively small.

The observed values of the ozone densities are far less than predicted by the Chapman reactions only, indicating that pure oxygen chemistry is not enough to explain stratospheric ozone. The resolution of this problem lies with catalytic chemistry, primarily involving minor constituents of odd hydrogen, nitrogen, and chlorine. The usual notation for these "odd" molecules is HO_x, NO_x, and ClO_x. In this notation x represents 0, 1, or 2.

Once formed, odd hydrogen, nitrogen, and chlorine (and some other molecules) destroy odd oxygen with no destruction of themselves. Some sample reactions include

$$OH + O_3 \rightarrow HO_2 + O_2 + 1.68\,eV,$$

$$HO_2 + O_3 \rightarrow OH + 2O_2 + 1.34\,eV$$

(8.108)

and

$$NO + O_3 \rightarrow NO_2 + O_2 + 2.06\,eV,$$

$$NO_2 + O \rightarrow NO + O_2 + 2.00\,eV.$$

(8.109)

Chlorine reacts immediately with ozone and destroys it. This, for instance, can be done in the following simple catalytic cycle:

$$Cl + O_3 \rightarrow ClO + O_2 + 1.68\,eV,$$

$$ClO + O \rightarrow Cl + O_2 + 2.39\,eV.$$

(8.110)

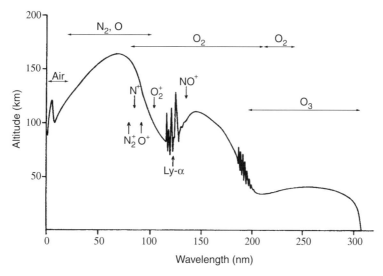

Figure 8.6 Altitude at which the intensity of solar radiation drops to $1/e$ of its value outside Earth's atmosphere for vertical incidence. (From *Herzberg*.[6])

Chlorine is mainly lost in the stratosphere in the form of inactive HCl (hydrogen chloride). This gets transported downward into the troposphere where it is removed by rain. The principal sources of Cl in the stratosphere are *halogenated methanes*. In these molecules halogen atoms take the place of one or more hydrogen in methane (CH_4).   The most important of these molecules are *trichlorofluoromethane* ($CFCl_3$) and *dichlorofluoromethane* (CF_2Cl_2). The former is mainly used in sprays; the latter is used in sprays, air conditioners, and refrigerators. These *fluorocarbons*, as they are often called, are practically insoluble, so that even the oceans are not an effective sink. They cannot be photodissocated in the troposphere (the short wavelength UV radiation is absorbed above the troposphere), but upon diffusing into the stratosphere they are dissociated by

$$CFCl_3 + h\nu(\lambda < 226 \text{ nm}) \rightarrow CFCl_2 + Cl,$$
$$CF_2Cl_2 + h\nu(\lambda < 215 \text{ nm}) \rightarrow CF_2Cl + Cl. \tag{8.111}$$

So why are we so concerned about stratospheric ozone? The answer is quite simple. The O_3 molecules in the stratosphere absorb most of the solar UV radiation between 200 and 300 nm and prevents this wavelength from reaching the ground. Stratospheric ozone is the only absorber in this wavelength range, and therefore it provides the only shield against this potentially very harmful UV radiation. This can be seen in Figure 8.6, which shows the altitude range and absorbing constituents for the 0 to 300 nm wavelength range.

[6] Herzberg, L., "Solar optical radiation and its role in upper atmospheric processes," in *Physics of the Earth's Upper Atmosphere*, ed. C. O. Hines, I. Paghis, T. R. Hartz, J. A. Fejer, p. 31, Prentice Hall, Englewood Cliffs, NJ, 1965.

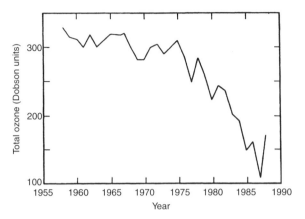

Figure 8.7 The decrease of total stratospheric ozone over
the Antarctic during the months of September–November.
(From *Farman et al.*[7]) and modified by *Aikin.*[8]

It is important to point out that there are two photodissociation continuum regions
for O_2 that are important in atmospheric physics. The stronger one, the so-called
Schumann–Runge continuum, lies in the wavelengths interval between 100 and 175 nm.
The absorption of this band takes place in the lower thermosphere, and it is practically
extinguished well above the stratosphere.

In the past decade a large decrease of stratospheric ozone was observed over the
Antarctic during the months of September to November (local spring). This seasonal
decrease for the past thirty years is shown in Figure 8.7. During these months, the
stratosphere is very cold due to the long absence of sunlight. In addition, there is a
long lasting stratospheric vortex in this region that breaks up only toward the end of
this period.

For years the origin of the Antarctic ozone hole was a subject of various competing
theories. It was quickly recognized that homogeneous catalytic chemistry (when gas
molecules interact with each other) involving chlorine could not explain the observed
dramatic decrease in stratospheric ozone content.

The solution to the puzzle lies in so-called *heterogeneous* reactions involving both
gaseous and solid materials. During the Antarctic winter, this region is the coldest part
of the terrestrial stratosphere. At temperatures near 195 K polar stratospheric clouds
(PSC) form containing a very large number of ice crystals. These ice crystals are
the solid components that help to convert large numbers of inert chlorine-containing
molecules into species that can undergo rapid photolysis in the springtime when the
Sun rises.

Gas-phase HCl penetrates into the ice crystals, while $ClONO_2$ (chlorine nitrate)
and N_2O_5 (dinitrogen pentoxide) react with the HCl/ice mixture, converting chlorine

[7] Farman, J. C., Gardiner, B. G., and Shanklin, J. D., "Large losses of total ozone in Antarctica
     reveal seasonal ClO_x/NO_x interaction," *Nature*, **315**, 207–210, 1985.
[8] Aikin, A. C., "Spring polar ozone behavior," *Planet. Space Sci.*, **40**, 7, 1992.

reservoir species to photochemically active gas-phase Cl_2, $ClNO_2$ (nitryl chloride), and HOCl (hypoclorous acid). These new species accumulate in the atmosphere until sunrise.

At sunrise Cl_2, $ClNO_2$, and HOCl are photolyzed, releasing Cl atoms, which destroy ozone. Eventually the polar air mass warms up, the polar stratospheric clouds evaporate, and the Antarctic ozone hole dissipates during the November–December time period.

8.8.2   Thermosphere

At the mesopause the atmosphere is still well mixed and the major molecular species are N_2 and O_2. Above the mesopause mixing gradually becomes slower and the various atmospheric constituents develop different altitude dependencies. It is in this region where the mixing ratios radically change and atomic species become dominant. Here we briefly outline the most important reactions contributing to the formation of atomic nitrogen and oxygen.

The strong Schumann–Runge continuum at $\lambda < 175$ nm produces the following dissociation of molecular oxygen:

$$O_2 + h\nu \rightarrow O(^3D) + O(^1D), \tag{8.112}$$

with an approximate rate of $J_2 \approx 4 \times 10^{-6} s^{-1}$. Association of two oxygen atoms primarily occurs in the presence of a third particle via the reaction

$$O + O + M \rightarrow O_2 + M, \tag{8.113}$$

with $k_{11} = 2.76 \times 10^{-34} \exp(710/T)$ cm^6/s.

Because there is no N_2 continuum where solar UV radiation is strong enough to dissociate it, ion–neutral reactions play a major role in forming odd nitrogen. N atoms are primarily formed through photoionization ($\lambda < 79.6$ nm):

$$\begin{aligned} N_2 + h\nu &\rightarrow N_2^+ + e^-, \\ N_2^+ + O &\rightarrow NO^+ + N. \end{aligned} \tag{8.114}$$

An alternative way to produce N involves oxygen ions:

$$O^+ + N_2 \rightarrow NO^+ + N. \tag{8.115}$$

The NO^+ dissociatively recombines and produces excited nitrogen:

$$NO^+ + e^- \rightarrow N(^2D) + O. \tag{8.116}$$

Finally, the metastable $N(^2D)$ can react with molecular oxygen to produce neutral nitric oxide:

$$N(^2D) + O_2 \rightarrow NO + O(^3P). \tag{8.117}$$

NO plays a very important role in the formation of the lower ionosphere.

8.9   Problems

Problem 8.1 Derive the dispersion relation for waves in an incompressible rotating fluid (Poincaré waves). Calculate the group and phase velocities for these waves (Hint: Start from Eqs. 8.21 and take the curl of the momentum equation.)

Problem 8.2 The kinetic energy carried by a plane wave is proportional to $\rho_m u^2$. Derive the altitude dependence of the wave energy in an isothermal hydrostatic atmosphere for vertically propagating waves.

Problem 8.3 A spacecraft approaches an Earth-sized planet of unknown mass. On-board instruments indicate that the planet has a dense, isothermal atmosphere at temperature $T = 300$ K, composed of pure water vapor. With the help of its "Maze & Blue" atmospheric remote sensing package, the spacecraft also measures near-surface gravity–acoustic waves that propagate upward at $45°$ with respect to the vertical direction. The spacecraft observes no waves between 4.25 mHz and 12 mHz.

1. What is the specific heat ratio of the atmosphere?
2. What is the mass of the planet relative to the mass of Earth?

Problem 8.4 What is the density distribution of a minor constituent in an isothermal atmosphere when the flux is constant and the eddy diffusion dominates? Assume that the eddy diffusion coefficient varies as

$$K(z) = K_0 \exp\left(\frac{z - z_0}{L}\right),$$

where

1. $L = H/2$,
2. $L = H$

(H is the major constituent scale height).

Problem 8.5 Consider the formation of NO_2 during the day by

$$HNO_3 + h\nu \rightarrow OH + NO_2 \tag{8.118}$$

and its removal, day and night, by

$$NO_2 + OH + M \rightarrow HNO_3 + M. \tag{8.119}$$

The rate coefficient, β, for reaction (8.119) is constant. The production rate, $J(t)$, in the 24-hour day of the Arctic summer is approximated by

$$J(t) = \frac{1}{2} J_{max}[1 - \cos \omega t], \tag{8.120}$$

where t is measured from midnight and $\omega = 2\pi/86,400$ s^{-1}. Regard all substances except NO_2 as constant. Take $\beta = 5 \times 10^{-30}$ cm^6 s^{-1}, $[M] = 10^{18}$ cm^{-3}, $[OH] = 10^7$ cm^{-3}, $[HNO_3] = 10^{10}$ cm^{-3}, and $J_{max} = 5 \times 10^{-5}$ s^{-1}.

1. How long after noon does $[NO_2]$ reach a maximum?
2. What is ratio of the maximum to minimum $[NO_2]$?

Problem 8.6 In the atmosphere of a very slowly rotating planet (the orbital and rotational periods of the planet are equal) the densities of the neutral gas species X, X_2, and X_3 are controlled by the following fast (much faster than the orbital period) chemical processes:

$$X_2 + h\nu \rightarrow X + X \qquad (J_1),$$
$$X_3 + X + M \rightarrow 2X_2 + M \quad (k_1),$$
$$X_2 + X \rightarrow X_3 \qquad\qquad (k_2),$$

where the density of species M is constant, $[M] = C$. Assume photochemical equilibrium and derive $[X]$ and the $[X_2]/[X_3]$ ratio at the subsolar point of the atmosphere.

Chapter 9

Airglow and Aurora

Humanity has always been fascinated by atmospheric sources of light. The first records of auroras date back thousands of years to biblical, Greek, and Chinese documents. The name *aurora borealis* (latin for northern dawn) was coined by the French mathematician and astronomer, P. Gassendi, who described a spectacular event observed in southern France on September 12, 1621. Airglow was discovered in 1901 by Newcomb who explained it as light from stars too faint to be seen individually. It was not before the 1930s that scientists realized that the source of the faint "light of the night sky" must be zodiacal light and atmospheric luminescence. Figure 9.1 shows a photograph of a spectacular aurora.

Both aurora and airglow are caused by excitation of atmospheric species followed by subsequent radiation of photons. They are, however, quite different in terms of excitation mechanisms, temporal and spatial characteristics, intensity, and dominant emissions.

Airglow is the amorphous, faint optical radiation continuously emitted in wavelengths from the far UV to near infrared (but excluding thermal emissions in the long wavelength infrared). The Earth's airglow mainly originates from discrete atomic and molecular transitions (an exception is a weak continuum in the green). Airglow is mainly caused by three fundamental processes: direct scattering of sunlight, emissions associated with ionization and recombination, and radiation associated with neutral photochemistry.

The aurora is mainly caused by excitation due to precipitating electrons and ions. Auroras typically are associated with high geomagnetic latitudes, where magnetospheric and solar wind electrons can readily access the upper atmosphere.

Figure 9.1 A spectacular aurora photographed in Finland. (Courtesy of Jyrki Manninen, Sodankylä Geophysical Observatory.)

9.1   Measuring Atmospheric Emissions: The Rayleigh

The basic unit of energy emission from auroras and airglows is the rayleigh (R), named after the fourth Lord Rayleigh who was a pioneer in airglow studies.

Let us assume that the *specific intensity* of the radiation crossing a given surface is $I_\nu(\cos\Theta, \Phi)$ (the specific intensity is the energy flux emitted per unit area, per unit time, per unit solid angle, per unit frequency). Let us also assume that there is an observer at a distance r from the emitting surface with an instrument that detects photons coming from the emitting surface (see Figure 9.2). Let us denote the small surface area of the detector by dS, and denote the projection of the area of the emitting surface normal to the line of sight of the observer by A.

The photon energy flux at frequency ν originating from a small surface element with unit projected area and arriving at the detector in unit time is $I_\nu dS/r^2$. If the emitting surface has a uniform intensity (the emission in a given direction is the same from all surface elements), then the total energy flux at wavelength ν detected by the instrument is $I_\nu A dS/r^2$. Because $A = r^2 d\Omega$, where $d\Omega$ is the solid angle occupied by the emitting surface as seen by the detector, the detected energy per unit frequency per unit time is $I_\nu(dS/r^2)r^2 d\Omega = I_\nu dS d\Omega$. The *specific surface brightness* is defined as detected energy per unit detector area, per unit time, per unit frequency, coming from unit solid angle. This means that the specific surface brightness of the emitting surface is simply I_ν. Thus the surface brightness for an extended source is independent of the distance of the observer and is identical to the intensity emitted by the surface.

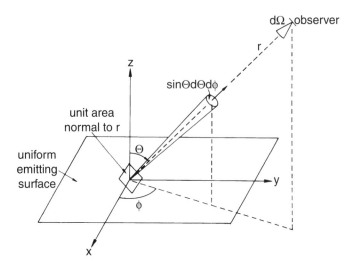

Figure 9.2 Specific surface brightness as seen by an observer.

Now consider a plane-parallel atmosphere that emits radiation everywhere at a rate independent of horizontal position but dependent on altitude, z. Let us denote the directional emission rate by $\varepsilon_\nu(z, \Theta, \phi)$ (measured in units of $\text{erg cm}^{-3}\,\text{s}^{-1}\,\text{sr}^{-1}\,\text{Hz}^{-1}$). Each volume element along the line of sight contributes to the total specific intensity measured by the observer. If there is no loss due to absorption or scattering, the observer (assumed to be above the emitting atmosphere) will measure

$$I_\nu(\Theta, \phi) = \int_0^\infty \varepsilon_\nu(z, \Theta, \phi) \frac{dz}{\cos \Theta}. \tag{9.1}$$

In optical aeronomy one is usually interested in the total emission over a spectral line or group of lines. Also, we usually count the number of photons (which is a direct measure of the number of aeronomical reactions). The number of photons emitted in a given direction is

$$\mathcal{I}(\Theta, \phi) = \int_{\text{line}} I_\nu(\Theta, \phi) \frac{d\nu}{h\nu} = \int_{\text{line}} \int_0^\infty \varepsilon_\nu(z, \Theta, \phi) \frac{dz}{\cos \Theta} \frac{d\nu}{h\nu}. \tag{9.2}$$

For most aeronomical reactions the differential volume emission rate is isotropic, $\varepsilon_\nu(z, \Theta, \phi) = \varepsilon_\nu(z)$. In this case the total photon emission rate in all directions from a column of unit cross section along the line of sight is

$$\mathcal{I}_t = \frac{4\pi}{\cos \Theta} \int_{\text{line}} d\nu \int_0^\infty dz \frac{\varepsilon_\nu(z)}{h\nu}. \tag{9.3}$$

The rayleigh unit (R) is the measure of this omnidirectional emission rate in a column of unit cross section along the line of sight with $1\,\text{R} = 10^6\,\text{photons cm}^{-2}\,\text{s}^{-1}$.

9.2   Atomic and Molecular Spectra

Atoms and molecules have many ways to store energy in their electronic configuration. These internal degrees of freedom are very important for the understanding of the physics of the airglow and aurora. In this section we give a brief "pedestrian's guide" to the symbols and terminology used in describing atomic and molecular spectra, and then in the next section we discuss specific aeronomical processes.

The detailed understanding of atoms and molecules involves a heavy dose of quantum physics and it goes well beyond the scope of this book. Here we will use concepts and results from quantum mechanics in a "recipe" manner, without detailed explanation of the physics behind them.

The modern theory of atomic structure has its origin in the Bohr model of the hydrogen atom. Bohr made two fundamental assumptions: (1) the angular momentum of the bound electron must be an integer times $\hbar = h/2\pi$ (h is the Planck constant, $h = 6.6261 \times 10^{-34}$ Js) and (2) when an electron jumps from one allowed state to another, the frequency of the emitted (or absorbed) radiation is related to the energy difference between the two states by $\Delta\epsilon = h\nu$. Using these assumptions, he calculated the possible electronic transitions in the electrostatic field of a positive nucleus and obtained the following expression:

$$\Delta\epsilon = \frac{m_e e^4}{2\hbar} \left(\frac{1}{n_1} - \frac{1}{n_2} \right) \tag{9.4}$$

where e and m_e are the charge and mass of the electron, and n_1 and n_2 are integers ($n_1 < n_2$). This formula is also called the *Rydberg formula*. The Rydberg formula successfully predicted the emissions of the hydrogen atom. It describes the four main series of atomic hydrogen spectra:

- *Lyman series* for $n_1 = 1$ and $n_2 = 2, 3, 4, \ldots$ in UV;
- *Balmer series* for $n_1 = 2$ and $n_2 = 3, 4, 5, \ldots$ in visible light;
- *Paschen series* for $n_1 = 3$ and $n_2 = 4, 5, 6, \ldots$ in infrared;
- *Brackett series* for $n_1 = 4$ and $n_2 = 5, 6, 7, \ldots$ in far infrared.

Although the "planetary" atom picture of Bohr was able to describe the hydrogen atom, it failed in the case of multielectron atoms and molecules. Modern quantum theory does not assign electrons to "planetary" orbits. Instead, it defines a complex wave function, ψ, for each particular energy level. The function $|\psi|^2$ expresses the probability of finding an electron in the vicinity of a given point at a given time. The behavior of the electron is completely characterized by this wave function, which is the solution of the *Schrödinger equation*. Unfortunately, as it most often happens in physics, exact solutions of the Schrödinger equation can only be found for the simplest atoms. For most atoms and molecules we have to live with approximate solutions.

9.2.1   Ground States of Atoms

One of the consequences of quantum theory is Heisenberg's uncertainty principle: Certain physical quantities (spatial location and particle momentum among them) cannot be simultaneously measured accurately. For instance, the product of the uncertainties of a location coordinate (Δx) and the corresponding momentum component (Δp_x) is always equal to or larger than $\hbar$:

$$\Delta x \Delta p_x \geq \hbar. \tag{9.5}$$

Although the use of the Schrödinger equation is necessary for the computation of atomic and molecular structures, it is too complicated for our pedestrian discussion. Instead we will use an empirical description of atomic and molecular structures that is based on more rigorous theory.

In general, four quantum numbers, n, l, m_l, and m_s, are used to describe each electron's state in the atom or molecule. The *total quantum number*, n $(1, 2, 3, \ldots)$, characterizes the electron shell (in essence the average distance from the nucleus). The various shells are designated by capital letters according to the following scheme:

value of n:	1	2	3	4	5	6	7	...
shell symbol:	K	L	M	N	O	P	Q	...

With the aid of quantum mechanics, one can find that the magnitude of the orbital angular momentum of an electron can only have the following discrete values:

$$|\mathbf{L}| = \sqrt{l(l+1)}\,\hbar, \tag{9.6}$$

where the quantum number l can have values of $0, 1, 2, 3, \ldots, n - 1$. The value of l is also symbolized by lowercase letters:

value of l:	0	1	2	3	4	5	...
symbol:	s	p	d	f	g	h	...

Within a given shell different values of l are possible: Each value of l defines a *subshell* (which is basically equivalent to a "classical" orbit). It is interesting to note that the symbols denoting subshells originate from early observations of atomic spectra. Emissions associated with $l = 0$ were associated with narrow spectral lines, and therefore they were called the *sharp* series and eventually denoted by the letter s. Other symbols stand for *principal series* (p), *diffuse series* (d), and *fundamental series* (f).

With the help of quantum mechanics, one can show that only the magnitude and one component of the orbital angular momentum vector (or any angular momentum vector for that matter) are well defined. The total angular momentum vector can be visualized as a vector precessing around the axis of the known component. The known component of $\mathbf{L}$ is also quantized and it only can assume values of $m_l\hbar$ ($m_l = -l, -l+1, \ldots, -1, 0, 1, \ldots, l-1, l$).

In addition to the orbital angular momentum, the electron also possesses an intrinsic angular momentum, called the *spin*. The known component of the electron spin vector is $m_s\hbar$, where the spin quantum number can assume only two values: $m_s = -1/2$ and $m_s = 1/2$.

The lowest energy (or ground-state) configuration for atoms containing many electrons can be explained using Pauli's exclusion principle and the atomic shell model. Pauli's exclusion principle states that in an atom (or molecule) no two electrons can have the same quantum numbers (n, l, m_l, m_s). Thus the maximum number of electrons on any given subshell is $2(2l+1)$, and the maximum number of electrons on any given shell is $2n^2$ (see Table 9.1). Noble gases (which are chemically inert and difficult to ionize) occur when a p subshell is completely filled (with the notable exception of helium). This means that noble gases are at $Z = 2, 10, 18, 36, 54$, and 86 (He, Ne, A, Kr, Xe, and Rn).

Ground states of atoms are denoted by the subshell configuration. For instance, the ground state of the oxygen atom ($Z = 8$) is $1s^2 2s^2 2p^4$. In this notation the subshell levels (such as $1s, 2s, 2p$, etc.) are followed by a superscript denoting the number of electrons on that particular subshell.

9.2.2   Atomic Excited States

Describing excited many-electron atomic states is quite difficult, because one not only has to take into account the Coulomb interaction between the electrons and the nucleus but also the Coulomb interaction between individual electrons, interactions between orbital angular momentum and spin vectors, and the interaction between the spin vectors of different electrons.

Table 9.1. *Maximum number of electrons on various shells and subshells.*

n	Shell	l	Subshell	Max. no. subshell electrons	Max. no. shell electrons
1	K	0	1s	2	2
2	L	0	2s	2	
		1	2p	6	8
3	M	0	3s	2	
		1	3p	6	
		2	3d	10	18
4	N	0	4s	2	
		1	4p	6	
		2	4d	10	
		3	4f	14	32

For light and medium-heavy atoms a scheme called *LS coupling* provides a relatively simple recipe for understanding excited atomic states. In this scheme the atom's total orbital angular momentum and total spin vectors are the vector sums of the individual $\mathbf{L}_i$ and $\mathbf{s}_i$ vectors of the individual electrons:

$$\mathbf{L} = \sum_i \mathbf{L}_i,$$
$$\mathbf{S} = \sum_i \mathbf{s}_i. \tag{9.7}$$

The atom's total angular momentum is now given by

$$\mathbf{J} = \mathbf{L} + \mathbf{S}. \tag{9.8}$$

The magnitudes of these three atomic angular momentum vectors are quantized as follows:

$$|\mathbf{L}|^2 = L(L+1)\hbar^2,$$
$$|\mathbf{S}|^2 = S(S+1)\hbar^2, \tag{9.9}$$
$$|\mathbf{J}|^2 = J(J+1)\hbar^2.$$

The known component quantum numbers of these vectors are quantized according to

$$M_\mathrm{L} = \sum_i (m_\mathrm{l})_i,$$
$$M_\mathrm{S} = \sum_i (m_\mathrm{s})_i, \tag{9.10}$$
$$M_\mathrm{J} = M_\mathrm{L} + M_\mathrm{S}.$$

Knowing M_L, M_S, and M_J, we can infer L, S, and J because we know that the component quantum number can take the potential values of $(-L \leq M_L \leq L)$, $(-S \leq M_S \leq S)$, and $(-J \leq M_J \leq J)$.

Excited states of atoms are characterized by the three quantum numbers, L, S, and J. The particular values of L are represented by the following spectroscopic notations:

value of L:	0	1	2	3	4	5	...
symbol:	S	P	D	F	G	H	...

Atomic states are specified by the letter symbol characterizing L with the value of $2S+1$ as a pre superscript (called multiplicity) and the value of J as a postsubscript. States with $S = 0$ ($2S+1 = 1$) are called singlet states; those with $S = 1/2$ ($2S+1 = 2$) are called doublet states. The multiplicities corresponding to $S = 1, 3/2, 2, 5/2, 3$ describe triplets, quartets, quintets, sextets, and septets, respectively.

Filled subshells are always in the $L = 0$, $S = 0$ (and consequently $J = 0$) ground state, which is denoted by $^1S_0$.

As an example let us consider the lowest electronic states of the oxygen atom. In its ground state the O atom has two filled subshells, $1s$ and $2s$. There are four electrons on the $2p$ subshell. The closed subshells do not contribute to L, S, or J. On the $2p$ subshell ($l = 1$) two electrons are characterized by $m_l = 0$ and $m_s = \pm 1/2$. The remaining two electrons can have $m_l = -1$ or $m_l = 1$ values; therefore we can have $M_L = -2, 0$, or 2. The total spin quantum number of these two electrons can be $M_S = -1, 0$, or 1. The potential electronic states of these last two electrons can be characterized[1] by $L = 0$, $S = 0$ (^1S), $L = 1$, $S = 1$ (^3P), or $L = 2$, $S = 0$ (^1D). The lowest energy state (ground state) of oxygen is $^3P_2$.

It is possible to excite an atom into energy levels above ground state. In returning to its ground state the atom will emit radiation with a corresponding line spectrum. For strong transitions the following selection rules apply:

$$\Delta J = 0, \pm 1 \quad \text{(but } J = 0 \to J = 0 \text{ not allowed)}$$
$$\Delta L = 0, \pm 1$$
$$\Delta S = 0$$
$$\Delta M_J = 0, \pm 1 \quad \text{(but if } \Delta J = 0, M_J = 0 \to M_J = 0 \text{ not allowed)}$$

These are called *electric dipole transitions* and are the strongest spectral lines. There are other, weaker transitions as well (such as electric quadrupole or magnetic dipole transitions).

In discussions of spectral observations Roman numerals indicate the state of ionization: $OI = O$, $OII = O^+$, etc. Square brackets in spectroscopy indicate that the transition is forbidden by the electric dipole selection rules (so these usually are electric quadrupole or magnetic dipole transitions).

Figure 9.3 shows the lowest energy electronic states of the oxygen atom along with the lifetimes and wavelengths of the various spectroscopic transitions. There are some very important lines shown in this diagram. The 5,577 Å green line is produced by the ^1D–^1S electric quadrupole transition; the ^3P–^1D magnetic dipole transition results in a triplet of 6,300 Å, 6,364 Å, and 6,392 Å red lines (however the 6,392 Å is too weak to be observed). Both of these transitions contribute to airglow and aurora.

9.2.3   Molecular Structure

The underlying principles of molecular structure are the same as those of atomic structure. However, the computational difficulties are even greater because of the presence of more than one nucleus. This complication destroys the approximate spherical symmetry of the central Coulomb force.

---

[1]   Note that in this case the two electrons have identical values of n, l, and m_l; consequently they must have opposite spins.

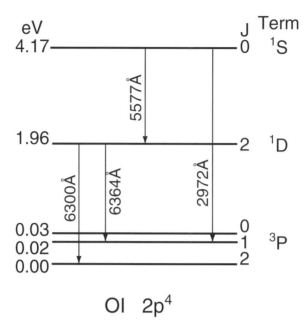

Figure 9.3 The low-energy portion of the atomic oxygen energy level diagram. [From *Chambarlain & Hunten (1987)*.]

Diatomic molecules retain some of the simplicity because they are symmetric about the internucleus axis. However, as a result of the presence of two heavy nuclei excerting attracting forces on each electron, the electronic orbital angular momentum is no longer a conserved quantity. Fortunately, the projection of the total angular momentum to the internuclear axis, Λ (the analogy of m_l), is conserved, and it therefore provides a good way to classify the electronic states of the diatomic molecule.

By analogy with the nomenclature of atomic spectra, we use the Greek characters Σ ($\Lambda = 0$), Π ($\Lambda = 1$), Δ ($\Lambda = 2$) to denote excited electronic states.

Under certain conditions diatomic molecules may also have reflection symmetries. If the molecule is *homonuclear* (such as N_2 or O_2), the molecule is symmetric over a reflection with respect to a plane perpendicular to the internucleus axis. In this case the wave function can either change sign or remain unchanged when reflected with respect to the symmetry plane. If the wave function does not change sign, it is called *ungerade* and is denoted by a subscript u. If the wave function does change sign, it is called *gerade* and is denoted by a subscript g. In case of axially symmetric homonuclear molecules there is another reflection symmetry with respect to any plane containing the internuclear axis. There are again two possibilities for the wave function: It is either symmetric for this reflection (denoted by a superscript +) or antisymmetric (denoted by $-$).

The projection of the electron spin to the internuclear axis is also a conserved quantity. The value of this quantum number is denoted the same way as in the case of atoms (with a presuperscript representing the multiplet number). An example of the notation is $^3\Sigma_g^+$, which indicates a combined spin of 1, $\Lambda = 0$, and a wave function that is antisymmetric for one and symmetric for the other reflection. It is also customary

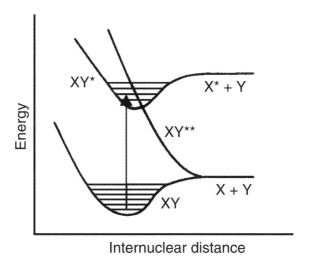

Figure 9.4 Potential energy curves for a diatomic molecule.
[From *Brasseur & Solomon* (*1984*).]

to indicate something about the electronic excitation of the molecule. A prefix X is used if the molecule is in ground state, whereas a letter near the beginning of the alphabet denotes excited states. Examples are the ground and an excited state of the N_2 molecule, $N_2(X^1\Sigma_g^+)$ and $N_2(C^3\Pi_u)$.

In addition to electronic excitation the molecule can also have rotational and vibrational excitations. Figure 9.4 shows two potential energy curves for a hypothetical diatomic molecule. When the two atoms come together, their potential energy follows one of these curves. Near their minimum energy value the curves are nearly parabolic (nearly harmonic), but at higher energies the nonlinear terms dominate. Vibrational levels are represented by horizontal lines in the potential wells. Dissociation can occur by first a collisional excitation to a higher energy vibrational state (XY^*) followed by dissociation to $X^* + Y$. Figure 9.5 shows an approximate potential energy diagram for the O_2 molecule.

9.3   Airglow

The principal sources of Earth's airglow are the absorption of electromagnetic radiation and chemical reactions. Excitation by charged particles is also important; however, interaction with energetic charged particles plays a somewhat secondary role.

Earth's night airglow is dominated by the green and red forbidden lines of neutral atomic oxygen, [OI], and by the yellow resonance doublet of sodium, NaI. Figure 9.6 shows the dominant part of the visible nightglow spectrum, from 540 nm to 680 nm.

There is a very large number of emission features in the airglow spectrum corresponding to a variety of excitation processes. Here we restrict our discussion to only a

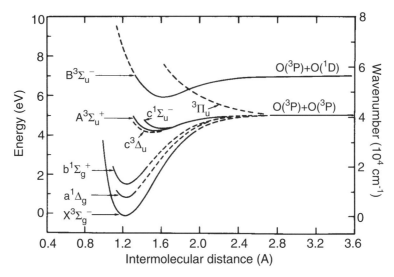

Figure 9.5 Potential energy diagram for the principal states of the O_2 molecule. [From *Brasseur & Solomon (1984)*.]

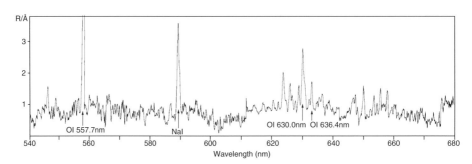

Figure 9.6 Earth's night airglow spectrum from 540 to 680 nm. (From *Broadfoot & Kendall* [2]).

few, representative examples. In particular, we will focus our attention on the atomic oxygen green and red emission lines, mainly because these lines are very strong in airglow spectra, because they can be generated by several typical processes, and because the physics of the green and red lines is well understood. First, let us discuss several fundamental emission mechanisms.

Resonance scattering is the absorption and emission of photons without a change in photon energy (or wavelength). *Fluorescence* is the absorption of a photon of energy ϵ_1 followed by the emission of a less energetic photon with $\epsilon_2 < \epsilon_1$.

The excess energy resulting from exothermic chemical reactions can increase the kinetic energy of the particles and result in vibrational, rotational, or electronic excitation of one or more particles. The excess energy that is stored in internal degrees of freedom will usually be emitted as electromagnetic radiation.

[2] Broadfoot, A. L., and Kendall, K. R., "The airglow spectrum, 3,100–10,000 Å," *J. Geophys. Res.*, **73**, 426, 1968.

Charged particles (mainly electrons) can cause excitation of atmospheric constituents via several processes. Fast particles can transfer energy to atoms and molecules on impact and result in excitation, dissociation, ionization, recombination, and charge exchange.

The main source of the atomic oxygen green line [$O(^1S)$–$O(^1D)$] is the so-called Barth mechanism:

$$O + O + M \rightarrow O_2^* + M,$$
$$O_2^* + O(^3P) \rightarrow O_2 + O(^1S), \tag{9.11}$$

where O_2^* is identified as the vibrationally excited $c^1\Sigma_u^-$ state, which is the upper state of the Herzberg II system.

Additional (but weaker) oxygen green line emission arises as a result of recombination of O_2^+ molecular ions. This recombination produces both $O(^1D)$ and $O(^1S)$ states and therefore contributes to both red line and green line emissions:

$$O_2^+ + e \rightarrow O^*(^1D, \, ^1S) + O. \tag{9.12}$$

About 90% of the excited oxygen atoms produced by this process are in the ^1D state, while about 10% is in ^1S.

The [OI] red line is produced by a variety of processes. Some of the most important reactions are the following:

$$e^* + O(^3P) \rightarrow e^* + O(^1D),$$
$$e^* + O_2 \rightarrow e^* + O + O(^1D),$$
$$O_2^+ + e \rightarrow O + O(^1D), \tag{9.13}$$
$$O_2 + h\nu \rightarrow O(^1D) + O,$$

where e^* denotes energetic electrons. The excited oxygen atom returns to its ground by producing red line emission.

The sodium NaI emission is believed to be the result of the following two-step process:

$$Na + O_3 \rightarrow NaO + O_2,$$
$$NaO + O \rightarrow Na(^2P) + O_2. \tag{9.14}$$

This line shows strong seasonal variation.

9.4   Aurora

The principal source of auroral phenomena is the dissipation of energy carried by precipitating magnetospheric particles (mainly keV electrons). In addition to occasionally spectacular optical emissions, auroral activity includes a number of other upper atmospheric phenomena that are all related to the precipitation of magnetospheric energetic particles. These include:

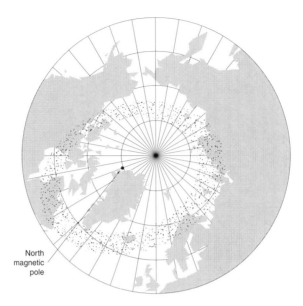

North
magnetic
pole

Figure 9.7 The auroral oval in the northern hemisphere for
average geomagnetic conditions.

- auroral radio absorption – the absorption of radio signals due to the increased
  state of ionization in the auroral region;
- auroral X rays – these are generated by the interaction of the precipitating energetic particles with the atmosphere and are detected by high-altitude balloons;
- magnetic disturbances – these are generated by the enhanced electric currents in
  the auroral region and are regularly detected by ground-based magnetometers;
- Very Low Frequency (VLF) and Ultra Low Frequency (ULF) radio emissions –
  these are generated in the magnetosphere by wave–particle interaction.

Auroral activity is generally related to solar activity. Auroral phenomena are highly
structured both in space and time, whereas their occurrence pattern is essentially zonal.
Most of the time auroral activity is centered in the *auroral oval*, which, in a first
approximation, is fixed with respect to the Sun (see Figure 9.7). The occurrence rates
rapidly fall off on both the poleward and equatorward sides of the auroral oval.

In this section we limit our discussion to the optical phenomena related to auroral
activity. Other aspects of auroral physics, such as the acceleration of precipitating
particles, current systems, or interaction with the magnetosphere and the solar wind,
will be discussed later in connection to the physics of the magnetosphere.

There are significant similarities between airglow and auroral emission spectra. This
is not surprising, because both phenomena are related to the radiative deexcitation
of ions, atoms, and molecules in the upper atmosphere. The major differences are
the consequences of the deep penetration of the energetic electrons into the lower
thermosphere and the resulting aeronomical processes.

The auroral emission features that historically received most of the attention are N_2 and N_2^+ lines together with the green and red lines of atomic oxygen. The N_2 band systems are excited by electron collisions. Excited levels leading to the first and second positive systems can be produced by the following reactions:

$$e + N_2\left(X^1\Sigma_g^+\right) \rightarrow e + N_2\left(A^3\Sigma_u^+\right),$$

$$e + N_2\left(X^1\Sigma_g^+\right) \rightarrow e + N_2\left(B^3\Pi_g\right), \qquad (9.15)$$

$$e + N_2\left(X^1\Sigma_g^+\right) \rightarrow e + N_2(C^3\Pi_u).$$

The excitation of N_2 bands may be a major factor in the slowing down of electrons in the 10–20 eV range. This process may also affect the population of $O(^1D)$, because it is efficiently created by collisions with low energy electrons.

The strongest N_2^+ lines are the first negative bands with emissions at 391, 428, and 471 nm. Because these lines are in the visible part of the electromagnetic spectrum, they were observed for decades by ground-based spectrometers. The excited N_2^+ ions are created by electron impact:

$$e + N_2\left(X^1\Sigma_g^+\right) \rightarrow N_2^+\left(B^2\Sigma_u^+\right) + 2e,$$

$$e + N_2\left(X^1\Sigma_g^+\right) \rightarrow N_2^+(A^2\Pi_u) + 2e. \qquad (9.16)$$

On the average, the first negative bands of N_2^+ (at 391, 428, and 471 nm) emit one photon per approximately 25, 75, and 300 nitrogen ions formed. It takes about 35 eV to produce an N_2^+ ion by electron impact, and therefore the generation of each 391 nm photon takes about 875 eV. The N_2^+ emission rate thus can be used to estimate the total energy flux of precipitating electrons. A typical auroral spectrum is shown in Figure 9.8.

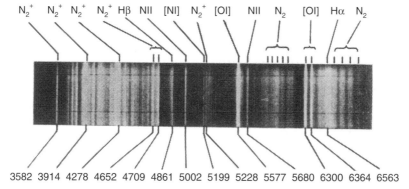

Figure 9.8 Low-dispersion spectrum of a green aurora. [From *Chambarlain* (*1961*).]

9.5   Auroral Electrons

Most bright auroras are generated by precipitating energetic electrons moving downward along "open" geomagnetic field lines. Open magnetic field lines are not connected to the conjugate hemisphere; instead they are connected to field lines in the distant magnetosphere or in the solar wind. The configuration of the geomagnetic field lines will be discussed in greater detail later in Chapter 14. The average energy of the primary electrons entering into the upper atmosphere from the magnetosphere is of the order of 5 to 10 keV. These electrons not only excite atoms and molecules in the atmosphere and thus create optical emissions, but also produce ionization, which gives rise to various radio wave absorption and reflection effects. The higher energy part of the precipitating electron fluxes can also generate X-ray and VLF radio emissions.

The energetic primary auroral electrons ionize the gases in the upper atmosphere and produce secondary electrons. Electrons are also produced in the atmosphere by photoionization. Both of these electron populations have similar energy spectra and they are produced by a distributed source inside the atmosphere (as opposed to the primary precipitating magnetospheric electron population, which is more energetic and only has an external source). Electrons created inside the atmosphere are generally referred to as *photoelectrons* or as *suprathermal electrons*.

In order to calculate auroral optical emissions and other related phenomena (such as radio absorption, X-ray emission, magnetic disturbances, or VLF emissions) one must have a large body of knowledge about the atomic and molecular processes (such as excitation and ionization cross sections, excited state lifetimes, elastic scattering cross sections, etc.), as well as detailed information about the precipitating electron distribution and the composition and physical state of the upper atmosphere. Some of these processes have been briefly discussed earlier in this book. Detailed information can be found in a series of aeronomy books; a representative sample is listed in the bibliography.

The field-aligned flux of energetic electrons in the upper atmosphere is described with the help of Eq. (7.39). However, in the present case the right-hand side of the equation must account for the complex set of aeronomic interactions of the electrons:[3]

$$\sqrt{\frac{m}{2\varepsilon}}\frac{\partial \Phi}{\partial t} + \mu\frac{\partial \Phi}{\partial s} + qE_{\parallel}\mu\varepsilon\frac{\partial}{\partial \varepsilon}\left(\frac{\Phi}{\varepsilon}\right) + \left(q\frac{E_{\parallel}}{\varepsilon} - \frac{1}{B}\frac{dB}{ds}\right)\frac{1-\mu^2}{2}\frac{\partial \Phi}{\partial \mu}$$

$$= S_{ee} + \sum_{i}\left(S_{ei} + S_{ei}^* + S_{ei}^-\right) + \sum_{\alpha}\left(S_{e\alpha} + S_{e\alpha}^* + S_{e\alpha}^+\right) + Q. \qquad (9.17)$$

Here S_{ee} describes the rate of change of the energetic electron flux due to collisions with thermal electrons, S_{ei} and $S_{e\alpha}$ account for the effects of elastic collisions with various types of ions and neutrals, S_{ei}^* and $S_{e\alpha}^*$ represent the rate of change due to excitation of ions and neutrals, S_{ei}^- is the electron loss rate due to recombination with ions, $S_{e\alpha}^+$ is the electron source rate due to electron impact ionization of neutrals, and

[3] Khazanov, G. V., Neubert, T., and Gefan, G. D., "A unified theory of ionosphere–plasmasphere transport of suprathermal electrons," *IEEE Transactions on Plasma Science*, **22**, 187, 1994.

finally Q is the electron production rate by processes other than electron impact (such as photoionization).

In general the source terms in Eq. (9.17) are quite complicated and difficult to compute. However, George Khazanov and his colleagues have published relatively easy to use approximations for the elastic collision terms:[4]

$$S_{ee} + \sum_i S_{ei} = An_e \left\{ \frac{\partial}{\partial \varepsilon} \left(\frac{\Phi}{\varepsilon} \right) + \frac{1}{2\varepsilon^2} \frac{\partial}{\partial \mu} \left[(1 - \mu^2) \frac{\partial \Phi}{\partial \mu} \right] \right\} \qquad (9.18)$$

$$S_{e\alpha} = \frac{n_\alpha \sigma_\alpha}{2} \frac{\partial}{\partial \mu} \left[(1 - \mu^2) \frac{\partial \Phi}{\partial \mu} \right], \qquad (9.19)$$

where n_e is the density of the thermal electrons, n_α is the density of the neutral particle species, α, σ_α is the momentum transfer elastic collision cross section for electron–α collisions, and finally the constant A has an approximate value of $A \approx 2.6 \times 10^{-12} \, \text{eV}^2\text{cm}^2$. It is interesting to note that elastic scattering on thermal plasma particles results in energy loss and pitch-angle diffusion, whereas elastic scattering on neutral particles primarily results in pitch-angle diffusion.

Khazanov and his colleagues also approximated the inelastic collision terms[4] in Eq. (9.17). Using the continuous energy loss approximation (i.e., assuming that the energy loss of a suprathermal electron in a single inelastic collision is small compared to the total energy of the electron), they concluded that S_{ei}^* and S_{ei}^- can generally be neglected, whereas $S_{e\alpha}^*$ and $S_{e\alpha}^+$ can be approximated the following way:

$$S_{e\alpha}^* = n_\alpha \sum_j \varepsilon_{\alpha j}^* \frac{\partial}{\partial \varepsilon} \left[\sigma_{\alpha j}^*(\varepsilon) \Phi \right], \qquad (9.20)$$

$$S_{e\alpha}^+ = n_\alpha \left\{ \frac{\partial}{\partial \varepsilon} \left[\sigma_\alpha^+(\varepsilon) \left(\varepsilon_\alpha^+ + \langle \varepsilon \rangle_\alpha \right) \Phi \right] + \int_{2\varepsilon + \varepsilon_\alpha^+}^{\infty} d\varepsilon \sigma_{e\alpha}(\varepsilon', \varepsilon) \Phi_0(\varepsilon') \right\}, \qquad (9.21)$$

where $\sigma_{\alpha j}^*$ is the total scattering cross section bringing the neutral particle to its j-th excited state ($\varepsilon_{\alpha j}^*$ is the threshold energy of this excited state), ε_α^+ is the ionization energy, and $\sigma_{e\alpha}(\varepsilon', \varepsilon)$ is the differential ionization cross section by an electron with energy ε' resulting in a secondary electron with energy ε. The other quantities are the total ionization cross section, σ_α^+, the average energy of secondary electrons, $\langle \varepsilon \rangle_\alpha$, and the pitch angle–averaged flux function, Φ_0. These quantities are defined in the following way:

$$\Phi_0(\varepsilon, s) = \frac{1}{2} \int_{-1}^{1} d\mu \Phi(\varepsilon, \mu, s),$$

$$\sigma_\alpha^+(\varepsilon) = \int_0^{(\varepsilon - \varepsilon_\alpha^+)/2} d\varepsilon_s \sigma_{e\alpha}(\varepsilon, \varepsilon_s), \qquad (9.22)$$

$$\langle \varepsilon \rangle_\alpha = \frac{1}{\sigma_\alpha^+} \int_0^{(\varepsilon - \varepsilon_\alpha^+)/2} d\varepsilon_s \varepsilon_s \sigma_{e\alpha}(\varepsilon, \varepsilon_s).$$

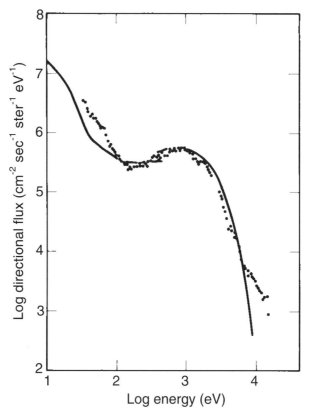

Figure 9.9 Comparison of an observed and calculated energy
spectrum of auroral electrons. (From *Evans.*[4])

Model calculations of auroral electrons and their interaction with the upper atmosphere are quite complicated. In general, these calculations involve not only electron transport theory, but also detailed information about atomic and molecular processes. These models are quite complex, and they represent great challenges. An observed auroral electron spectrum and the result of an appropriate model calculation are shown in Figure 9.9. The peak near 1 keV represents the flux of primary precipitating electrons; the large flux of <100 eV secondary electrons was produced inside the atmosphere.

Useful byproducts of the auroral electron precitipation calculation are the altitude profiles of various atomic and molecular excited states. These profiles can be used to calculate the various optical emissions resulting from the aurora.

9.6   Problems

Problem 9.1 In the Bohr model the angular momentum of bound electrons is $n\hbar$, where n is a positive integer. Assume that the electrons move around the positively

---

[4] Evans, D., "Precipitating electron fluxes formed by a magnetic field aligned potential difference," *J. Geophys. Res.*, **79**, 2853, 1974.

charged nucleus on circular orbits (the charge of the nucleus is Ze, where Z is the number of protons in the nucleus). Calculate the orbital velocity of an electron on a shell characterized by quantum number n.

Problem 9.2 The orbital velocity of electrons naturally cannot exceed the speed of light. In the Bohr model what is the heaviest noble gas that can exist in nature? (Hint: Use the result of the previous problem.)

Problem 9.3 What is the ground-state shell configuration of atomic nitrogen ($Z = 7$)?

Problem 9.4 The lowest energy state of the nitrogen atom is $^4P_{3/2}$. Give the possible values of l, m_l, and m_s of all electrons on "open" subshells.

Problem 9.5 The threshold of the $O_2 + h\nu \rightarrow O(^1D) + O(^3P)$ reaction is 174.9 nm. Solar radiation at 1,700 Å dissociates an O_2 molecule leaving one of the oxygen atoms in the ^1D state and the other one in the ground state. What will be the total kinetic energy given to the two oxygen atoms?

Problem 9.6 Assume that monoenergetic 3 keV electrons precipitate parallel to the geomagnetic field and produce an auroral arc of 1,000 km in east–west extent and a north–south width of 10 km, with a homogeneous surface brightness of 5 k rayleigh in the 391.4 nm band.

1. Estimate the electron flux needed to produce this arc.
2. Calculate the corresponding net downward energy in watts per square meter for this electron flux.

Chapter 10

The Ionosphere

There are ions and electrons at all altitudes of the terrestrial atmosphere. Below about 60 km thermal charged particles (which have comparable energies to the neutral gas constituents) do not play any significant role in determining the chemical or physical properties of the atmosphere. Above ≈ 60 km, however, the presence of electrons and ions becomes increasingly important. This region of the upper atmosphere is called the *ionosphere*. Note that the ionosphere overlaps with the upper mesosphere, the thermosphere, and the geocorona.

The typical vertical structure of the ionosphere is shown in Figure 10.1 (*Hargreaves 1992*). Inspection of Figure 10.1 reveals that the ionosphere exhibits a strong diurnal variation and it also varies with the solar cycle. The identification of the atmospheric layers is usually related to inflection points in the vertical density profile: The main regions are local minimums. The primary ionospheric regions are the following:

- D region (≈ 60–90 km, peaks around 90 km);
- E region (≈ 90–140 km, peaks around 110 km);
- F_1 region (≈ 140–200 km, peaks around 200 km);
- F_2 region (≈ 200–500 km, peaks around 300 km);
- Topside ionosphere (above the F_2 region).

It can be seen that the D and F_1 regions disappear at night, while the E and F_2 regions become much weaker. The topside ionosphere can also be considered to be the extension of the magnetosphere.

Figure 10.2 shows the typical composition of the dayside ionosphere under solar minimum conditions. It can be seen that at low altitudes the major ions are O_2^+ and

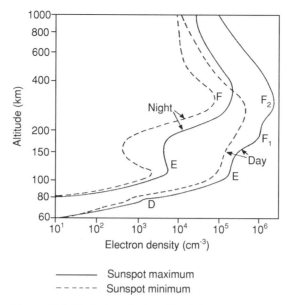

————— Sunspot maximum
– – – – – Sunspot minimum

Figure 10.1 Typical vertical profile of electron density in the midlatitude ionosphere. [From *Hargreaves* (*1992*).]

NO^+, near the F_2 peak it changes to O^+, and in the topside ionosphere eventually H^+ becomes dominant. For all practical purposes, the ionosphere can be considered quasineutral (the net charge is practically zero in every volume element containing a significant number of charged particles).

The ionosphere is formed by the ionization of three primary atmospheric constituents: N_2, O_2, and O. The primary ionization mechanism is photoionization by extreme ultraviolet (EUV) and X-ray radiation. In some regions ionization by precipitating magnetospheric particles or cosmic rays may also be important. The primary ionization process is sometimes followed by a series of chemical reactions, which produce other ions while preserving the total number of electrons. Finally, recombination removes free charges and transforms the ions to neutral particles.

10.1 Ionization Profile

Consider electromagnetic radiation entering an atmosphere and being gradually weakened by various atomic and molecular interactions (such as ionization). Let us denote the photon flux per unit frequency by Φ_ν. The change of flux, $d\Phi_\nu$, due to absorption by the neutral gas in an infinitesimal distance, ds, is the following:

$$d\Phi_\nu = -n\sigma_\nu \Phi_\nu ds, \tag{10.1}$$

where $n(z)$ is the altitude-dependent neutral gas concentration, σ_ν is the frequency-dependent photoabsorption cross section, and ds is the path length element in the

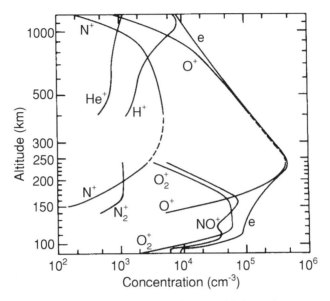

Figure 10.2 Typical composition of the dayside ionosphere at solar minimum. (From *Johnson*.[1])

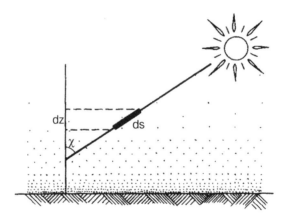

Figure 10.3 The relationship between dz and ds.

direction of the optical radiation (see Figure 10.3). Equation (10.1) assumes that there are no local sources or sinks of radiation (such as resonant scattering). This condition is usually satisfied in the dayside upper atmosphere for the ionizing solar UV radiation.

It is obvious from Figure 10.3 that $ds = \sec\chi \, dz$ (where χ is the zenith angle of the incoming solar radiation). Equation (10.1) can be readily solved to obtain the altitude

---

[1] Johnson, C. Y., "Ionospheric composition and density from 90 to 1200 kilometers at solar minimum," *J. Geophys. Res.*, **71**, 330, 1966.

dependence of the solar radiation flux:

$$\Phi_\nu(z) = \Phi_{\nu_\infty} \exp\left(-\sec\chi \int_z^\infty \sigma_\nu n(z')\, dz'\right),\tag{10.2}$$

where Φ_{ν_∞} is the incident photon intensity per unit frequency. The argument of the exponential function in Eq. (10.2) is called the *optical depth* and is represented by the symbol τ_ν:

$$\tau_\nu = \sec\chi \int_z^\infty \sigma_\nu n(z')\, dz'.\tag{10.3}$$

Radiation entering the upper atmosphere is usually attenuated by several neutral constituents, each with its own absorption cross section. One can easily generalize Eq. (10.3) and obtain the optical depth for a multispecies atmosphere:

$$\tau_\nu = \sec\chi \sum_t \int_z^\infty \sigma_{\nu t} n_t(z')\, dz'.\tag{10.4}$$

Now the attenuation of solar radiation due to atmospheric absorption can be written simply as

$$\Phi_\nu(z) = \Phi_{\nu\infty} \exp\left(-\tau_\nu\right).\tag{10.5}$$

Next, we evaluate the optical depth for the thermosphere. We derived the density profiles for thermospheric constituents assuming that molecular diffusion dominates (see Eq. 8.61). Assuming a single dominant thermospheric species, one can substitute this equation into Eq. (10.3) to get

$$\tau_\nu = \sec\chi \int_z^\infty \sigma_\nu n_0 \left(\frac{T_0}{T}\right)^{1+\alpha_T} \exp\left[-\frac{z'-z_0}{H}\right] dz'.\tag{10.6}$$

Here the subscript 0 again refers to values at a reference altitude z_0. For the sake of simplicity, we can assume that the upper thermosphere is isothermal (this is not such a bad assumption) and that σ_ν is independent of altitude. In this case we obtain

$$\tau_\nu = \sec\chi \int_z^\infty \sigma_\nu n_0 \exp\left[-\frac{z'-z_0}{H}\right] dz' = \sec\chi\, \sigma_\nu n(z) H\tag{10.7}$$

or in the case of a multispecies atmosphere

$$\tau_\nu = \sec\chi \sum_t \sigma_{\nu t} n_t(z) H_t.\tag{10.8}$$

Inspection of Eq. (10.7) reveals that the optical depth increases exponentially with decreasing altitude. Because the attenuation of the solar radiation is proportional to

$\exp(-\tau_\nu)$, it is therefore obvious that the solar radiation at a particular frequency is rapidly extinguished once it reaches the $\tau \approx 1$ level.

In the thermosphere solar radiation is absorbed mainly via ionization processes. Let us assume that $\sigma_\nu \approx \sigma_{\nu i}$ and calculate the electron production rate profile. In this approximation each absorbed photon creates a new electron-ion pair, and therefore the electron production in a ray-path element, ds, is the following:

$$S_i\, ds = n\sigma_{\nu i}\Phi_\nu(z)\, ds, \tag{10.9}$$

where S_i is the total electron production rate (S_i is measured in particles $\mathrm{cm}^{-3}\,\mathrm{s}^{-1}$). Substituting the altitude dependences of the concentration and the photon flux (Eqs. 8.61 and 10.5), we obtain

$$S_i = n_0\sigma_{\nu i}\Phi_{\nu\infty}\exp\left(-\frac{z-z_0}{H} - \sec\chi\,\sigma_{\nu i}Hn_0\exp\left[-\frac{z-z_0}{H}\right]\right). \tag{10.10}$$

The altitude of maximum ionization can be obtained by solving the following equation:

$$\frac{dS_i}{dz} = 0. \tag{10.11}$$

This equation leads to the following expression:

$$\sec\chi\,\sigma_{\nu i}Hn(z_{max}) = \tau(z_{max}) = 1, \tag{10.12}$$

where z_{max} is the altitude of maximum ionization. It can be seen that ionization is maximum at the altitude where the optical depth is unity. We can choose $z_0 = z_{max}|_{\chi=0}$ (the altitude of maximum ionization for perpendicular solar radiation), and obtain the following expression for the ionization rate profile:

$$S_i = S_0\exp\left(1 - \frac{z}{H} - \sec\chi\exp\left[-\frac{z}{H}\right]\right), \tag{10.13}$$

where

$$S_0 = \frac{\Phi_{\nu\infty}}{H}e^{-1}. \tag{10.14}$$

The maximum ionization rate for a given incidence angle is given by

$$S_{max} = S_0\cos\chi. \tag{10.15}$$

Equation (10.13) is the classical Chapman formula. It predicts the ionization profile due to the absorption of solar radiation in an isothermal atmosphere. The Chapman ionization function is shown in Figure 10.4.

Using the Chapman formula for ionization, it is possible to determine a simple approximation for the altitude profile of electron density. To do so one has to assume that the main loss process is ion–electron recombination with a recombination coefficient of α. In this very simple approximation we assume that the total recombination rate is αn_e^2, since an electron and an ion must come together in order to recombine, and

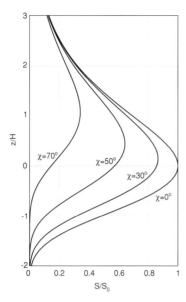

Figure 10.4 The normalized Chapman ionization function.

we assume that the ion and electron concentrations are equal. Neglecting transport processes, one can assume local equilibrium between production and loss:

$$S_i = \alpha n_e^2, \tag{10.16}$$

giving

$$n_e = \sqrt{\frac{S_0}{\alpha}} \exp\left(\frac{1}{2} - \frac{z}{2H} - \frac{\sec \chi}{2} \exp\left[-\frac{z}{H}\right]\right). \tag{10.17}$$

This is the vertical profile of the electron density in a simple *Chapman layer*.

Although the Chapman formulas provide a convenient analytical tool to describe ionospheric altitude profiles, the method is overly simplified and has to be used with great care. Usually one has to use complicated numerical techniques to obtain realistic altitude profiles taking into account various production, loss, and transport processes. The Chapman formalism provides a good first step in understanding ionospheric profiles, but usually one has to go beyond this simple method. Nevertheless, the simple Chapman theory gives a good general understanding of the D, E, and F regions of the ionosphere. In particular, the E and F_1 regions are essentially Chapman layers, whereas the D and F_2 regions are somewhat complicated.

10.2 Ion Composition and Chemistry

To be ionized a neutral atom or molecule must absorb a photon with energy above the ionization potential. Table 10.1 lists the first ionization potential and the corresponding wavelength of photons capable of ionizing the given neutral species. Inspection of Table 10.1 immediately reveals that the X-ray (0.1 to 17 nm) and EUV (17 to 175 nm) parts of the solar spectrum are major sources of ionization in the ionosphere. In this section we will discuss some of the reactions playing a dominant role in the formation of the various layers of the ionosphere.

10.2.1 The D Region

The D region is the most complex and least understood region of the terrestrial ionosphere. The chemistry is extremely complex, and it involves negative ions, hydrated ions, and electrons. The primary source of ionization in the D region is ionization by solar X rays (in the 0.2 to 0.8 nm range) and Lyman-α ($\lambda =121.5$ nm) ionization of the NO molecule. In addition, ionization by precipitating magnetospheric particles (mainly electrons) and by galactic cosmic rays may also play important roles.

The solar X rays ionize all ions, and consequently the most important primary ions are formed from the dominant constituents of the neutral atmosphere, N_2 and O_2. In addition, the Lyman-α line ionizes the minor neutral constituent, NO.

The initial positive ions are N_2^+, O_2^+, NO^+, and to a lesser extent O^+. The unstable N_2^+ ion is rapidly converted to O_2^+ by the following charge exchange reaction:

$$N_2^+ + O_2 \rightarrow O_2^+ + N_2. \tag{10.18}$$

This process leaves O_2^+ and NO^+ as the major positive ions. In the lower D region (near the mesopause), however, hydrated positive ions are dominant. These hydrated ions include H_3O^+, $H_5O_2^+$, as well as $H^+(H_2O)_n$ clusters.

Table 10.1. *First ionization potentials of the main atmospheric constituents.*

Species	Ionization potential (eV)	Maximum wavelength (nm)
NO	9.25	134.0
O_2	12.08	102.7
H_2O	12.60	98.5
O_3	12.80	97.0
H	13.59	91.2
O	13.61	91.1
CO_2	13.79	89.9
N	14.54	85.3
H_2	15.41	80.4
N_2	15.58	79.6
Ar	15.75	78.7
Ne	21.56	57.5
He	24.58	50.4

The negative ion composition in the D region is similarly complex and fairly poorly known. The first step in the formation of negative ions is believed to be the following electron attachment process:

$$e + O_2 + M \rightarrow O_2^- + M. \tag{10.19}$$

This reaction is followed by further reactions that eventually form more complex negative ions, such as CO_3^-, NO_2^-, and NO_3^-, the last of which is the most abundant negative ion in the D region. The negative ion chemistry is further complicated by the formation of hydrated negative ions.

10.2.2 The E Region

The E region is essentially a Chapman layer that is formed by the 80 to 102.7 nm part of the EUV spectrum. The main initial ion in this region is O_2^+, with some production of N_2^+ and O^+. The N_2^+ ions are rapidly transformed to other ions by the charge exchange reaction (10.18) and by the following other reactions:

$$\begin{aligned} N_2^+ + O &\rightarrow NO^+ + N, \\ N_2^+ + O &\rightarrow O^+ + N_2. \end{aligned} \tag{10.20}$$

Oxygen ions are removed by the following reactions:

$$\begin{aligned} O^+ + N_2 &\rightarrow NO^+ + N, \\ O^+ + O_2 &\rightarrow O_2^+ + O. \end{aligned} \tag{10.21}$$

The end result of these reactions is that the major ions in the dayside E region are O_2^+ and NO^+ (under average conditions the NO^+ concentration is somewhat larger than the concentration of O_2^+). The total ion density (or electron density) profile in the E layer is basically consistent with a Chapman layer.

Figure 10.5 shows the daytime positive ion composition measured *in situ* by a rocket-borne ion mass spectrometer. (The presence of the H_2O^+ ion is probably due to contamination by the rocket.) The total electron density was obtained from an ionogram. Inspection of Figure 10.5 shows that during the day the dominant ion in the lower F_1 region is NO^+, whereas above ≈ 170 km the O^+ concentration rapidly exceeds all other ions.

10.2.3 The F_1 Region

The F_1 region is also essentially a Chapman layer. The ionizing solar flux is in the <91 nm EUV spectral region. It is basically absorbed in this layer and thus does not penetrate into the E region. The principal initial ion in the F_1 region is O^+, with some contribution from N_2^+.

The radiative recombination of the oxygen ion is extremely slow, $k = 3 \times 10^{-12}$ cm^{-3} s^{-1}, and therefore O^+ recombines by a two-step process. First, the atom–ion

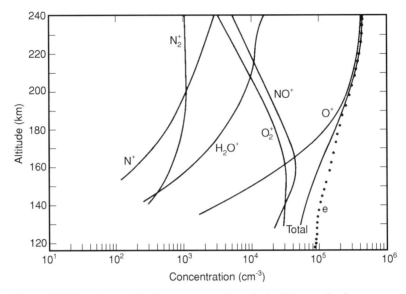

Figure 10.5 Daytime positive ion concentrations obtained by a rocket-borne mass spectrometer. (From *Holmes et al.*[2])

interchange (10.21) takes place. This is followed by dissociative recombination of O_2^+ and NO^+.

In the lower F_1 region (around 140 km) the lifetime of the O^+ is quite short (a few seconds), and therefore the NO^+ molecular ion becomes dominant. Above approximately 180 km, O^+ rapidly takes over and becomes the major ion.

10.2.4   The F_2 Peak Region

In the F_2 region the major ion is O^+ with a density peak in the 200 to 400 km range. This region clearly cannot be a Chapman layer, since the atmosphere above the F_1 region is optically thin to most ionizing radiation. The formation of the F_2 peak is caused by an interplay between ion sources, sinks, and ambipolar diffusion.

In the F_2 region the dominant ionization source is the photoionization of atomic oxygen:

$$O + h\nu \rightarrow O^+ + e. \tag{10.22}$$

The photoionization coefficient of oxygen is about $I_i \approx 10^{-7} \text{ s}^{-1}$, and the ion production rate is

$$S_i = I_i n_O = I_i n_{O0} \exp\left(-\frac{z - z_0}{H_O}\right), \tag{10.23}$$

where n_{O0} is the neutral oxygen density at the reference altitude, z_0.

[2] Holmes, J. C., Johnson, C. Y., and Young, J. M., "Ionospheric chemistry," *Space Research*, **5**, 756, 1965.

The oxygen ions are lost by a two-step process involving N_2 and O_2. The first step is an atom–ion interchange (see reaction 10.21):

$$O^+ + O_2 \rightarrow O_2^+ + O \qquad k_{O_2} \approx 2 \times 10^{-11}\,\text{cm}^{-3}\,\text{s}^{-1},$$
$$O^+ + N_2 \rightarrow NO^+ + N \qquad k_{N_2} \approx 1 \times 10^{-12}\,\text{cm}^{-3}\,\text{s}^{-1}. \qquad (10.24)$$

This interchange process is followed by very rapid dissociative recombination:

$$O_2^+ + e \rightarrow O + O \qquad k_{O_2^+} \approx 3 \times 10^{-7}\,\text{cm}^{-3}\,\text{s}^{-1},$$
$$NO^+ + e \rightarrow N + O \qquad k_{NO^+} \approx 3 \times 10^{-7}\,\text{cm}^{-3}\,\text{s}^{-1}. \qquad (10.25)$$

The dissociative recombination processes are so fast that it is perfectly adequate to consider reactions (10.24) when calculating the loss rate:

$$L_e n_e = k_{O_2} n_e n_{O_2} + k_{N_2} n_e n_{N_2} = n_e (k_{O_2} n_{O_2} + k_{N_2} n_{N_2}) \qquad (10.26)$$

The net electron (or ion) production rate can be obtained by subtracting the loss rate from the production rate:

$$P_i = S_i - L_e n_e = I_i n_{O0} \exp\left(-\frac{z - z_0}{H_O}\right)$$
$$- n_e \left[k_{O_2} n_{O_2 0} \exp\left(-\frac{z - z_0}{H_{O_2 0}}\right) + k_{N_2} n_{N_2 0} \exp\left(-\frac{z - z_0}{H_{N_2 0}}\right) \right]. \qquad (10.27)$$

Under local photochemical equilibrium conditions, there is no net production, $P_i = 0$. Assuming that the thermospheric species are in thermodynamic equilibrium (all species have the same temperature), one can obtain the following electron density profile:

$$n_e = \frac{I_i n_{O0} \exp\left(-\frac{z - z_0}{H_O}\right)}{k_{O_2} n_{O_2 0} \exp\left(-\frac{z - z_0}{H_{O_2 0}}\right) + k_{N_2} n_{N_2 0} \exp\left(-\frac{z - z_0}{H_{N_2 0}}\right)}. \qquad (10.28)$$

It is easy to see that this electron density profile indefinitely increases with increasing altitude. When all temperatures are equal the scale heights are inversely proportional to the particle mass, and therefore $H_O = 1.75\,H_{N_2} = 2 H_{O_2}$. When the altitude increases, the denominator decreases much faster than the numerator, and thus the electron density increases with altitude.

This solution clearly contradicts the observations, and therefore we immediately conclude that the F_2 region cannot be in local photochemical equilibrium. Transport phenomena must be taken into account to resolve this problem. This will be done in the next sections.

10.3   Gyration-Dominated Plasma Transport

Let us consider the transport of a charged particle species in the presence of a strong magnetic field. We start with the multispecies Navier–Stokes equations (4.34), but neglect the effects of viscosity, and assume that only elastic collisions have a significant effect on the transport process (ionization and recombination are also going on, but they are only taken into account in the continuity equation, and their effects on momentum and energy transport are neglected). We also assume that the charged particle species have no internal degrees of freedom (i.e., they are monatomic). In this case the continuity, momentum, and energy equations for charged particles become:

$$\frac{\partial m_s n_s}{\partial t} + \nabla \cdot (m_s n_s \, \mathbf{u}_s) = m_s S_s - L_s m_s n_s,$$

$$m_s n_s \frac{\partial \mathbf{u}_s}{\partial t} + m_s n_s \, (\mathbf{u}_s \cdot \nabla) \mathbf{u}_s + \nabla p_s - m_s n_s \mathbf{g} - q_s n_s (\mathbf{E} + \mathbf{u}_s \times \mathbf{B})$$

$$= \sum_t m_s n_s \bar{\nu}_{st} (\mathbf{u}_t - \mathbf{u}_s), \tag{10.29}$$

$$\frac{3}{2} \frac{\partial p_s}{\partial t} + \frac{3}{2} (\mathbf{u}_s \cdot \nabla) p_s + \frac{5}{2} p_s (\nabla \cdot \mathbf{u}_s) + \nabla \cdot \mathbf{h}_s$$

$$= \sum_t \frac{m_s n_s \bar{\nu}_{st}}{m_s + m_t} [3k(T_t - T_s) + m_t (\mathbf{u}_t - \mathbf{u}_s)^2],$$

where the summation runs over all species, neutral and charged alike.

Consider a very strong magnetic field, $B \to \infty$. If the perpendicular component of the flow velocity is not zero, the dominant terms in the momentum equation can only be the gyration and the collision terms (all other terms are limited by other processes). In this case the momentum equation becomes

$$\pm \Omega_{cs} u_{s\perp} = \sum_t \bar{\nu}_{st} |(\mathbf{u}_t - \mathbf{u}_s)_\perp|, \tag{10.30}$$

where Ω_{cs} is the gyrofrequency of particles s and the $\pm$ sign refers to the sign of the particle charge. This means that if we place a partially ionized gas in a strong magnetic field (such as the terrestrial magnetic field in the ionosphere) the plasma transport perpendicular to the magnetic field is limited by the collisional momentum transfer. Above approximately 200 km the gyrofrequency is much larger than the collision frequency. Therefore charged particle species cannot have significant perpendicular velocities. Thus to a good approximation one can write

$$\mathbf{u}_s = u_s \mathbf{b}, \tag{10.31}$$

where $\mathbf{b} = \mathbf{B}/B$ is the unit vector along the magnetic field line.

Since above $\approx$200 km most of the ionospheric transport takes place along magnetic field lines, let us derive transport equations for field-aligned transport. First, let us substitute Eq. (10.31) into the continuity equation:

$$\frac{\partial m_s n_s}{\partial t} + m_s n_s u_s (\nabla \cdot \mathbf{b}) + (\mathbf{b} \cdot \nabla)(m_s n_s u_s) = m_s S_s - L_s m_s n_s. \tag{10.32}$$

We know that $(\mathbf{b} \cdot \nabla) = \partial/\partial s$ is the gradient operator along the field line. Also,

$$(\nabla \cdot \mathbf{b}) = \left(\nabla \cdot \frac{\mathbf{B}}{B} \right) = -\frac{1}{B} \frac{\partial B}{\partial s}. \tag{10.33}$$

Substituting Eq. (10.33) into Eq. (10.32) yields the following:

$$\frac{\partial m_s n_s}{\partial t} + B \frac{\partial}{\partial s}\left(\frac{m_s n_s}{B} u_s\right) = m_s S_s - L_s m_s n_s. \tag{10.34}$$

The magnetic field is time independent to a very good approximation, and therefore one obtains the following form of the continuity equation for field-aligned transport in the presence of a strong magnetic field:

$$\frac{\partial}{\partial t}\left(\frac{m_s n_s}{B}\right) + \frac{\partial}{\partial s}\left(\frac{m_s n_s}{B} u_s\right) = \frac{m_s S_s}{B} - L_s \frac{m_s n_s}{B}. \tag{10.35}$$

This is a very interesting result. Equation (10.35) shows that the quantity $1/B$ behaves like the cross sectional area of a variable cross-section tube. This is not surprising, because we know that $1/B$ is proportional to the area of a magnetic flux tube. In other words, the continuity equation is a conservation law for the charged particle density per unit magnetic flux tube.

Next, we multiply the momentum equation by $\mathbf{b}$:

$$m_s n_s \frac{\partial u_s}{\partial t} + m_s n_s u_s \frac{\partial u_s}{\partial s} + m_s n_s u_s^2 \mathbf{b} \cdot \frac{\partial \mathbf{b}}{\partial s} + \frac{\partial p_s}{\partial s} - m_s n_s g_\| - q_s n_s E_\|$$
$$= \sum_t m_s n_s \bar{\nu}_{st}(u_{t\|} - u_s). \tag{10.36}$$

Knowing that

$$\mathbf{b} \cdot \frac{\partial \mathbf{b}}{\partial s} = \frac{1}{2}\frac{\partial(\mathbf{b} \cdot \mathbf{b})}{\partial s} = 0 \tag{10.37}$$

gives us the following form of the field-aligned momentum equation:

$$m_s n_s \frac{\partial u_s}{\partial t} + m_s n_s u_s \frac{\partial u_s}{\partial s} + \frac{\partial p_s}{\partial s} - m_s n_s g_\| - q_s n_s E_\|$$
$$= \sum_t m_s n_s \bar{\nu}_{st}(u_{t\|} - u_s). \tag{10.38}$$

Here the subscript $\|$ refers to vector components parallel to the magnetic field line.

Finally, one can substitute the field-aligned transport requirement into the energy equation to obtain

$$\frac{3}{2}\frac{\partial p_s}{\partial t} + B\frac{3}{2}\frac{\partial}{\partial s}\left(u_s \frac{p_s}{B}\right) + p_s \frac{\partial u_s}{\partial s} - u_s \frac{p_s}{B}\frac{\partial B}{\partial s} + \nabla \cdot \mathbf{h_s}$$
$$= \sum_t \frac{m_s n_s \bar{\nu}_{st}}{m_s + m_t}[3k(T_t - T_s) + m_t(\mathbf{u}_t - \mathbf{u}_s)^2]. \tag{10.39}$$

It is also assumed that the heat flow is primarily along the magnetic field lines, $\mathbf{h}_s = h_s \mathbf{b}$.

With this assumption the energy equation becomes

$$\frac{3}{2}\frac{\partial}{\partial t}\left(\frac{p_s}{B}\right) + \frac{3}{2}\frac{\partial}{\partial s}\left(u_s\frac{p_s}{B}\right) + p_s\frac{\partial}{\partial s}\left(\frac{u_s}{B}\right) + \frac{\partial}{\partial s}\left(\frac{h_s}{B}\right)$$

$$= \frac{m_s n_s}{B}\sum_t \frac{\bar{\nu}_{st}}{m_s + m_t}[3k(T_t - T_s) + m_t(\mathbf{u}_t - \mathbf{u}_s)^2]. \tag{10.40}$$

Equations (10.35), (10.38), and (10.40) represent the continuity, momentum, and energy equations describing the transport of a charged species in the presence of strong magnetic fields.

10.4 Ambipolar Electric Field and Diffusion

Let us consider the field-aligned momentum equation (10.38) in the ionosphere and assume steady-state conditions. Let us start with electrons:

$$m_e n_e u_e\frac{du_e}{ds} + \frac{dp_e}{ds} - m_e n_e g_\| + e n_e E_\| = \sum_t m_e n_e \bar{\nu}_{et}(u_{t\|} - u_e), \tag{10.41}$$

where e is the magnitude of the electron charge.

In the ionosphere one can make several additional assumptions. First, because of the very small mass of electrons, both the inertial term and the gravitational term are much smaller than the pressure gradient force term, and therefore these terms can be neglected. Second, collisions also play only a minor role in electron dynamics, and so the electron collision term can be neglected, leaving

$$\frac{dp_e}{ds} + e n_e E_\| = 0. \tag{10.42}$$

Thus that the electron pressure gradient is balanced by an electric field in the ionosphere. This electric field is called the *ambipolar electric field*. The physical mechanism responsible for the generation of the ambipolar electric field is that the very light electrons are not gravitationally bound in the ionosphere and collisional effects are also negligible above ≈ 200 km. In this region the electrons could be driven away from the ionosphere along the magnetic field lines by the electron pressure gradient force. However, the heavier ions are bound by gravity to Earth, and they cannot move with the light electrons. When the electrons start to move away, the ions and electrons are slightly separated and a significant electric field is generated. The magnitude of this ambipolar electric field (sometimes also called the polarization electric field) is such that it "pulls back" the electrons and "pulls up" the ions with equal force.

For the sake of simplicity, let us assume that there is one dominant ion (which is singly ionized) in the partially ionized mixture. To a first approximation one can neglect field-aligned currents, and therefore the electron and ion concentrations and velocities must be the same. This means that ion–electron collisions do not contribute to the collision term. To a first approximation one can also neglect neutral winds and assume that all neutral species have zero velocity. In this approximation the ion momentum equation becomes the following:

$$m_i n_i u_i \frac{du_i}{ds} + \frac{dp_i}{ds} - m_i n_i g_\parallel - e n_i E_\parallel = -\bar{\nu}_{in} m_i n_i u_i, \tag{10.43}$$

where the subscript "i" here refers to ions. Also, we introduced the total ion–neutral collision frequency:

$$\bar{\nu}_{in} = \sum_{t=\text{neutrals}} \bar{\nu}_{it}. \tag{10.44}$$

In the F_2 region the ion velocities are relatively small. Therefore one can neglect the ion inertial term (it is quadratic in the velocity). This assumption leads to the following expression for the ion velocity:

$$u_i = -\frac{1}{\bar{\nu}_{in} m_i n_i} \left[\frac{dp_i}{ds} - m_i n_i g_\parallel - e n_i E_\parallel \right]. \tag{10.45}$$

One can also substitute the ambipolar electric field expression (Eq. 10.42) into the ion momentum equation and recognize that the plasma must be quasineutral ($n_e = n_i$). In this case Eq. (10.45) becomes

$$u_i = -\frac{1}{\bar{\nu}_{in} m_i n_i} \left[\frac{d(p_i + p_e)}{ds} - m_i n_i g_\parallel \right]. \tag{10.46}$$

Equation (10.46) describes the ambipolar diffusion of ions due to the presence of the ambipolar electric field. This is the slow response of the ions to the fact that the electrons are not gravitationally bound and could run away from the ionosphere if they were not "tied" to the ions.

One can readily obtain the diffusive equilibrium solution characterized by $u_i = 0$. Also, one has to take into account the relation between the altitude increment, dz, and the field line length element, ds. Let us introduce the *magnetic dip angle*, ϑ, which is defined as the angle between the horizontal direction and the magnetic field line. In this case $ds = dz / \sin \vartheta$. The parallel component of the gravitational acceleration is $g_\parallel = -g \sin \vartheta$. Now the density profile in diffusive equilibrium equation is given by

$$\sin \vartheta \frac{d(p_i + p_e)}{dz} + m_i n_i g \sin \vartheta = 0. \tag{10.47}$$

If the ion and electron temperatures are constant, this equation yields the following differential equation for n_i:

$$n_i = -\frac{k(T_e + T_i)}{m_i g} \frac{dn_i}{dz}. \tag{10.48}$$

The solution is an isothermal hydrostatic equilibrium

$$n_i = n_{i0} \exp\left(-\frac{z - z_0}{H_i} \right), \tag{10.49}$$

where the ion scale height is

$$H_i = \frac{k(T_e + T_i)}{m_i g} = \frac{kT_i}{m_i g} + \frac{kT_e}{m_i g}. \tag{10.50}$$

This is a very interesting result. The ion scale height is the sum of two different scale heights. One is determined by the ion gas only; in effect, this term is the scale height of an ion gas in diffusive equilibrium. However, the effect of electrons adds an extra term to the ion scale height. This second term is due to the ambipolar electric field generated by the electron pressure gradient, which pulls the ions upward, and therefore increases its scale height. If the ion and electron temperatures are equal, the ion scale height is twice as large as the corresponding scale height of a neutral gas with the same molecular mass and temperature. In other words, the ambipolar electric field reduces the "effective" mass (attracted by the gravitational field) of the ions by half and "creates" an equal "effective" electron mass (as if the electrons had a "gravitational" mass of $m_i/2$).

10.5 Diffusive Equilibrium in the F_2 Region

Let us return to the F_2 region and examine the altitude variation of ion density. The continuity equation of the ions can be written in the following form (see Eqs. 10.35, 10.27, and 10.46):

$$\frac{\partial}{\partial t}\left(\frac{m_i n_i}{B}\right) + \frac{\partial}{\partial s}\left(\frac{m_i n_i}{B} u_i\right) = I_i \frac{m_i n_{O0}}{B} \exp\left(-\frac{z - z_0}{H_O}\right)$$
$$- \frac{m_i n_i}{B}\left[k_{O_2} n_{O_2 0} \exp\left(-\frac{z - z_0}{H_{O_2 0}}\right) + k_{N_2} n_{N_2 0} \exp\left(-\frac{z - z_0}{H_{N_2 0}}\right)\right]. \tag{10.51}$$

Let us assume steady-state conditions and apply our diffusive velocity approximation given by Eq. (10.46). Also, in the F_2 region the magnetic field is nearly constant. With these simplifications Eq. (10.51) becomes the following:

$$-\frac{d}{ds}\left(\frac{1}{\bar{\nu}_{in}}\left[\frac{d(p_i + p_e)}{ds} - m_i n_i g_\parallel\right]\right) = I_i m_i n_{O0} \exp\left(-\frac{z - z_0}{H_O}\right)$$
$$- m_i n_i\left[k_{O_2} n_{O_2 0} \exp\left(-\frac{z - z_0}{H_{O_2 0}}\right) + k_{N_2} n_{N_2 0} \exp\left(-\frac{z - z_0}{H_{N_2 0}}\right)\right]. \tag{10.52}$$

For the sake of additional simplicity, we assume that the electron and ion temperatures are constant in the F_2 region. This assumption is not quite true, but it helps to bring us some physical insight into the problem. One also has to take into account the relation between the altitude increment, dz, and the field line length element, ds. This relation is $ds = dz/\sin\vartheta$, where ϑ is the magnetic dip angle. Noting that the parallel component of the gravitational acceleration is $g_\parallel = -g\sin\vartheta$, Eq. (10.52) can now be written as

$$-\sin^2\vartheta \frac{d}{dz}\left(\frac{k(T_i + T_e)}{\bar{\nu}_{in}}\left[\frac{dn_i}{dz} + \frac{m_i g}{k(T_i + T_e)}n_i\right]\right) = I_i m_i n_{00} \exp\left(-\frac{z - z_0}{H_O}\right)$$
$$- m_i n_i \left[k_{O_2} n_{O_2 0} \exp\left(-\frac{z - z_0}{H_{O_2 0}}\right) + k_{N_2} n_{N_2 0} \exp\left(-\frac{z - z_0}{H_{N_2 0}}\right)\right]. \quad (10.53)$$

Equation (10.53) can be further simplified by recalling that the dominant neutral in the F2 region is atomic oxygen, whereas the dominant ion is O^+. It was also assumed earlier that all neutral constituents have the same temperature, T_n. Additional simplification can be achieved if we make the assumption that the ion temperature is equal to the neutral temperature ($T_i = T_n$), the ratio of the electron and ion temperatures is χ_e, and the N_2 scale height is close to the O_2 scale height ($H_{N_2} \approx H_{O_2} = H_n/2$, where H_n is the neutral oxygen scale height). In this case we get the following:

$$\sin^2\vartheta \frac{d}{dz}\left((1 + \chi_e)H_n \frac{g}{\bar{\nu}_{in}}\left[\frac{dn_i}{dz} + \frac{1}{(1 + \chi_e)H_n}n_i\right]\right)$$
$$+ S_0 \exp\left(-\frac{z - z_0}{H_n}\right) - n_i L_0 \exp\left(-2\frac{z - z_0}{H_n}\right) = 0, \quad (10.54)$$

where we introduced two constants:

$$S_0 = I_i n_{00}, \quad (10.55)$$
$$L_0 = [k_{O_2} n_{O_2 0} + k_{N_2} n_{N_2 0}].$$

It was already mentioned that O is the dominant neutral in the F2 layer. This means that the major contribution to the ion–neutral collision frequency comes from momentum transfer collisions between O and O^+. Because $\nu_{in} \propto n_n$, the altitude dependence of the collision frequency can be written as

$$\bar{\nu}_{in} = \bar{\nu}_0 \exp\left(-\frac{z - z_0}{H_n}\right). \quad (10.56)$$

Substituting expression (10.55) into Eq. (10.54) yields the following:

$$\frac{d^2 n_i}{dz^2} + \frac{2 + \chi_e}{(1 + \chi_e)H_n}\frac{dn_i}{dz} + \frac{n_i}{(1 + \chi_e)H_n^2}\left[1 - \frac{\nu_0 L_0 H_n}{g \sin^2\vartheta}\exp\left(-3\frac{z - z_0}{H_n}\right)\right]$$
$$+ \frac{\nu_0 S_0}{(1 + \chi_e)H_n g \sin^2\vartheta}\exp\left(-2\frac{z - z_0}{H_n}\right) = 0. \quad (10.57)$$

Inspection of Eq. (10.57) reveals a great deal about the ion concentration in the F2 region. However, a word of caution is in order before we discuss the physical interpretation of this equation. It is clear that Eq. (10.57) breaks down in the equatorial ionosphere, where the magnetic field lines are nearly horizontal. In this region $\vartheta \approx 0$, and therefore Eq. (10.57) predicts photochemical equilibrium (the production and loss terms are multiplied by the same large number and therefore they must balance). In this region a more sophisticated approach must be used to explain the observed properties of the F_2 region (this approach takes into account reduced diffusion across magnetic field lines). This complicated derivation goes beyond the scope of this book.

Equation (10.57) is valid in the middle- and high-latitude regions of the ionosphere. Usually, one has to solve it numerically, because in its general form it is too complicated for meaningful analytic solutions. However, we can investigate the asymptotic behavior of Eq. (10.57) at high altitudes and choose physically meaningful upper boundary conditions. It is worth pointing out that Eq. (10.57) is a second-order ordinary differential equation, which requires two boundary conditions. These boundary conditions can be the ion density at the bottom of the F_2 region and the ion density at $z \rightarrow \infty$.

As $z \rightarrow \infty$ Eq. (10.57) becomes the following:

$$\frac{d^2 n_i}{dz^2} + \frac{2 + \chi_e}{(1 + \chi_e) H_n} \frac{dn_i}{dz} + \frac{n_i}{(1 + \chi_e) H_n^2} = 0. \tag{10.58}$$

This asymptotic equation has two solutions:

$$n_{iE} \approx \exp\left(-\frac{z - z_0}{(1 + \chi_e) H_n}\right) = \exp\left(-\frac{z - z_0}{H_i}\right),$$

$$n_{iD} \approx \exp\left(-\frac{z - z_0}{H_n}\right). \tag{10.59}$$

Solution n_{iE} corresponds to the diffusive equilibrium solution for ions (see Eq. 10.49), whereas n_{iD} follows the neutral density profile. The corresponding asymptotic diffusion velocities are the following (cf. Eq. 10.46):

$$u_{iE} = 0,$$

$$u_{iD} = \frac{g}{\nu_0} \chi_e \sin \vartheta \exp\left(\frac{z - z_0}{H_n}\right). \tag{10.60}$$

The diffusion velocity in the second case is upward and it increases exponentially with altitude. Clearly, sooner or later our low-speed approximation will break down and a more sophisticated approach must be used. This will be done when we discuss the polar wind. The asymptotic diffusive flux in the diffusive case is constant:

$$\Phi_D = n_{iD} u_{iD} = n_D \frac{g}{\nu_0} \chi_e \sin \vartheta \tag{10.61}$$

Observations indicate that under typical conditions the midlatitude ionosphere exhibits diffusive equilibrium-type ion density profiles above the F_2 peak. In the polar ionosphere, however, the diffusion solution is observed.

A pair of full numerical solutions to Eq. (10.57) is shown in Figure 10.6 [see *Banks & Kockarts (1973)*]. The model describes the main characteristics of the F_2 peak quite well.

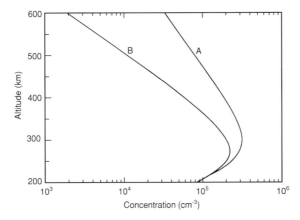

Figure 10.6 The F_2 region O^+ density showing diffusive equilibrium (A) and upward diffusion (B) solutions. In the calculation a daytime 1,000 K model thermosphere was used with vertical magnetic field. [From *Banks & Kockarts* (*1973*).]

10.6   The Topside Ionosphere and Plasmasphere

A very important feature of the topside ionosphere (above 500 km) is the increasing importance of H^+ ions. At high altitudes (above approximately 1,000 km), the light hydrogen ion becomes dominant (with the exception of the polar wind, which will be discussed in the next section).

In the topside ionosphere hydrogen ions are produced by the accidentally resonant charge exchange reaction with oxygen:

$$O^+(^4S) + H(^2S) \rightleftharpoons O(^3P) + H^+ + \Delta\epsilon. \tag{10.62}$$

Because $\Delta\epsilon$ is small relative to thermal energies in the thermosphere, reaction (10.62) is quite fast in both directions. In effect, this reaction is both the major source and major sink of hydrogen ions in the topside ionosphere. The momentum transfer cross section of reaction (10.62) is about $\sigma_e \approx 2 \times 10^{-15}$ cm^2. The production and loss rates can now be written as

$$S_{H^+} = m_H \sigma_e \sqrt{\frac{8kT_n}{\pi m_H}} n(O^+)\, n(H),$$

$$\tag{10.63}$$

$$L_{H^+} = m_H \sigma_e \sqrt{\frac{8kT_n}{\pi m_H}}\, n(O) n(H^+).$$

Here the relative velocity between the two particles is approximated by the average thermal speed of the light ion.

For photochemical equilibrium conditions, production balances loss at all altitudes. This gives the following ion density ratio:

$$\frac{n(\text{H}^+)}{n(\text{O}^+)} = \frac{n(\text{H})}{n(\text{O})}\sqrt{\frac{T_n}{T_i}} = \sqrt{\frac{T_n}{T_i}}\exp\left(\frac{15}{16}\frac{z-z_0}{H_O}\right) \tag{10.64}$$

This means that even in photochemical equilibrium the $[\text{H}^+]/[\text{O}^+]$ ratio increases exponentially with altitude and eventually H^+ becomes the major ion.

Helium ions are produced in the upper ionosphere through photoionization by solar radiation with $\lambda < 50.4$ nm. The principal loss mechanisms of He^+ involve charge transfer with O_2 and N_2. However, in the topside ionosphere the O_2 and N_2 densities are negligible, and therefore there is no significant chemical loss process for He^+. Due to the lack of loss mechanisms, early calculations resulted in a gross overestimate of the He^+ densities in the topside ionosphere. The missing He^+ loss process was resolved by Axford,[3] who suggested the *polar wind* as a possible loss mechanism for helium ions.

Let us consider the topside ionosphere above $\sim$1,000 km, where H^+ is the dominant ion and production and loss processes are negligible. Furthermore, we would like to find a low-speed, near-equilibrium solution for the density profile in the topside ionosphere. In this case the steady-state continuity equation becomes the following (see Eq. 10.35):

$$\frac{d}{ds}\left(\frac{n(\text{H}^+)}{B}u_i\right) = 0, \tag{10.65}$$

where we assumed that the ion and electron densities are equal, and we use the diffusion velocity given by Eq. (10.46). Substituting the diffusion velocity into Eq. (10.65) yields the following differential equation for the H^+ electron density:

$$\frac{d}{ds}\left(\frac{1}{B}\exp\left(\frac{\sin\vartheta}{H_n}(s-s_0)\right)\left[\frac{dn(\text{H}^+)}{ds} + \frac{\sin\vartheta}{H_p}n(\text{H}^+)\right]\right) = 0. \tag{10.66}$$

Here we assumed the topside ionosphere to be isothermal and used the altitude dependence of the ion–neutral collision frequency. Other symbols are: $\vartheta =$ magnetic dip angle, $s =$ distance along magnetic field line, and $H_n = kT_n/gm_\text{H}$ and $H_p = k(T_e + T_i)/gm_\text{H}$ are the neutral and plasma scale heights. This equation is basically analogous to Eq. (10.58) with the exception that now we cannot neglect the change of the magnetic field magnitude with altitude. After some manipulation Eq. (10.66) yields the following:

$$\frac{d^2n(\text{H}^+)}{ds^2} + \frac{1}{H_p}\left(1 + \sin\vartheta\frac{H_p}{H_n} + \frac{H_p}{H_B}\right)\frac{dn(\text{H}^+)}{ds}$$

$$+ \frac{1}{H_p^2}\left(\sin\vartheta\frac{H_p}{H_n} + \frac{H_p}{H_B}\right)n(\text{H}^+) = 0, \tag{10.67}$$

[3] Axford, W. I., "The polar wind," *J. Geophys. Res.*, **73**, 6855, 1968.

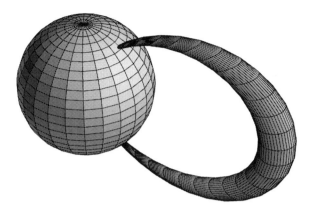

Figure 10.7 Closed magnetic flux tubes in the plasmasphere.

where we introduced the "magnetic scale height" of

$$\frac{1}{H_B} = -\frac{1}{B}\frac{dB}{ds}.$$
(10.68)

If the magnetic scale height is assumed to be slowly varying along the field line (which is a reasonable first approximation), Eq. (10.67) has two solutions. The first solution corresponds to diffusive equilibrium with the plasma scale height:

$$n_E(\mathrm{H}^+) = n_0 \exp\left(-\frac{s - s_0}{H_p}\right).$$
(10.69)

The other solution corresponds to a diffusive flux along the magnetic field line:

$$n_D(\mathrm{H}^+) = n_0 \exp\left[-\left(\sin\vartheta\,\frac{H_p}{H_n} + \frac{H_p}{H_B}\right)\frac{s - s_0}{H_p}\right].$$
(10.70)

In the midlatitude topside ionosphere the magnetic field lines are closed dipole field lines that connect two conjugate hemispheres (see Figure 10.7). For these field lines the diffusive equilibrium solution must hold. Therefore Eq. (10.69) describes the density distribution in the topside ionosphere and in the plasmasphere (the *plasmasphere* is the region of corotating closed field lines filled with plasma of ionospheric origin).

10.7   The Polar Wind

In this section we will discuss the behavior of ionospheric plasma in the high-latitude ionosphere. The high-latitude ionosphere is defined as the region where the geomagnetic field lines are "open", that is, they are not connected to a conjugate ionosphere. The configuration and dynamics of open geomagnetic field lines will be discussed in Section (14.4).

The effects of open magnetic field lines on light ions (mainly H^+ and He^+) have been observed for a long time: These ions have greatly reduced densities in the high-latitude ionosphere. As we will see in this section, this reduced density is due to the high-speed escape of these light ions along open geomagnetic field lines from the ionosphere to the magnetosphere.

Our previous results and discussions have already paved the road toward the understanding of the polar wind. Consider the topside ionosphere (above the F_2 peak) with O^+ as the dominant ion and a light ion (H^+) as a minor constituent. To a very good approximation the magnetic field can be considered vertical, so the magnetic dip angle is $\vartheta \approx 90°$.

For the sake of simplicity, we assume that all temperatures are different, $T_n < T_i < T_e$, but they vary only slowly with altitude. In this case the major ion (O^+) is near diffusive equilibrium with a scale height of

$$H_i = \frac{k(T_e + T_i)}{gm_O}.$$ (10.71)

The electron density is given by

$$n_e = n_0 \exp\left(-\frac{z - z_0}{H_i}\right).$$ (10.72)

The ambipolar electric field is determined by the electron pressure gradient (see Eq. 10.42):

$$E_\parallel = -\frac{1}{en_e}\frac{dp_e}{dz} = \frac{kT_e}{eH_i} = \beta_T \frac{gm_O}{e}.$$ (10.73)

This is an upward electric field that makes the oxygen ions "lighter" by a factor of $\beta_T = T_e/(T_e + T_i)$.

The minor ion continuity and momentum equations are the following (see Eqs. 10.35 and 10.38):

$$\frac{\partial}{\partial t}\left(\frac{n(H^+)}{B}\right) + \frac{\partial}{\partial z}\left(\frac{n(H^+)}{B}u(H^+)\right) = 0$$ (10.74)

and

$$m_H n(H^+)\frac{\partial u(H^+)}{\partial t} + m_H n(H^+)u(H^+)\frac{\partial u(H^+)}{\partial z} + \frac{\partial p(H^+)}{\partial z}$$
$$+ m_H n(H^+)g - en(H^+)E_\parallel = m_H n(H^+)\bar{\nu}_{HO}[u(O^+) - u(H^+)].$$ (10.75)

For the sake of simplicity, we neglected the effects of production and loss terms in these equations. In the momentum equation the collision term describes the drag force caused by Coulomb collisions between the major and minor ion species ($\bar{\nu}_{HO}$ is the H^+–O^+ momentum transfer collision frequency).

Let us consider steady-state solutions of these equations and recognize that $u(O^+) \approx 0$. In this case Eqs. (10.74) and (10.75) can be combined to obtain

$$\frac{1}{u(H^+)}\frac{du(H^+)}{dz} = \frac{g\left(\beta_T\frac{m_O}{m_H}-1\right) - \frac{1}{H_B}a^2(H^+) - \bar{\nu}_{HO}u(H^+)}{u^2(H^+) - a^2(H^+)}, \tag{10.76}$$

where $a^2(H^+) = kT_i/m_H$ is the square of the H^+ characteristic speed, and we substituted Eq. (10.73) for the electric field.

Next, we express Eq. (10.76) in terms of the dimensionless quantities $M(H^+) = u(H^+)/a(H^+)$ and $\zeta = z/H_B$, where $M(H^+)$ is the field-aligned Mach number (slightly different from the acoustic Mach number) and ζ is a normalized distance. This gives

$$\frac{1}{M(H^+)}\frac{dM(H^+)}{d\zeta} = \frac{\frac{gH_B}{a^2(H^+)}\left(\beta_T\frac{m_O}{m_H}-1\right) - 1 - \frac{H_B\bar{\nu}_{HO}}{a(H^+)}M(H^+)}{M^2(H^+) - 1}. \tag{10.77}$$

Near the base of the topside ionosphere (≈ 400 km) the constants can be evaluated to yield the following expression:

$$\left.\frac{1}{M(H^+)}\frac{dM(H^+)}{d\zeta}\right|_{\zeta_0} \approx \frac{10^2 - 10^4 M(H^+)}{M^2(H^+) - 1}. \tag{10.78}$$

At ζ_0 the typical light ion diffusion Mach number is around 0.1. Therefore dM/dz is positive at the base of the topside ionosphere. This means that the light ions are accelerated upward in this region.

At high altitudes the collision term vanishes and Eq. (10.77) becomes the following:

$$\left.\frac{1}{M(H^+)}\frac{dM(H^+)}{d\zeta}\right|_{\zeta > 0.2} \approx \frac{10^2}{M^2(H^+) - 1}. \tag{10.79}$$

One can see that this equation predicts an accelerating flow if the flow is supersonic ($M > 1$), whereas it predicts deceleration if the flow is subsonic ($M < 1$). If the flow is decelerating at high altitudes, it eventually should reach diffusive equilibrium; however, observations clearly show that light ions do not follow diffusive equilibrium in the high-latitude topside ionosphere. This solution, therefore, although mathematically admissible, is physically incorrect.

Figure 10.8 shows the mathematically admissible solutions of Eq. (10.77) for hydrogen ions moving in a background of an O^+, O, H, O_2, and N_2 gas mixture. We know that the physical solution (which matches the observed altitude profiles) must start with subsonic velocities at low altitudes and end up with low densities (and consequently high velocities) at high altitudes. The only solution satisfying these criteria is the transonic solution (A) shown in Figure 10.8. Interestingly Eq. (10.77) resembles the famous Parker equation establishing the existence of the solar wind. This is not an

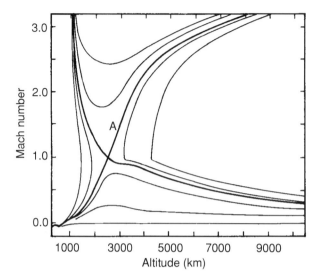

Figure 10.8 The mathematically admissible minor ion (H^+)
Mach number profiles. The exospheric temperature of the
stationary background gas is 1,000 K. (From *Banks & Holzer.*[4])

accident, because the "inventor" of the polar wind concept[5] was greatly influenced by
Parker's fundamental work.

The physics in the polar wind is quite simple. In the presence of a heavy major ion
(O^+) the upward-pointing electron pressure gradient creates an ambipolar electric field
(see Eq. 10.73). The combination of gravity and the ambipolar electric field results in a
net upward acceleration for all positive particles with mass less than half the mass of the
major ion (in this case all positive ions lighter than 8 amu will be accelerated upward).
In contrast, at the base of the topside ionosphere the frictional drag force between the
nearly stationary major ion and the upward-moving light ions represents a large but
rapidly decreasing (with altitude) downward force. The result of the combined force
is that the light ions (primarily H^+) gradually accelerate as they move upward through
regions of decreasing resistance. The light ion fluid will reach sonic velocity at the
point where friction and upward electric field accelerations balance: At this point both
the numerator and the denominator of Eq. (10.76) vanish, but $du/dz > 0$ (this can be
shown using L'Hospital's rule).

Detailed calculations of the multicomponent polar wind are quite complicated and
numerically challenging. They involve the simultaneous solution of the multispecies
continuity, momentum, and energy equations from about 200 km altitude to several
Earth radii. Figure 10.9 shows the result of such a calculation involving O^+, H^+,
He^+, and electrons as dynamic species. The model considers the appropriate sources

[4] Banks, P. M., and Holzer, T. E., "Features of plasma transport in the upper atmosphere,"
  J. Geophys. Res., **74**, 6304, 1969.
[5] Axford, W. I., "The polar wind," *J. Geophys. Res.*, **73**, 6855, 1968.

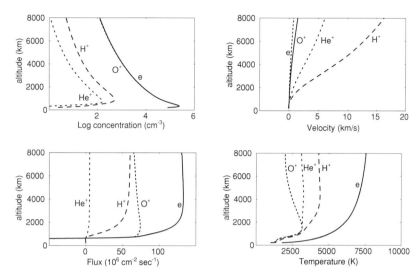

Figure 10.9 Calculated altitude profiles of density, velocity, particle flux, and temperature in the polar wind. (From *Gombosi et al.*[6])

and sinks of mass, momentum, and energy at all altitude regions for summer solar maximum conditions.

10.8 Ionospheric Energetics

The main source of energy in the terrestrial ionosphere is the extreme ultraviolet (EUV) radiaton of the Sun. The energy of most ionizing photons exceeds the threshold energy of ionization. It is this excess energy that maintains through various processes the thermal structure of the ionosphere.

The first step of energy deposition from the EUV radiation to the ionosphere is the process of photoionization. Because the mass of the photoelectron is several orders of magnitude smaller than the mass of the freshly ionized ion, most of the excess energy of the photoionization process is initially carried by the photoelectrons. The magnitude of this excess energy is typically tens of electron volts. Typical electron thermal energies in the ionosphere are <1 eV, and therefore photoelectrons represent a suprathermal electron population in the ionosphere.

In the ionosphere the most effective energy loss mechanisms for the initial photoelectrons involve heavy particles (neutrals and ions): Above approximately 50 eV the dominant energy loss is due to ionization and optically allowed electronic excitations of neutral atoms and molecules (these excitations eventually result in optical line emissions), around 20 eV most energy loss goes to the excitation of metastable

[6] Gombosi, T. I., Kerr, L. K., Nagy, A. F., and Cannata, R. W., "Helium in the polar wind," *Adv. Space Research*, **12** (6), 183, 1992.

molecular levels, and below about 5 eV photoelectrons lose energy by exciting vibrational levels of the N_2 molecule. Finally, below approximately 2 eV, Coulomb collisions with thermal electrons become the primary energy loss mechanism.

The transport and energy loss of photoelectrons is described by the kinetic equation considered earlier in connection with auroral electrons (Eq. 9.17). In general this equation is quite complicated and one can only solve it numerically. However, much physical insight might be gained by considering a simple approximation that can be solved analytically. In this derivation we will follow the method of Banks and Nagy.[7]

Let us consider the kinetic equation of suprathermal electron transport (Eq. 9.17), assume steady-state conditions, and neglect the effects of field-aligned electric fields and the variation of the magnetic field magnitude. One can also neglect the effects of electron–ion inelastic collisions. Inside the ionosphere these assumptions are quite justified. In this approximation Eq. (9.17) becomes the following:

$$\mu \frac{\partial \Phi}{\partial s} = S_{ee} + \sum_i S_{ei} + \sum_\alpha (S_{e\alpha} + S_{e\alpha}^* + S_{e\alpha}^+) + Q \qquad (10.80)$$

Here again $\Phi = \Phi(\varepsilon, \mu, s)$ is the electron flux, while S_{ee} is the change of rate of the energetic electron flux due to collisions with thermal electrons, S_{ei} and $S_{e\alpha}$ account for the effects of elastic collisions with various types of ions and neutrals, $S_{e\alpha}^*$ represents the rate of change due to excitation of neutrals, $S_{e\alpha}^+$ is the electron source rate due to electron impact ionization of neutrals, and finally Q is the electron production rate by photoionization.

One can substitute Khazanov's approximate collision terms[8] (Eqs. 9.18 and 9.19) to obtain the following:

$$\mu \frac{\partial \Phi}{\partial s} = \left(\frac{An_e}{2\varepsilon^2} + \sum_\alpha \frac{n_\alpha \sigma_\alpha}{2} \right) \frac{\partial}{\partial \mu} \left[(1 - \mu^2) \frac{\partial \Phi}{\partial \mu} \right] + An_e \frac{\partial}{\partial \varepsilon} \left(\frac{\Phi}{\varepsilon} \right)$$

$$+ \sum_\alpha n_\alpha \frac{\partial}{\partial \varepsilon} \left[\left(\sigma_\alpha^{in}(\varepsilon) \varepsilon_\alpha^{in} \right) \Phi \right] + \sum_\alpha n_\alpha q_\alpha + Q, \qquad (10.81)$$

where the numerical constant A has the value of $A \approx 2.6 \times 10^{-12}$ eV^{2}cm^2 and the inelastic collision term and secondary electron production rate are defined as

$$\sigma_\alpha^{in}(\varepsilon) \varepsilon_\alpha^{in} = \sum_j \varepsilon_{\alpha j}^* \sigma_{\alpha j}^*(\varepsilon) + \sigma_\alpha^+(\varepsilon) \left(\varepsilon_\alpha^+ + \langle \varepsilon \rangle_\alpha \right),$$

$$\qquad (10.82)$$

$$q_\alpha(\varepsilon) = \int_{2\varepsilon + \varepsilon_\alpha^+}^{\infty} d\varepsilon \, \sigma_{e\alpha}(\varepsilon', \varepsilon) \Phi_0(\varepsilon').$$

[the various symbols were defined in connection with Eqs. (9.18)–(9.21)].

[7] Banks, P. M., and Nagy, A. F., "Concerning the influence of elastic scattering upon photoelectron transport and escape," *J. Geophys. Res.*, **75**, 1902, 1970.
[8] Khazanov, G. V., Neubert, T., and Gefan, G. D., "A unified theory of ionosphere–plasmasphere transport of suprathermal electrons," *IEEE Transactions on Plasma Science*, **22**, 187, 1994.

The first term on the right-hand side of Eq. (10.81) describes pitch-angle diffusion, the second term describes heating of the thermal plasma, the third term is the energy loss rate due to inelastic collisions with neutral particles, the fourth term describes the production of secondary electrons, and the last term is the photoelectron production rate.

Photoelectrons primarily propagate along magnetic field lines. Therefore as a first approximation it is sufficient to know the upward and downward field-aligned electron fluxes in the ionosphere. These fluxes are defined the following way:

$$
\Phi^+(\varepsilon, s) = \int_0^1 d\mu \Phi(\varepsilon, \mu, s),
$$

$$
\Phi^-(\varepsilon, s) = \int_{-1}^0 d\mu \Phi(\varepsilon, \mu, s). \tag{10.83}
$$

Next, we average Eq. (10.81) for upward- and downward-moving electrons. This opeartion yields the following two equations:

$$
\frac{\partial \langle \mu \Phi \rangle_+}{\partial s} = \left(\frac{An_e}{2\varepsilon^2} + \sum_\alpha \frac{n_\alpha \sigma_\alpha}{2} \right) \left[-\frac{\partial \Phi}{\partial \mu} \right]_{\mu=0}
$$

$$
+ \frac{\partial}{\partial \varepsilon} \left[\left(\frac{An_e}{\varepsilon} + \sum_\alpha n_\alpha \sigma_\alpha^{in} \varepsilon_\alpha^{in} \right) \Phi^+ \right] + \sum_\alpha n_\alpha q_\alpha + Q, \tag{10.84}
$$

$$
\frac{\partial \langle \mu \Phi \rangle_-}{\partial s} = \left(\frac{An_e}{2\varepsilon^2} + \sum_\alpha \frac{n_\alpha \sigma_\alpha}{2} \right) \left[-\frac{\partial \Phi}{\partial \mu} \right]_{\mu=0}
$$

$$
+ \frac{\partial}{\partial \varepsilon} \left[\left(\frac{An_e}{\varepsilon} + \sum_\alpha n_\alpha \sigma_\alpha^{in} \varepsilon_\alpha^{in} \right) \Phi^- \right] + \sum_\alpha n_\alpha q_\alpha + Q, \tag{10.85}
$$

where $\langle \rangle_\pm$ represents the average for upward- and downward-moving particles. When deriving Eqs. (10.84) and (10.85), we assumed that newly born electrons start moving up and down along the field line with equal probability.

Next, let us examine the individual terms in Eqs. (10.84) and (10.85). The left-hand sides consist of field-aligned derivatives of the first pitch-angle moments of upward- and downward-moving particles. It can be shown that in the ionosphere the upward and downward fluxes are not very far from isotropy. Therefore $\langle \mu \Phi \rangle_\pm \approx \pm \langle |\mu| \rangle \Phi^\pm$, where $\langle |\mu| \rangle$ is the average magnitude of the electron pitch angle. It can also be shown that $\partial \Phi / \partial \mu$ at $\mu = 0$ is the net flux of backscattered particles, because at $\mu = 0$ the particle changes direction ($\mu = 0$ corresponds to $90°$ pitch angle). Then one can write

$$
\left[\frac{\partial \Phi}{\partial \mu} \right]_{\mu=0} = \mathcal{P}_e(\Phi^+ - \Phi^-), \tag{10.86}
$$

where $\mathcal{P}_e$ is the photoelectron backscatter propability for elastic collisions. Now we

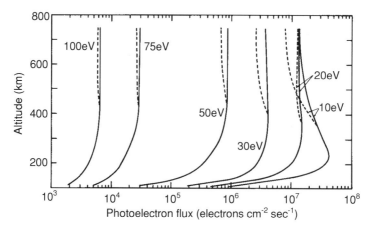

Figure 10.10 Calculated photoelectron fluxes for a sunlit thermosphere
($T = 1250$ K) and overhead Sun. Solid and dashed curves represent Φ^+
and Φ^-, respectively. [From *Banks & Kockarts* (*1973*).]

get the following *two-stream transport equations* for photoelectrons:

$$\langle|\mu|\rangle \frac{\partial \Phi^+}{\partial s} = -\mathcal{R}\Phi^+ + \mathcal{R}\Phi^- + \frac{\partial}{\partial \varepsilon}\left[\mathcal{L}\Phi^+\right] + \sum_\alpha n_\alpha q_\alpha + Q, \qquad (10.87)$$

$$-\langle|\mu|\rangle \frac{\partial \Phi^-}{\partial s} = -\mathcal{R}\Phi^- + \mathcal{R}\Phi^+ + \frac{\partial}{\partial \varepsilon}\left[\mathcal{L}\Phi^-\right] + \sum_\alpha n_\alpha q_\alpha + Q, \qquad (10.88)$$

where

$$\mathcal{R} = \left(\frac{An_e}{2\varepsilon^2} + \sum_\alpha \frac{n_\alpha \sigma_\alpha}{2}\right)\mathcal{P}_e,$$

$$\mathcal{L} = \frac{An_e}{\varepsilon} + \sum_\alpha n_\alpha \sigma_\alpha^{in} \varepsilon_\alpha^{in}. \qquad (10.89)$$

In general Eqs. (10.87) and (10.88) can only be solved numerically for any given
model of an upper atmosphere. Two-stream transport models have given us valuable
insight into the energetics of atmospheres of other planets (primarily Venus, Mars,
and Jupiter). The solution not only provides us with the fluxes of photoelectrons in the
upper atmosphere, but it also results in heating rates for the thermal plasma, excitation
rates of various optical line emissions, and, last but not least, ionization rates.

Figure 10.10 shows altitude profiles of photoelectron fluxes calculated with the
two-stream model. It can be seen that a significant flux of low-energy photoelectrons
escape from the ionosphere along magnetic field lines. In the midlatitude ionosphere
(where the magnetic field lines close in the opposite hemisphere) a large fraction
of this photoelectron flux passes through the plasmasphere and eventually enters the
conjugate ionosphere, where it acts as a source of heating and ionization. In the high-
latitude ionosphere these photoelectrons escape into the magnetosphere along the open
geomagnetic field lines.

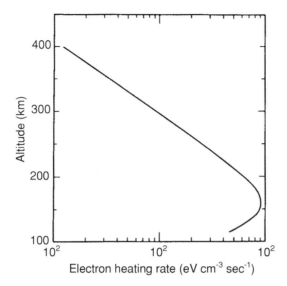

Figure 10.11 Calculated heating rate profile for thermal electrons. [From *Banks & Kockarts (1973)*.]

Once we know the photoelectron fluxes as a function of altitude and electron energy, the heating rate of thermal electrons and ions can also be calculated. Figure 10.11 shows an example for the altitude profile of the thermal electron heating rate due to photoelectrons in the midlatitude ionosphere. Notice that the photoelectron heating rate peaks at fairly low altitudes and drops exponentially above the peak. This means that below the F_2 peak local photoelectron heating of thermal electrons and ions is locally balanced by heat transfer to the dense neutral atmosphere. However, above the F_2 peak heat conduction becomes increasingly important. In this region the electron temperature is basically controlled by a downward electron heat flux originating in the magnetosphere (for open magnetic field lines) or carried by photoelectrons arriving from the conjugate hemisphere. A typical midlatitude ionospheric temperature profile is shown in Figure 10.12.

10.9   Ionospheric Conductivities and Currents

The dense regions of the ionosphere (the D, E, and F regions) contain significant concentrations of free electrons and ions. The presence of mobile charges makes the ionosphere highly conducting. A natural consequence of the high conductivity is that electric currents can be generated in the ionosphere by various physical processes. These current systems can be quite complicated, because not only is the ionosphere a conducting medium, but it is also collisional, and it is penetrated by a strong magnetic field.

The ionospheric current system can be described with the help of a few simplified equations. The current is given by the generalized Ohm's law (see Eq. 4.74):

$$\mathbf{j} = \bar{\sigma} \cdot (\mathbf{E} + \mathbf{u} \times \mathbf{B}). \qquad (10.90)$$

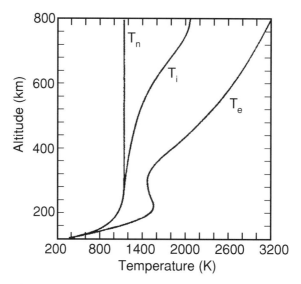

Figure 10.12 Calculated ionospheric temperature profiles. (From *Roble.*[9])

The conductivity tensor was discussed earlier in the book (see Section 4.3.2). In a "magnetic" coordinate system (where the z axis points along the local magnetic field line) the conductivity tensor can be given in the form (see Eq. 4.77):

$$\bar{\sigma} = \begin{pmatrix} \bar{\sigma}_P & -\bar{\sigma}_H & 0 \\ \bar{\sigma}_H & \bar{\sigma}_P & 0 \\ 0 & 0 & \bar{\sigma}_0 \end{pmatrix}, \tag{10.91}$$

where $\bar{\sigma}_0$, $\bar{\sigma}_P$, and $\bar{\sigma}_H$ are the *specific conductivity*, *Pedersen conductivity*, and *Hall conductivity*. In general the conductivity components include the sum of all collisional effects:

$$\bar{\sigma}_0 = e^2 n_e \left(\frac{1}{\bar{\nu}_e m_e} + \frac{1}{\bar{\nu}_i m_i} \right),$$

$$\bar{\sigma}_P = e^2 n_e \left(\frac{\bar{\nu}_e}{\bar{\nu}_e^2 + \Omega_e^2} \frac{1}{m_e} + \frac{\bar{\nu}_i}{\bar{\nu}_i^2 + \Omega_i^2} \frac{1}{m_i} \right), \tag{10.92}$$

$$\bar{\sigma}_H = -e^2 n_e \left(\frac{\Omega_e}{\bar{\nu}_e^2 + \Omega_e^2} \frac{1}{m_e} + \frac{\Omega_i}{\bar{\nu}_i^2 + \Omega_i^2} \frac{1}{m_i} \right),$$

where $\bar{\nu}_e$ and $\bar{\nu}_i$ are the total electron and ion momentum transfer collision frequencies, and Ω_e and Ω_i are the electron and ion gyrofrequencies.

Figure 10.13 shows the altitude variation of the three conductivity components. It is interesting to see that above a few hundred kilometers the specific conductivity σ_0 becomes nearly independent of altitude, because the n_e nearly cancels with the density factor in the collision frequencies. The near constancy of the specific conductivity is an

---

[9] Roble, R. G., "The calculated and observed diurnal variation of the ionosphere over Millstone Hill on 23–24 March 1970," *Planet. Space Sci.*, **23**, 1017, 1975.

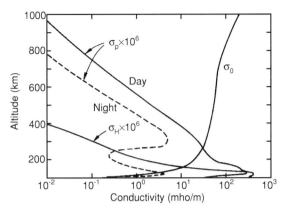

Figure 10.13 Typical conductivity values in the
midlatitude daytime ionosphere. [From *Kelley (1989)*.]

important factor in understanding ionospheric current systems. The Hall conductivity
falls off very rapidly, and it is important only in the D and E regions.

The high conductivity parallel to the Earth's magnetic field has very important
consequences for ionospheric and magnetospheric physics. If we take the $\sigma_0 \to \infty$
limit, it would prohibit any potential drop along geomagnetic field lines: In this case
all field lines would be electric equipotentials, and the potential difference between
two field lines would remain constant over large distances. This means that electric
fields generated in the ionosphere would map along the magnetic field lines into the
magnetosphere, and, conversely, magnetospheric or solar wind potential differences
would map into the ionosphere along field lines.

Consider the midlatitude ionosphere and neglect all transport phenomena ($\mathbf{u} = \mathbf{0}$).
Now the current density (given by Eq. 10.90) can be expressed with the help of the
conductivity components. To do this we must recognize that the Pedersen conductivity
creates electric currents that are perpendicular to the magnetic field and parallel to the
electric field, Hall currents are perpendicular both to the electric and magnetic fields,
and field-aligned currents flow along the magnetic field. Now the current density can
be written as

$$\mathbf{j} = \bar{\sigma}_P \, \mathbf{E}_\perp - \bar{\sigma}_H \, (\mathbf{E} \times \mathbf{b}) + \bar{\sigma}_0 \, \mathbf{E}_\parallel, \tag{10.93}$$

where $\mathbf{b}$ is again the unit vector along the magnetic field line, and the subscripts $\parallel$ and
$\perp$ refer to vector components with respect to the magnetic field.

Because space plasmas are generally quasineutral, the conservation of charge implies that the current density must be divergenceless. In other words,

$$\nabla \cdot \mathbf{j} = \nabla_\perp \cdot (\bar{\sigma}_\perp \, \mathbf{E}_\perp) + \nabla_\parallel \cdot (\bar{\sigma}_0 \, \mathbf{E}_\parallel) = 0, \tag{10.94}$$

where the 2×2 conductivity tensor, $\bar{\sigma}_\perp$, is the following:

$$\bar{\sigma}_\perp = \begin{pmatrix} \bar{\sigma}_P & -\bar{\sigma}_H \\ \bar{\sigma}_H & \bar{\sigma}_P \end{pmatrix}. \tag{10.95}$$

Equation (10.94) can also be written as

$$\nabla_\perp \cdot (\bar{\sigma}_\perp \mathbf{E}_\perp) = -\frac{\partial j_\parallel}{\partial s}, \tag{10.96}$$

where s is again the distance along the field line. This equation can be integrated along the magnetic field line from the bottom of the ionosphere ($s = s_0$ to ∞):

$$j_\parallel(\infty) - j_\parallel(s_0) = -\int_{s_0}^{\infty} \nabla_\perp \cdot (\bar{\sigma}_\perp \mathbf{E}_\perp)\, ds. \tag{10.97}$$

At the bottom of the ionosphere the field-aligned current must vanish, since there are no mobile charges to conduct it downward. Also, because the field lines are nearly equipotentials, the electric field is constant along the field line, and therefore Eq. (10.97) can be written as

$$j_\parallel(\infty) = \nabla_\perp \cdot \left(\left[\int_\infty^{s_0} \bar{\sigma}_\perp\, ds \right] \mathbf{E}_\perp \right) = \nabla_\perp \cdot (\Sigma_\perp \mathbf{E}_\perp), \tag{10.98}$$

where $\Sigma_\perp$ is the perpendicular height-integrated conductivity tensor:

$$\Sigma_\perp = \begin{pmatrix} \Sigma_P & -\Sigma_H \\ \Sigma_H & \Sigma_P \end{pmatrix}. \tag{10.99}$$

Equation (10.98) means that the field-aligned currents coming from the solar wind or magnetosphere close through the dense regions of the ionosphere. This electric circuit will be discussed in more detail later in Section 14.5.

10.10 Problems

Problem 10.1 A major surprise from early planetary probes was the detection of a nighttime ionosphere on Venus. Venus has a very long night (60 days) and without an additional source of ionization no nighttime ionosphere could exist. The first Venus orbiters (*Venera-9* and *-10*) detected a flux of 3×10^8 cm^{-2} s^{-1} of precipitating low-energy (30 eV) electrons. Assume that the nighttime atmosphere is isothermal at $T = 150$ K, that it is primarily composed of CO_2 molecules, and that the neutral density is 2×10^9 cm^{-3} at $z_0 = 140$ km. Assume that each precipitating electron results in one ionization and then it is lost (the ionization cross section is $\sigma_{imp} = 3 \times 10^{-16}$ cm^2). Neglect transport effects and assume that the recombination rate is $\alpha = 4 \times 10^{-7}$ cm^3 s^{-1}.

1. Plot the altitude profile of electron density.
2. What is the peak electron density?
3. What is the altitude of the electron density peak?

Problem 10.2 A planet has an atmosphere consisting of atomic and molecular oxygen, whose densities at the surface of the planet are 6.84×10^{11} and 4.68×10^{12} cm^{-3}, respectively. Assume a neutral gas temperature of 1,000 K and a gravitational acceleration of 10 m/s^2, both constant with altitude. The solar photon flux reaching the top of the atmosphere is 1.0×10^9 photons/cm^2/s/Å.

1. Calculate the atomic and molecular oxygen densities at 100 km assuming that they are in diffusive equilibrium.
2. Calculate the solar photon flux at 100 km, given that the absorption cross sections are 2×10^{-18} cm^2 for both species, independent of wavelength (assume an overhead sun, $\chi = 0°$).
3. Calculate the O^+ and O_2^+ ionization rates at 100 km, given that the ionization cross section is 2×10^{-18} cm^2 for both atomic and molecular oxygen.

Part III

Sun–Earth Connection

Chapter 11

The Sun

The Sun is an ordinary star of spectral type G2V with magnitude of 4.8. However, it is the only star we have in our immediate vicinity and it is the source of most of the energy that controls physical phenomena in our space environment. The Sun is also a living, dynamic star with varying activity as demonstrated in Figure 11.1. Changes in solar activity result in many important phenomena in the space environment, ranging from flares, to coronal mass ejections, to geomagnetic storms. The fundamental physical properties of the Sun are given in Table 11.1.

The Sun consists primarily of hydrogen (90%) and helium (10%). Elements such as C, N, and O constitute about 0.1% of its mass. The interior can be divided into four zones (see Figure 11.2):

1. *The core.* This is the high density, high temperature region at the center of the Sun, where thermonuclear energy production takes place. The core extends from the center to about $R_\odot/4$ (1/64-th of the Sun's volume), but it contains about half of the solar mass. Practically all of the Sun's energy production takes place in this region.

2. *The radiative zone.* The energy produced in the core is transported through the core and the radiative zone by gamma ray diffusion. The gamma rays are scattered, absorbed, and reemitted many times before they reach the outer edge of the radiative zone.

3. *The convection zone.* This zone is located in the uppermost 30% of the solar interior. In this region the solar material is convectively unstable, because the radial temperature gradients are large. When a blob of plasma (with a typical size of $\sim 1,000$ km near the solar surface) is displaced upward from its equilibrium location, it enters into a colder and consequently higher density region.

Table 11.1. *Some properties of the Sun*

Quantity	Value	Units
Mass ($M_\odot$)	1.99×10^{30}	kg
Radius ($R_\odot$)	6.96×10^{8}	m
Mean density ($\rho_\odot$)	1.41×10^{3}	kg m^{-3}
Surface gravitational acceleration ($g_\odot$)	274	m s^{-2}
Escape velocity at surface (v_∞)	618	km s^{-1}
Effective blackbody temperature	5,770	K
Luminosity ($L_\odot$)	3.83×10^{26}	W
Radiative flux density ($\mathcal{F}_\odot$)	6.28×10^{7}	W/m^{-2}
Solar constant ($f_\odot$ = energy flux at 1 AU)	1.36×10^{3}	W/m^{-2}
Equatorial rotation period	26	days
Inclination of Sun's equator to ecliptic	7	deg
Mean distance from Earth (= 1 AU)	1.5×10^{11}	m

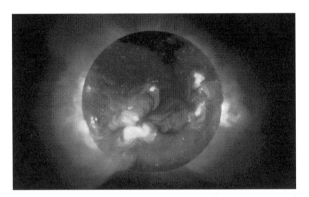

Figure 11.1 An X-ray image of the Sun made by the Soft X-ray Telescope (SXT) onboard the *Yohkoh* spacecraft. (Courtesy of ISAS and NASA.)

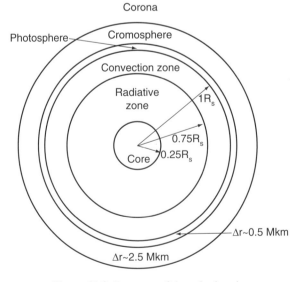

Figure 11.2 Structure of the solar interior.

The buoyancy force pushes the blob even further upward. The resulting *granulation* covers the solar surface. The colder material "sinks" downward near the dark edges of the granules.

4. *The atmosphere.* The solar atmosphere consists of four layers. The lowest is the thin and dense (~ 500 km thick, with densities of about 10^{23} particles per m^3) *photosphere* that emits most of the sunlight. The equivalent *blackbody temperature* of the photosphere is 5,770 K.[1] The next layer is the *chromosphere* (thickness $\sim 3,000$ km, density $\sim 10^{17}$ m^{-3}), where the temperature increases from 4,200 K to $\sim 10^4$ K. The chromosphere is the source region of several transition lines (such as H-α and some UV lines) that are very important in the terrestrial upper atmosphere. The chromosphere is followed by a very narrow *transition layer* where the temperature increases from $\sim 10^4$ K to $\sim 10^6$ K. This transition layer is one of the most interesting and least understood layers of the solar atmosphere. The uppermost layer of the solar atmosphere is the *solar corona*. This region extends into the interplanetary space where it becomes the *solar wind*.

In the following sections we will examine regions of the solar interior and atmosphere in some detail.

11.1   Thermonuclear Energy Generation in the Core

The prevailing view is that the core of the Sun is a very efficient fusion reactor burning hydrogen fuel at temperatures $\approx 1.5 \times 10^7$ K and producing He nuclei. The average kinetic energy of a particle at this very high temperature is about 2 keV, which is well above all ionization energies. This means that all particles are either electrons or completely ionized bare nuclei. The ~ 2 keV average energy also means that there is a sufficient number of energetic particles in the high energy tail of the distribution to overcome the repulsive electrostatic forces between hydrogen nuclei. The penetration probability is of the order of 10^{-9} for the reaction between two protons that is required to initiate the proton–proton chain.

The chain of thermonuclear reactions starts with the fusion of two protons:

$$^1\text{H} + {}^1\text{H} \rightarrow {}^2\text{D} + e^+ + \nu + 1.442\,\text{MeV}. \tag{11.1}$$

The superscript before the symbol refers to the total number of nucleons in the atomic nucleus, while the letter symbol denotes a bare nucleus (positively charged). Also, the symbol ν represents a neutrino and not a photon. The excess energy in the process is due to the small mass difference between two protons and a deuterium nucleus, and it is mainly carried by the neutrino.

---

[1]   The equivalent temperature, T_e, is the temperature of a blackbody with the same radiative flux density, $\mathcal{F}_\odot = \sigma_{SM} T_e^4$, where $\sigma_{SM} = 5.67 \times 10^{-8}$ W m^{-2} K^{-4}.

In the p–p chain this reaction is by far the slowest, so it is reaction (11.1) that controls the energy production rate in the core, and consequently, it controls the temperature and luminosity of the Sun.

In the next step of the chain reaction, a deuterium and a proton react to produce a ^{3}He nucleus:

$$^2D + {}^1H \rightarrow {}^3He + \gamma + 5.493 \text{ MeV}. \tag{11.2}$$

Finally, two ^{3}He nuclei come together and produce a ^{4}He and two protons:

$$^3He + {}^3He \rightarrow {}^4He + {}^1H + {}^1H + 12.859 \text{ MeV}. \tag{11.3}$$

The net result of reactions (11.1), (11.2), and (11.3) is the following:

$$4{}^1H \rightarrow {}^4He + 2e^+ + 2\nu + 26.73 \text{ MeV}. \tag{11.4}$$

The net energy produced by this "hydrogen cycle" is 26.73 MeV, which is the equivalent of the mass difference between four hydrogen nuclei and a helium nucleus (0.029 atomic mass units). It is this energy that fuels the Sun, sustains life, and drives most physical processes in the solar system.

11.2 Internal Structure

11.2.1 Pressure Balance

The radial momentum balance inside the Sun can be approximated by a hydrostatic equilibrium. In this approximation we assume steady-state conditions and the flow velocity is neglected. Inside the Sun the magnetic field plays a relatively insignificant role; therefore the forces acting on charged particles are also neglected. Finally, the Sun's interior is treated as a well-mixed single fluid. In this approximation the single-fluid momentum equation (Eq. 4.56) can be written in the following form:

$$\nabla p = \rho_m \mathbf{g}. \tag{11.5}$$

It should be noted that the gravitational acceleration, $\mathbf{g}$, is not created by the mass of the entire star. According to Newton's theorem, the gravitational field of a spherical mass distribution at a radial distance r is the same as the gravitational field of the enclosed mass concentrated to the center of the sphere. Mathematically speaking this means that

$$\mathbf{g} = -G\frac{M(r)}{r^2}\mathbf{e}_r, \tag{11.6}$$

where $\mathbf{e}_r$ is the unit vector in the radial direction, G is the universal gravitational constant ($G = 6.6726 \times 10^{-11}$ m^3 s^{-2} kg^{-1}), and the enclosed mass is given by

$$M(r) = \int_0^r 4\pi r'^2 \rho_m(r')\,dr'. \tag{11.7}$$

This equation can also be written as

$$\frac{dM(r)}{dr} = 4\pi r^2 \rho_m(r). \tag{11.8}$$

Now the radial component of the pressure balance equation (Eq. 11.5) can be written in the form

$$\frac{dp}{dr} = -G\frac{M(r)}{4\pi r^4}\frac{dM(r)}{dr}. \tag{11.9}$$

Assuming thermodynamic equilibrium between the various ionized gas components, the pressure and density in the solar interior can be related by the perfect gas law:

$$p = \left(\sum_s n_s\right)kT \approx (n_e + n_p + n_\alpha)kT = \rho_m \frac{k}{\bar{m}}T, \tag{11.10}$$

where n_p and n_α refer to the concentration of fully ionized hydrogen (protons) and helium atoms (alpha particles), while $\bar{m}$ is the mean molecular mass of the fully ionized plasma. Because the solar plasma is quasineutral, the electron concentration is

$$n_e = \sum_{s=\text{ions}} Z_s n_s \approx n_p + 2n_\alpha, \tag{11.11}$$

where Z_s is the number of protons in the given nucleus. In a good approximation $n_\alpha/n_p \approx 0.08$ throughout the solar interior, and therefore the mean molecular mass is

$$\bar{m} = \frac{\sum_s m_s n_s}{\sum_s n_s} \approx \frac{m_e n_e + m_p n_p + m_\alpha n_\alpha}{n_e + n_p + n_\alpha} \approx 0.6 m_p. \tag{11.12}$$

With these results the pressure balance equation can be written in the following form:

$$\frac{kT}{\bar{m}}\frac{d\rho_m}{dr} + \frac{k\rho_m}{\bar{m}}\frac{dT}{dr} = -G\frac{M(r)}{4\pi r^4}\frac{dM(r)}{dr}. \tag{11.13}$$

Equations (11.8) and (11.13) give differential equations for $M(r)$ and ρ_m. However, to solve the problem one still needs an equation for the temperature gradient in the solar interior. This can be obtained by considering the transport of energy in the interior of the Sun.

11.2.2 Energy Transport in the Solar Interior

In equilibrium the radial energy flow through the surface of an inside sphere of radius r, $L(r)$, must be equal to the total energy production inside the sphere. This means that the outward energy flow can be expressed as

$$L(r) = 4\pi \int_0^r r'^2 \rho_m(r') \epsilon_n(r') \, dr', \tag{11.14}$$

where ϵ_n is the rate of nuclear energy production per unit mass. Because most of the energy production takes place well inside the core, outside about $0.3 R_\odot$ the energy flow is constant, $L(r > 0.3 R_\odot) = L_\odot$.

Energy transport inside the Sun takes place mainly by diffusion. Diffusive energy transport occurs by radiative transfer in the inner zone and by convection in the outer part of the solar interior.

In the radiative zone energy is transported by the absorption and reradiation of photons. The energy density produced locally by a blackbody reradiation is proportional to T^4; therefore the energy flux can be approximated by

$$\Phi_\varepsilon = -D_\varepsilon \frac{dT^4}{dr}, \tag{11.15}$$

where the diffusion coefficient is proportional to the radiative mean free path of photons. The radiative mean free path can be written as $\lambda_\gamma \propto 1/\rho_m$ (just like the collisional mean free path of molecules). Therefore one can write

$$\Phi_\varepsilon \propto -\frac{T^3}{\rho_m} \frac{dT}{dr}. \tag{11.16}$$

However, we know that the energy flux is given by $L(r)/4\pi r^2$, so we can conclude that

$$\frac{L(r)}{4\pi r^2} = -\frac{1}{K_T} \frac{T^3}{\rho_m} \frac{dT}{dr}, \tag{11.17}$$

where K_T is a multiplication factor depending on the optical properties of the solar interior. Outside the core this relation gives the following equation for the temperature gradient:

$$\left(\frac{dT}{dr} \right)_{\text{rad}} = -K_T \frac{\rho_m}{T^3} \frac{L_\odot}{4\pi r^2}. \tag{11.18}$$

In the convective zone the radiative temperature gradient is too large and the plasma becomes convectively unstable. In this region another mode of energy transport – convective motion of macroscopic-size cells – becomes the dominant energy transport mechanism. One can obtain an estimate for the temperature gradient in the convective region by assuming that magnetic field effects can be neglected and that the motion of the cells is "slow" (adiabatic). In this case the temperature profile in the slightly unstable solar atmosphere is essentially given by the adiabatic temperature gradient (or *adiabatic lapse rate*), which was derived earlier in the book (see Eq. 8.14). In the solar interior one must use the local gravitational acceleration; therefore the adiabatic lapse rate becomes

$$\left(\frac{dT}{dr} \right)_{ad} = -\frac{2}{5} \frac{\bar{m}}{k} G \frac{M(r)}{r^2}. \tag{11.19}$$

If the radiative temperature gradient exceeds the adiabatic lapse rate, then the solar material becomes convectively unstable and convective energy transport takes over. This happens in the upper part of the solar interior. In this region the temperature gradient is given by expression (11.19). Deep inside the Sun the radiative temperature gradient is smaller than the adiabatic lapse rate: In this region radiative energy transport dominates (given by Eq. 11.18).

11.2.3   Radial Structure

To obtain the radial profiles of density, temperature, and energy transport, one has to integrate Equations (11.8) and (11.13) from the center of the Sun outward to the surface. The third equation needed for the solution defines the temperature gradient: This is given by (11.18) in the inner radiative zone and by Eq. (11.19) in the outer convective zone.

The boundary conditions are quite simple. At the center of the Sun ($r = 0$), we have $M(0) = 0$, $L(0) = 0$, $T_0 = 1.5 \times 10^7$ K, and $\rho_0 = 1.5 \times 10^5$ kg m^{-3}. The thermonuclear energy production rate, ϵ_n, is known only with $\sim 50\%$ accuracy, and it is given by a complicated integral.

The governing equations describing the interior structure of the Sun have been solved numerically using increasingly sophisticated methods and physical models. The results of a standard model are shown in Figure 11.3.

11.3   Solar Oscillations

The outer regions of the Sun contain fairly dense and hot gases. This medium is capable of periodic oscillations around an equilibrium solution. In the first approximation one can consider oscillations around local hydrostatic equilibrium. These oscillations are similar to the ones discussed in connection with acoustic–gravity waves in the terrestrial atmosphere. Since the hydrostatic scale height in the outer regions of the Sun, H, is much smaller than the solar radius, one can again use a local planar geometry to investigate these oscillations.

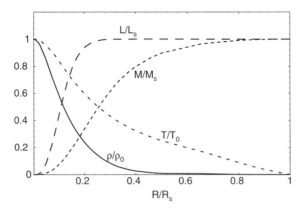

Figure 11.3 Radial variation of physical parameters inside
the Sun. (From *Bachcall & Ulrich*.[2])

Let us consider the transport equations in the solar interior (including the lower
regions of the solar atmosphere, such as the photosphere). We neglect magnetic field
effects and assume that small-amplitude periodic oscillations take place around a
spherically symmetric, hydrostatic equilibrium solution. The amplitudes of these os-
cillations are denoted by $\hat{\rho}$, $\hat{\mathbf{u}}$, and $\hat{p}$. In this case the density, velocity, and pressure
are approximated as

$$\rho_m = \hat{\rho}_0 \exp\left(-\frac{z}{H}\right) + \hat{\rho} \exp\left[-\frac{z}{2H} + i(\mathbf{k}\cdot\mathbf{r} - \omega t)\right],$$

$$\mathbf{u} = \hat{\mathbf{u}} \exp\left[\frac{z}{2H} + i(\mathbf{k}\cdot\mathbf{r} - \omega t)\right], \tag{11.20}$$

$$p = \hat{p}_0 \exp\left(-\frac{z}{H}\right) + \hat{p} \exp\left[-\frac{z}{2H} + i(\mathbf{k}\cdot\mathbf{r} - \omega t)\right],$$

where $\hat{\rho}_0$ and $\hat{p}_0$ are the hydrostatic density and pressure values at $z = 0$, $H = \hat{p}_0/\hat{\rho}_0 g$
is the hydrostatic scale height, ω is the frequency, and $\mathbf{k}$ is the wavenumber. It is again
assumed that $\hat{\rho}_0 \gg \hat{\rho}$ and $\hat{p}_0 \gg \hat{p}$.

These oscillations are identical to the ones discussed with respect to terrestrial
acoustic–gravity waves (see Section 8.3.4). The characteristic equations and the resulting dispersion relation are also the same. Solar gravity waves are low-frequency
oblique waves propagating with velocities well below the local sound speed. At high
frequencies the dispersion relation describes the well-known acoustic wave.

Helioseismology is the study of solar oscillations. It is possible because wave
motions are excited and can propagate in the Sun, and because these waves can be
observed at the surface. Resonant cavities inside the Sun organize internal gravity and
surface gravity waves into standing wave patterns named p, g, and f modes.

The characteristic frequency of solar oscillations, the gravity resonant frequency, is
approximately $\omega_g = 0.02$ s^{-1}. The oscillation period corresponding to this frequency

---

[2] Bachcall, J., and Ulrich, R., "Solar models, neutrino experiments, and helioseismology," *Rev.
Modern Phys.*, **60**, 297, 1988.

is $T_g = 2\pi/\omega_g \approx 5$ min. Observations show that the basic oscillation period of the Sun is about 5 minutes, which corresponds to standing radial acoustic waves near the solar surface. Gravity waves are below this frequency, whereas other acoustic waves are observed at higher frequencies. Helioseismology is one of our principal sources of information about the solar interior.

11.4   Generation of Solar Magnetic Fields

The solar magnetic field is generated by a *dynamo mechanism* operating in the convective zone by large-scale motions of the conducting fluid. The complete solar dynamo is very complicated and not fully understood. Therefore here we just briefly outline some of the basic elements leading to the generation of solar magnetic fields.

Solar dynamo theory is essentially based on the solution of the magnetic field induction equation for the magnetic field (Eq. 4.86) and the single-fluid momentum equation. The induction equation (also called the convection–diffusion equation) is the following:

$$\frac{\partial \mathbf{B}}{\partial t} = \nabla \times (\mathbf{u} \times \mathbf{B}) + \eta_m \nabla^2 \mathbf{B}, \tag{11.21}$$

where η_m is the magnetic viscosity given by Eq. (4.87).

The solar magnetic field is more or less axisymmetric, and therefore much theoretical work was done for axisymmetric dynamos. In solar dynamo theory a cylindrical coordinate system is used (where the R, z, and φ coordinates denote the radial distance in the plane perpendicular to the axis of symmetry, the distance along the axis of symmetry, and the azimuth angle in the perpendicular plane). The axisymmetric field can be written as the sum of a toroidal (i.e., azimuthal) and a poloidal component (the poloidal component is the sum of the radial and axial components):

$$\mathbf{B} = B_\varphi \mathbf{e}_\varphi + \mathbf{B}_p. \tag{11.22}$$

In the upper regions of the Sun the equatorial regions rotate faster than the polar regions. Such *differential rotation* can distort a purely poloidal field and create a toroidal magnetic field component. This effect may be demonstrated by considering an axisymmetric magnetic field and a flow given by

$$\mathbf{u} = u_\varphi \mathbf{e}_\varphi + \mathbf{u}_p. \tag{11.23}$$

Now the toroidal component of the induction equation (Eq. 11.21) can be written as

$$\frac{\partial B_\varphi}{\partial t} + R(\mathbf{u}_p \cdot \nabla)\left(\frac{B_\varphi}{R}\right) = R(\mathbf{B}_p \cdot \nabla)\left(\frac{u_\varphi}{R}\right) + \eta_m \left(\nabla^2 - \frac{1}{R^2}\right)B_\varphi. \tag{11.24}$$

Here $\mathbf{u}_p$ is the poloidal velocity component, and we made use of the $\nabla \cdot \mathbf{B} = 0$ condition and assumed that the solar plasma was incompressible ($\nabla \cdot \mathbf{u} = 0$). We also made use of the fact that axial symmetry means that all derivatives with respect to the azimuth angle must vanish. The left-hand side of Eq. (11.24) describes the time rate of change and the advection of the toroidal component, while the two terms on the right-hand side describe the generation of toroidal field by the poloidal component by the shear of the angular velocity (u_φ/R) and the ohmic dissipation due to resistivity.

The most important term in Eq. (11.24) is the first term on the right-hand side, which generates toroidal magnetic field from the poloidal component with the help of angular velocity shear. Such shear is present in the Sun owing to its differential rotation. This means that if there is poloidal field (such as a nontilted dipole) toroidal components will be generated as the Sun differentially rotates. This phenomenon will eventually result in the 22-year solar cycle.

The next question is how can one maintain a basic poloidal field inside the Sun. Let us examine the poloidal component of the induction equation (Eq. 11.21). An axisymmetric poloidal field can be expressed in terms of a vector potential:

$$\mathbf{B}_p = \nabla \times (A_p \mathbf{e}_\varphi). \tag{11.25}$$

Now Eq. (11.21) yields the following for the poloidal vector potential:

$$\frac{\partial A_p}{\partial t} + \left(\frac{\mathbf{u}_p}{R} \cdot \nabla\right)(R A_p) = \eta_m \left(\nabla^2 - \frac{1}{R^2}\right) A_p. \tag{11.26}$$

Equation (11.26) is bad news for solar dynamo theory. It contains only A_p, and therefore it shows that one cannot generate poloidal magnetic fields from the toroidal component. The situation is seemingly even more hopeless, since the presence of the diffusion term on the right-hand side indicates that the poloidal field slowly diffuses away because of the resistivity of the solar plasma.

Solar dynamo theory was saved by Parker[3] who pointed out that as rising blobs of plasma expand in the convection zone, they also tend to twist due to the Coriolis force. If they carry magnetic flux with them, the twist converts toroidal magnetic fields into poloidal ones. The rate of generation of the poloidal field is proportional to B_φ. Parker parameterized this global effect of many rising convection cells by introducing a toroidal electric field:

$$E_\varphi = \alpha B_\varphi. \tag{11.27}$$

This process is called the *alpha effect*. The constant of proportionally (α) has the units of velocity, and it is a measure of the mean rotational speed of eddies. With Parker's addition the induction equation for A_p becomes

$$\frac{\partial A_p}{\partial t} + \left(\frac{\mathbf{u}_p}{R} \cdot \nabla\right)(R A_p) = \alpha B_\varphi + \eta_m \left(\nabla^2 - \frac{1}{R^2}\right) A_p. \tag{11.28}$$

This equation makes it now possible to create a solar dynamo.

---

[3] Parker, E. N., "Hydromagnetic dynamo models," *Astrophys. J.*, **122**, 293, 1955.

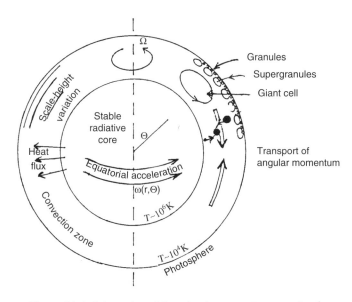

Figure 11.4 Schematics of the solar dynamo. (From *Moffatt*.[4])

The solar dynamo is shown schematically in Figure 11.4. The rotation of the Sun has an important double influence on the rising blobs in the convection zone: (i) Coriolis forces cause a deflection of rising blobs, which is believed to be responsible for the differential rotation of the Sun, and (ii) as the blobs rise, they also expand and tend to rotate more slowly (conserving their intrinsic angular momentum); therefore they twist and create the α effect. Thus the two ingredients of the α–ω dynamo are both consequences of the Coriolis force.

Equations (11.24) and (11.28) together can account for the interchange between toroidal and poloidal magnetic fields owing to the combined effects of differential rotation and the twisting of plasma blobs. Such a so-called α–ω dynamo can account for the observed solar cycle variation of sunspots.

11.5   The Sunspot Number and the Solar Cycle

The existence of sunspots has been known since the earliest days of astronomy going back at least two millennia. Sunspots are dark spots on the solar surface that can be as large as 10^5 km. They consist of a cold central umbra ($T \approx 4,100$ K) with very strong local magnetic fields (up to 0.3 T or 3×10^4 G) surrounded by a penumbra of light and dark radial filaments. The magnetic field is almost radial in the penumbra with a nearly radially outward flowing plasma ($u \approx 6$ km/s). A moderately large sunspot is shown in Figure 11.5.

[4] Moffatt, H. K., "Aspects of dynamo theory," in *Solar System Magnetic Fields*, ed. E. R. Priest, p. 172, D. Reidel, 1985.

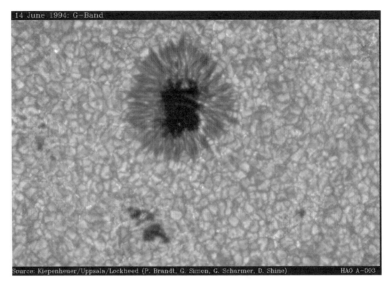

14 June 1994: G−Band

Source: Kiepenheuer/Uppsala/Lockheed (P. Brandt, G. Simon, G. Scharmer, D. Shine)   HAO A−093

Figure 11.5 A moderately large sunspot. (Courtesy of P. Brandt, Kiepenheuer Institut für Sonnenphysik, Freiburg, Germany; G. Scharmer, Uppsala, Sweden; and G. Simon, National Solar Observatory. Primary image processing by D. Shine, Lockheed Corporation.)

It was recognized in the middle of the nineteenth century that the number of sunspots exhibit a roughly 10 year periodicity. The time interval from a sunspot minimum (also called *solar minimum*) to the next sunspot minimum is called a *solar cycle*. Detailed analysis revealed that the average length of the solar cycle is about 11 years. The period from 1755 to 1766 has been chosen as solar cycle 1. Solar cycle 21 began in 1976, reached its maximum in 1979, and ended in 1986. Solar cycle 22 started in 1986, reached its maximum in 1991, and ended in 1996. The present solar cycle started in 1996 and is expected to reach its sunspot maximum in 2000 or 2001.

The main historic record of the Sun's magnetic activity comprises the daily observations of the number of sunspots on the disk. This simple index correlates relatively well with many features of solar activity, and therefore it is used as a proxy for the solar cycle.

The sunspot index is defined as

$$R = k(10g + f), \tag{11.29}$$

where f is the number of individual sunspots, g is the number of recognizable sunspot groups, and k is a correction factor accounting for the evolution in observing technology. Historical data are included with poorly known correction factors, and therefore they are accurate only within a factor of 2 or so.

The variation of the annual mean sunspot number is shown in Figure 11.6. It can be seen that the height of the cycles varies considerably and the solar minima also vary in depth. The cycle period also varies considerably from about 8 to 15 years with an average of about 11.1 years. The most remarkable feature seen in Figure 11.6 is

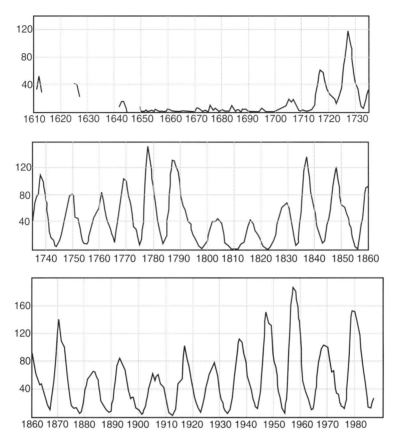

Figure 11.6 The variation of sunspot number with time from 1610 to 1987.
(From *Waldmeier*[5] and modified by *Eddy et al.*[6])

the greatly reduced sunspot activity between 1640 and 1700. This near disappearance of sunspots is called the *Maunder minimum*. The sunspot record indicates a general rise in the solar activity level (at least in the sunspot numbers) during the twentieth century. This long-term variation in the solar activity makes it even more difficult to separate changes in the Earth's environment generated by human activity from those generated by solar activity.

The remarkably regular 11-year variation of sunspot numbers is accompanied by a similarly regular variation in the latitude distribution of sunspots and in the polarity of their magnetic fields. The latitude band occupied by sunspots drifts toward the equator as the solar cycle progresses from minimum to maximum. This phenomenon is shown in Figure 11.7. The first sunspots of a new solar cycle usually appear around heliographic latitudes 25–30°, whereas the last sunspots are found about 20° closer to the solar equator.

[5] Waldmeier, M., "*The Sunspot Activity in the Years 1610–1960*," Schulthess, Zürich, 1961.
[6] Eddy, J. A., Gilman, P. A., and Trotter, D. E., "Solar rotation during the Maunder minimum," *Solar Phys.*, **46**, 3, 1976.

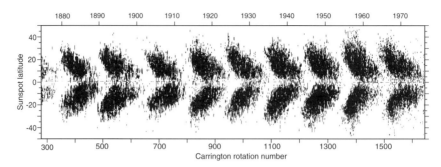

Figure 11.7 The "butterfly" diagram showing the latitude drift of sunspot occurrence. [From *Foukal* (*1989*).]

Sunspots first appear as small dark pores, which over a day or so gradually grow bigger and develop into sunspots. The effective temperature of the umbra is a few thousand kelvins lower than the blackbody temperature of the photosphere. Because of its lower effective temperature, a sunspot emits much fewer photons than the surrounding areas and therefore appears much darker. The lower effective temperature of sunspots is associated with strong local magnetic fields.

Sunspots are usually associated with strong local magnetic fields. Usually each sunspot (or sunspot group) is associated with a single magnetic polarity. These unipolar spots are accompanied by another sunspot or sunspot group of opposite polarity. These groups together form a bipolar sunspot group. Observations show that during any given solar cycle the polarity of the leading and trailing sunspots in a bipolar sunspot group is usually the same for a hemisphere. For instance, during solar cycle 21 the leading and trailing spots in the northern solar hemisphere were predominantly of north and south polarity, respectively. During the same solar cycle the leading spots were predominantly of south polarity in the southern hemisphere. The sunspot groups in the next solar cycle are of opposite polarities: In solar cycle 22 the leading spots of northern hemisphere bipolar groups were of south polarity. This 11-year reversal of sunspot group polarities defines a 22-year magnetic solar oscillation.

A simple conceptual model of the 22-year solar cycle was suggested by Babcock.[7] The Babcock model describes the solar cycle as a progression through five distinct stages.

The first stage starts at solar minimum, when the solar magnetic field is approximately a relatively weak axisymmetric dipole field (the surface field is about $1 \text{ G} = 10^{-4}$ T). The field lines emerge from the solar surface only above $\approx 55°$. This field is mainly poloidal, with negligible toroidal component.

In the second stage the submerged field is intensified by the solar dynamo. In this stage kinetic energy from the differential rotation of the Sun gets transferred

[7] Babcock, H. W., "The topology of the Sun's magnetic field and the 22-year cycle," *Astrophys. J.*, **133**, 572, 1961.

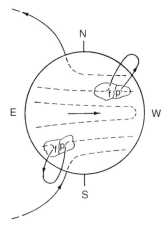

Figure 11.8 The formation of toroidal magnetic fields and bipolar regions. (From *Babcock*.[7])

to the energy of a newly generated toroidal magnetic field. The initially meridional field lines progressively distort into an increasingly east–west direction (see Figure 11.8).

Babcock was able to show that the time required to achieve a given level of field intensification increased toward the equator. When a certain level of intensification is achieved the field penetrates into the photosphere where the field line footpoints appear as sunspots with strong magnetic field (about 1 kG). Naturally, each emerging magnetic field line returns to the solar interior through a second sunspot. This scenario is shown in Figure 11.8. The letters "f" and "p" denote magnetic polarities following and preceding the direction of solar rotation. The emergence of the magnetic field through the photosphere is the third stage of the solar cycle.

In the fourth stage the Sun's poloidal field is neutralized and reversed. The "f" portions of the active regions tend to move toward higher latitudes, while the "p" regions migrate toward the equator. Eventually the "f" polarities of the active regions neutralize the existing poloidal field, while the "p" regions originating at the two hemispheres merge near the solar equator where their magnetic fields cancel each other. Further poleward migration of the "f" regions replace the old poloidal field with a new one with opposite magnetic polarity.

Finally, in stage five we find a reversed weak axisymmetric poloidal field about 11 years after the beginning of phase one.

11.6   The Solar Atmosphere

The solar atmosphere extends from the photosphere to very large heliocentric distances. Here we will discuss the near-solar regions of the atmosphere: the photosphere, the chromosphere, the transition region, and the lower corona. The expanding corona (solar wind) will be discussed later in this book.

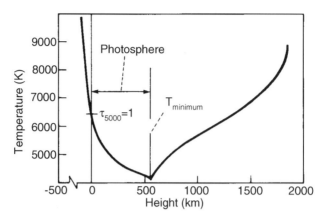

Figure 11.9 Temperature profile in the photosphere. (From *Gingerich et al.*[8])

11.6.1 The Photosphere

The photosphere is the apparent solar surface as seen by white-light observations. It is this narrow layer that emits 99% of the Sun's light and heat. Figure 11.9 shows the variation of temperature in the lower solar atmosphere. The solar surface is defined as the location where the optical depth of a $\lambda = 5{,}000$ Å photon is 1 (the probability that a photon can escape from this surface is $1/e$).

The photosphere is the lowest region of the solar atmosphere extending from the surface to the temperature minimum at around 500 km. Because the gas in the photosphere absorbs and reemits radiation approximately as a blackbody, the photosphere essentially emits a blackbody continuous spectrum. The blackbody temperature of the emission spectrum is about 5,762 K. Figure 11.10 shows the observed solar irradiance spectrum from UV to near infrared, where most of the radiation is emitted. The integral under the solar spectrum gives the *solar constant*.

The density in the photosphere is about 10^{14} cm^{-3}. The gas is mostly neutral with a very small fraction of ionization ($n_e/n_n \approx 10^{-4}$).

11.6.2 The Chromosphere

The chromosphere is a several thousand kilometer thick layer of glowing, transparent gas above the photosphere. The temperature increases from about 4,300 K to 10^4 K owing to the absorption of acoustic waves emerging from the convection zone (see Figure 11.9). In essence, in the chromosphere the dominant energy input is the absorption of mechanical energy, as opposed to the radiation energy input to the nearly blackbody photosphere.

[8] Gingerich, O., Noyes, R. W., and Kalkofen, W., "The Harvard-Smithonian reference atmosphere," *Solar Phys.*, **18**, 347, 1971.

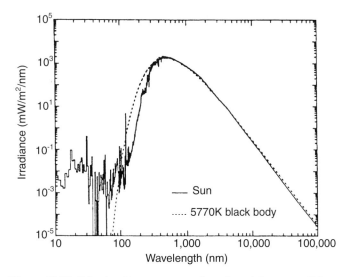

Figure 11.10 Solar irradiance spectrum for solar minimum conditions compared with the spectrum of a 5,770 K blackbody radiator. (From *Lean.*[9])

The density in the chromosphere is even lower than it is in the photosphere. It is nearly transparent to the visible continuum radiation, and it can only be observed in narrow atomic lines. The chromosphere is optically thick in several important atomic transition lines, such as the red H-α line (6,563 Å) or UV lines of species like CaII (i.e., Ca^+).

The heating mechanism responsible for the rise in chromospheric temperature is still disputed, but shock wave dissipation is the most likely candidate. It is believed that acoustic waves are generated by photospheric turbulence, and they propagate outward into regions of decreasing density. Here the sound waves steepen into shocks and dissipate their energy to the chromosphere and the corona.

11.6.3   The Transition Region

At the top of the chromosphere the temperature rapidly increases from about 10^4 K to over 10^6 K. This sharp increase takes place within a narrow region (see Figure 11.11) called the *transition region*. The physical mechanism leading to and maintaining such a huge temperature gradient is not understood at the present time and is one of the most exciting questions in solar system astrophysics.

11.6.4   The Corona

The corona is the outermost, most tenuous region of the solar atmosphere, extending to large distances and eventually becoming the solar wind. Historically the corona was only observed during total solar eclipses, when the Moon blocked the bright

---

[9] Lean, J., "Variations in the Sun's radiative output," *Rev. Geophys.*, **29**, 505, 1991.

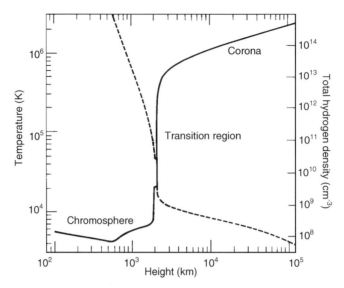

Figure 11.11 Calculated temperature (solid line) and density (dashed) profiles in the transition region. (From *Gabriel*.[10])

photosphere and the sky gets dark. Normally, the sky is much brighter than the corona (except very close to the Sun in some emission lines). Today sophisticated *corona-graphs* (which in effect produce artificial eclipses) make it possible to observe this very interesting region of the solar atmosphere on a daily basis. However, ground-based coronagraphs cannot see far above the limb, because of the bright sky. Space-based coronagraphs can see the corona much further, since there is no scattered radiation contaminating their field of view.

The corona is characterized by very high temperatures (a couple of million degrees) and by the presence of low density, fully ionized plasma. The heating mechanism of the corona is not understood and much more research is needed in this area.

Figure 11.12 shows an eclipse photograph taken on March 18, 1988, in the Philippines. With the solar disk completely obscured (and therefore the scattered photospheric light greatly reduced[11]) the corona becomes visible. The most common coronal structures seen on eclipse photographs are the *helmet streamers*: bright elongated structures, which are fairly wide near the solar surface, but taper off to a long, narrow spike. The base of a helmet streamer often contains a darker cavity, in which bright prominences can sometimes be seen. Streamers appear brighter than the rest of the corona, because they are higher density regions and the scattering of solar radiation is therefore enhanced. Near the poles the coronal intensity is generally depressed, particularly around solar maximum.

[10] Gabriel, A. H., "A magnetic model of the solar transition region," *Phil. Trans. Royal Soc. London, A*, **281**, 339, 1976.

[11] Photospheric light is normally scattered both in the corona (by free electrons) and in the atmosphere of Earth. During an eclipse, the photospheric light scattered in the terrestrial atmosphere is greatly reduced, and therefore the corona becomes visible.

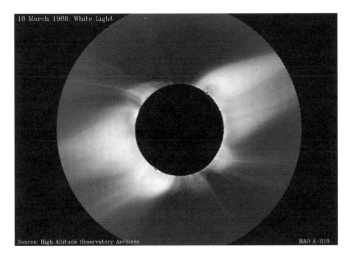

Figure 11.12 The solar corona during the solar eclipse of November 3, 1994. (Courtesy of the High Altitude Observatory.)

One of the major discoveries of the *Skylab* mission was the observation of extended dark coronal regions in X-ray solar images. These *coronal holes* are characterized by low density cold plasma (the temperature is about half a million degrees colder than in the bright coronal regions) and unipolar magnetic fields (the magnetic field lines emerging from coronal holes are directly connected to the magnetic field lines extending to the distant interplanetary space). In contrast, bright coronal regions are filled with hot, higher density plasma and they correlate with closed field line regions (the magnetic field lines emerging from regions return to the solar surface within a couple of solar radii).

Near solar minimum coronal holes cover about 20% of the solar surface. The polar coronal holes are essentially permanent features, whereas the lower latitude holes only last for several solar rotations. The low latitude holes are usually connected to a polar coronal hole.

Coronal holes are connected to open coronal field lines. The interplay between the inward pointing gravity and outward pointing pressure gradient force results in a rapid outward expansion of the coronal plasma along the open magnetic field lines (see the next chapter about the solar wind for details). At low heliolatitudes the direction of the coronal magnetic field is far from radial. Therefore the plasma cannot leave the vicinity of the Sun along magnetic field lines. At the base of low-latitude coronal holes, however, the magnetic field direction is not far from radial, and the expansion of the hot plasma can take place along open magnetic field lines without much resistance. This expansion greatly reduces the density of the plasma on open field lines, which appears as darker regions in the X-ray images. A coronal hole is shown in Figure 11.13. It should be noted that the development of coronal holes during the solar cycle is closely related to the evolution of the Sun's magnetic fields.

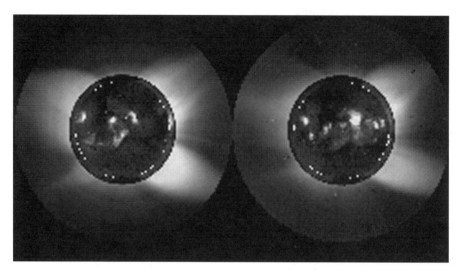

Figure 11.13 Combined soft X-ray and white-light coronagraph images of the Sun showing the close relation between coronal holes and low-density plasma regions. (Coronagraph data from the Mauna Loa Solar Observatory, operated in Hawaii by the High Altitude Observatory. Primary image processing by A. Lecinsky. X-ray data courtesy of the *Yohkoh* Science Team.)

11.7 Radiative Transfer in the Solar Atmosphere

In the solar atmosphere photons are both emitted and absorbed. In this section we briefly outline the transfer of radiation in the tenuous solar atmosphere. For the sake of mathematical simplicity, we assume that the thickness of the radiating layer (from where most of the observed solar spectrum originates) is small compared to the radius of the Sun. In this approximation the solar atmosphere can locally be approximated as a plane parallel layer. We choose a Cartesian coordinate system with the z axis pointing upward and the angle Θ characterizing the angle between the vertical direction and the direction pointing toward the observer.

We want to obtain an equation for the change of beam intensity (radiative energy per unit area per unit time per unit wavelength), I_λ, where λ is the wavelength of the photon. The monochromatic extinction coefficient per unit path length and the emission rate are denoted by $\kappa_\lambda(z)$ and $\epsilon_\lambda(z)$, respectively. The change of beam intensity along the line of sight in the Θ direction while the beam passes through an infinitesimal layer of thickness dz is the following:

$$I_\lambda(\Theta, z + dz) - I_\lambda(\Theta, z) = [\epsilon_\lambda(z) - \kappa_\lambda(z)I_\lambda(\Theta, z)]\frac{dz}{\cos\Theta}, \tag{11.30}$$

or

$$\cos\Theta\frac{dI_\lambda(\Theta, z)}{dz} = \epsilon_\lambda(z) - \kappa_\lambda(z)I_\lambda(\Theta, z). \tag{11.31}$$

The opacity of the infinitesimal layer is characterized by the *optical depth*, defined as

$$d\tau_\lambda = -\kappa_\lambda dz. \tag{11.32}$$

As a result of the negative sign in Eq. (11.32), the optical depth increases as we go deeper and deeper into the solar atmosphere. One can now rewrite Eq. (11.31) with the help of Eq. (11.32):

$$\cos\Theta \frac{dI_\lambda}{d\tau_\lambda} = I_\lambda - S_\lambda, \tag{11.33}$$

where we introduced the source function, $S_\lambda = \epsilon_\lambda/\kappa_\lambda$. This is the standard form of the radiative transfer equation in the solar atmosphere.

Equation (11.33) can be formally solved the following way. First, we multiply both sides of the equation by $\exp(-\tau_\lambda/\cos\Theta)/\cos\Theta$ and then integrate for τ_λ. This procedure leads to the following result:

$$I_\lambda(\tau_\lambda, \cos\Theta) = I_\lambda(\tau_{0\lambda}, \cos\Theta) \exp\left(-\frac{\tau_{0\lambda}}{\cos\Theta}\right)$$
$$+ \frac{1}{\cos\Theta} \int_{\tau_\lambda}^{\tau_{0\lambda}} d\tau_\lambda' S_\lambda(\tau_\lambda') \exp\left(-\frac{\tau_\lambda'}{\cos\Theta}\right), \tag{11.34}$$

where $\tau_{0\lambda}$ is the optical depth at some reference altitude. If we integrate from the observer ($\tau_\lambda = 0$) to deep into the Sun ($\tau_{0\lambda} = \infty$), then we get the total emergent intensity:

$$I_\lambda(0, \cos\Theta) = \frac{1}{\cos\Theta} \int_0^\infty d\tau_\lambda' S_\lambda(\tau_\lambda') \exp\left(-\frac{\tau_\lambda'}{\cos\Theta}\right). \tag{11.35}$$

11.7.1   Local Thermodynamic Equilibrium Approximation

In the most general case the calculation of the source function requires the knowledge of chemical composition and ionization states of the various components. However, the photospheric source function can be approximated as the radiation generated by a locally isothermal cavity with opaque gas walls. The radiation inside this "cavity" is in local thermodynamic equilibrium (LTE) with the gas walls. In this approximation each surface element of the cavity walls emits and absorbs about the same radiative power, independent of the material and geometry of the cavity. The intensity in the cavity is given by the *Planck function*:

$$B_\lambda(T) = \frac{2\pi hc^2}{\lambda^5} \frac{1}{e^{\frac{hc}{\lambda kT}} - 1}, \tag{11.36}$$

where h is *Planck's constant*. The integral of B_λ over wavelength yields the total radiation intensity in the cavity:

$$\int\limits_0^\infty d\lambda B_\lambda(T) = \frac{\sigma_{SB}}{\pi} T^4, \tag{11.37}$$

where σ_{SB} is the *Stefan–Boltzmann constant*.

Since $B_\lambda(T)$ is the intensity of radiation inside an adiabatic cavity, it characterizes both the emitted and absorbed radiation. If the "walls" of the cavity are perfectly absorbing, the cavity acts as a "blackbody," and the emission rate becomes the Planck function. The radiation from such a cavity is called *blackbody radiation*, and it is characterized by $B_\lambda(T)$. In the case of nearly blackbody cavities the local emission and absorption approximately balance each other, and one obtains in a good approximation

$$S_\lambda \approx B_\lambda(T). \tag{11.38}$$

In this case the equation of radiative transfer becomes

$$\cos\Theta \frac{dI_\lambda}{d\tau_\lambda} = I_\lambda - B_\lambda(T). \tag{11.39}$$

11.7.2 The Gray Atmosphere

The calculation of the total emergent intensity becomes greatly simplified if we assume the extinction coefficient, $\kappa_\lambda(z)$, to be independent of wavelength, λ (in other words $\kappa_\lambda(z) = \kappa(z)$). This assumption is called the *gray approximation*. Because gray approximation is quite good in the solar photosphere, it can be applied for the solar blackbody radiation.

Assuming that the LTE and gray approximations are both applicable, we obtain the following expression for the total emergent intensity per unit wavelength(see Eq. 11.35):

$$I_\lambda(0, \cos\Theta) = \frac{1}{\cos\Theta} \int\limits_0^\infty d\tau' B_\lambda(\tau') \exp\left(-\frac{\tau'}{\cos\Theta}\right). \tag{11.40}$$

In the next step we integrate Eq. (11.40) to obtain the wavelength-integrated total emergent intensity:

$$I(0, \cos\Theta) = \int\limits_0^\infty d\lambda I_\lambda(0, \cos\Theta)$$

$$= \frac{1}{\cos\Theta} \int\limits_0^\infty d\tau' \exp\left(-\frac{\tau'}{\cos\Theta}\right) \int\limits_0^\infty d\lambda B_\lambda(\tau'). \tag{11.41}$$

With the help of Eq. (11.37) this expression can be written as

$$I(0, \cos \Theta) = \frac{\sigma_{SB}}{\pi \cos \Theta} \int_0^\infty d\tau' T^4(\tau') \exp\left(-\frac{\tau'}{\cos \Theta}\right).$$ (11.42)

In general, evaluating the integral in Eq. (11.42) can be quite difficult. It is clear that in the gray approximation there is a one-to-one correspondence between altitude and optical depth. The exponential term in the integral will basically "emphasize" a relatively narrow temperature range near $\tau/\cos \Theta \approx 1$. The value of the integral can be characterized by a blackbody radiation of an "effective" photosphere with a temperature of T_{eff}. Assuming near isotropy, the integral can be approximated by the following expression:

$$I(0, \cos \Theta) \approx \frac{3\sigma_{SB} T_{eff}^4}{4\pi}\left(\cos \Theta + \frac{2}{3}\right),$$ (11.43)

where $T_{eff} = T(\tau = 1)$. Inspection of Eq. (11.43) reveals that the total emergent intensity depends on the angle between line of sight and the radial direction (Θ). The emergent intensity decreases by a factor of 2.5 as Θ changes from $\Theta = 0$ to $\Theta = \pi/2$. This effect is called *limb darkening*, and it is well established observationally. Limb darkening occurs because the radiation from the limb of the Sun must travel through much more material than the radiation originating from the central areas.

11.8 Flares and Coronal Mass Ejections

A *solar flare* is a localized explosive release of energy that appears as a sudden, short-lived brightening of an area in the chromosphere. Solar flares release their energy mainly in the form of electromagnetic radiation and energetic particles. Solar flares are classified according to their radiative energy release.

The "brilliance" of the flare is usually measured in two frequency bands: optical and X-ray. The optical flare classification is faint (F), normal (N), or brilliant (B). Although the optical emission in flares increases only a few percent, the X-ray emission may be enhanced by several orders of magnitude. The most common X-ray index is based on the peak energy flux of the flare in the 1 to 8 Å soft X-ray band measured by geosynchronous satellites. The first symbol characterizes the order of magnitude (C $= 10^{-3}$ ergs cm^{-2} s^{-1}, M $= 10^{-2}$ ergs cm^{-2} s^{-1}, X $= 10^{-1}$ ergs cm^{-2} s^{-1}), followed by the most significant digit of the actual peak flux. For instance, a peak flux of 6.3×10^{-2} ergs cm^{-2} s^{-1} is reported as an M6 soft X-ray flare. Figure 11.14 shows a solar flare observed in H$_\alpha$.

The total energy released from flares range from 10^{21} to 10^{25} J. The largest flares eject large amounts of energetic particles.[12] Since flares mostly occur in closed field

---

[12] However, CME-driven shocks typically accelerate more particles in the outer corona and in interplanetary space.

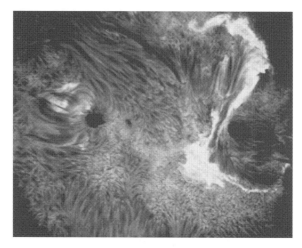

Figure 11.14 Solar flare in H_α. (Courtesy of the National
Solar Observatory/Sacramento Peak.)

line regions, their plasma emission is usually not very significant. However, flare-generated energetic particle events, their propagation through interplanetary space, and their interaction with the Earth's space environment pose very interesting questions with potential importance for human activities. These issues will be discussed in detail later in this book (see Chapter 13).

Observations made with white-light coronagraphs in the early 1970s demonstrated that large amounts of mass (10^{15} to 10^{16} g) are sporadically ejected from the Sun into the interplanetary medium. Such transient ejections of material are called *coronal mass ejections* (CMEs).

Figure 11.15 shows a series of white-light images capturing a large coronal mass ejection. In CMEs coronal material is ejected with speeds ranging from 50 km/s to as high as 2,000 km/s. Prior to the CME eruption, a helmet streamer was visible for a few days. As the helmet streamer swells in the initial stage of the ejection, a dark cavity comes into view.

Most CME events originate in closed magnetic field regions in the corona, where the magnetic field is strong enough to constrain the plasma from expanding outward (plasma transport across magnetic field lines is very limited). These closed field regions are found primarily in the coronal streamer belt near the solar magnetic equator. In three dimensions open field lines are possible within the streamer belt and are probably relatively common there. It is believed that most of the low-speed solar wind actually originates on open field lines in the streamer belt.

CMEs are commonly (but not always) observed in association with other forms of solar activity (such as eruptive prominences, radio bursts, etc.). The physical origin of CMEs is not clear. However, there is convincing evidence that CMEs are not generated by solar flares. Many CMEs are associated with long-duration (many hours-long) X-ray events, which are usually associated with restructuring of the solar corona beneath the CME.

Flares and coronal mass ejections are different aspects of solar activity that are not necessarily related. Flares are essentially photon output, whereas CMEs mainly

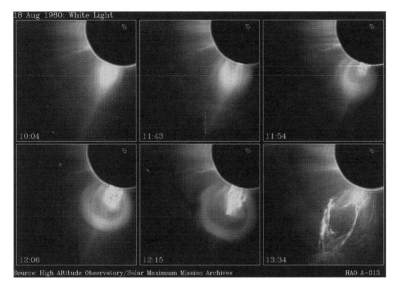

Figure 11.15 Coronal mass ejection (CME). (Coronagraph data from the *Solar Maximum Mission*, archived at the High Altitude Observatory. Primary image processing by J. Burkepile, High Altitude Observatory, National Center for Atmospheric Research, Boulder, CO.)

produce plasma.[13] CMEs and flares are very important sources of dynamical phenomena in the space environment. Interplanetary, magnetospheric, and ionospheric phenomena associated with flares and CMEs will be discussed later in this book (see Chapter 14). However, at this time neither the trigger mechanism for CMEs and solar flares nor the particle acceleration mechanisms are understood beyond a rudimentary level.

11.9   Problems

Problem 11.1 The temperature of the solar interior is 15 million K.

1. What is the average kinetic energy of a proton in joules and in eV?
2. What is the mean molecular speed of a proton, $\bar{v}$?
3. How close to each other can two protons get if both move with speed $\bar{v}$? How many proton radii is this?

Problem 11.2 Assume that all solar energy is produced by the chain reaction given in Section 11.1.

1. Throughout the whole Sun, how many times per second must this chain of reactions occur to account for the solar luminosity?
2. Assume that the solar interior is transparent to solar neutrinos. What is the flux of solar neutrinos at the Earth?

---

[13] Also, CME-driven shocks are the prime cause for major energetic particle events at Earth orbit.

Chapter 12

The Solar Wind

The solar wind is the extension of the solar corona to very large heliocentric distances. As we shall see later in this chapter the solar wind exists because of the huge pressure difference between the hot plasma at the base of the corona and the interstellar medium.

The existence of a *continuous* solar wind was first suggested by Ludwig Biermann[1] based on his studies of the acceleration of plasma structures in comet tails. The detailed mathematical theory of the solar wind was put forward by Eugene Parker.[2] The solar wind was first sporadically detected[3] by the Soviet space probes *Lunik 2* and *3*, but the first continuous observation of the solar wind was made with the *Mariner 2* spacecraft.[4]

In this chapter we shall describe the "classic" theory of the solar wind, which is based on the fluid approximation of coronal and interplanetary plasmas.

12.1 Hydrostatic Equilibrium – It Does Not Work

The simplest theoretical description of the solar corona is based on the assumption of a spherically symmetric, steady-state hydrostatic corona. Single-fluid equations can be applied since the gas is assumed to be a fully ionized, quasineutral

[1] Biermann, L., "Kometenschweife und Solare Korpuskularstrahlung," *Zeit. Astrophys.*, **29**, 274, 1951.

[2] Parker, E. N., "Dynamics of the interplanetary gas and magnetic fields," *Astrophys. J.*, **128**, 664, 1958.

[3] Gringauz, K. I., Bezrukikh, V. V., Ozerov, V. D., and Rybchinskii, R. E., "A study of interplanetary ionized gas, high energy electrons, and corpuscular radiation from the Sun by means of the three-electrode trap for charged particles on the second Soviet Cosmic Rocket," *Doklady Akademii Nauk SSSR*, **131**, 1301, 1960.

[4] Neugebauer, M, and Snyder, C. W., "*Mariner 2* observations of the solar wind, 1. Average properties," *J. Geophys. Res.*, **71**, 4469, 1966.

proton–electron plasma. The effects of magnetic field and heat conduction are ne-glected in this simple approximation. In this case the governing equations become the following (Eq. 4.89):

$$\frac{1}{r^2}\frac{d}{dr}(r^2\rho u) = 0,$$

$$\rho u\frac{du}{dr} + \frac{dp}{dr} + \rho G\frac{M_\odot}{r^2} = 0, \tag{12.1}$$

$$\frac{3}{2}u\frac{dp}{dr} + \frac{5}{2}p\frac{1}{r^2}\frac{d}{dr}(r^2 u) = 0,$$

where ρ, u, and p denote the mass density, radial velocity, and pressure, respectively.

The simplest hydrostatic solution of Eqs. (12.1) can be obtained by assuming a stationary solar atmosphere ($u = 0$). In this case the continuity and energy equations are automatically satisfied and the momentum equation simply expresses hydrostatic equilibrium in a spherical atmosphere:

$$\frac{dp_s}{dr} + \rho_s G\frac{M_\odot}{r^2} = 0, \tag{12.2}$$

where p_s and ρ_s are the pressure and density in the hydrostatic case.

In the early simple models of the solar corona an isothermal corona was assumed. In this case the pressure can be expressed as

$$p_s = n_p kT + n_e kT \approx \frac{2kT}{m_p}\rho_s \tag{12.3}$$

because we assumed that the coronal plasma is quasineutral and the electrons and ions have the same temperature. Substituting Eq. (12.3) into Eq. (12.2) yields the following differential equation for the pressure in hydrostatic equilibrium:

$$\frac{dp_s}{dr} + p_s\frac{Gm_p M_\odot}{2kT}\frac{1}{r^2} = 0. \tag{12.4}$$

Equation (12.4) can be readily solved to obtain

$$p_s = p_\odot \exp\left[\frac{m_p g_\odot}{2kT}R_\odot\left(\frac{R_\odot}{r} - 1\right)\right], \tag{12.5}$$

where $g_\odot$ is the gravitational acceleration at the base of the corona. This solution is the general form of the hydrostatic equilibrium solution for isothermal atmospheres. If $z = r - R_\odot \ll R_\odot$, this solution becomes the well-known barometric altitude formula:

$$\lim_{z \ll R_\odot} p_s \approx p_\odot \exp\left[-z\frac{m_p g_\odot}{2kT}\right]. \tag{12.6}$$

An interesting feature of the hydrostatic solution (Eq. 12.5) is the fact that the pressure does not vanish as $r \to \infty$:

$$p_\infty = \lim_{r/R_\odot \to \infty} p_s = p_\odot \exp\left(-\frac{m_p g_\odot}{2kT}R_\odot\right). \tag{12.7}$$

It is easy to confirm that for a $\approx 10^6$ K coronal temperature $p_\infty/p_\odot \approx 3 \times 10^{-4}$. However, the pressure at the base of the corona is a few tenths of a Pascal, whereas the estimated pressure of the interstellar medium is at least 10 orders of magnitude smaller. It is immediately obvious that $p_\infty \gg p_{\text{interstellar}}$ and therefore the hydrostatic solution cannot represent an equilibrium solution for the hot solar corona.

12.2   Coronal Expansion

The recognition that a hot, static corona cannot exist motivated the young Eugene Parker in the late 1950s to turn his attention to coronal solutions with nonzero radial velocities.[5] Today this kind of solution is very natural for us, but at that time it was quite revolutionary.

The physics behind the idea of an expanding solar corona is quite simple. The very low pressure of the interstellar medium acts as a "vacuum cleaner," which "sucks out" the gas from the solar corona.

Let us consider the simplest scenario and assume a steady-state spherically symmetric corona and neglect electromagnetic effects. In this case the fully ionized plasma is described by Eqs. (12.1).

The pressure gradient can be expressed from the energy equation:

$$\frac{dp}{dr} = -\frac{5}{3}\frac{p}{u}\frac{du}{dr} - \frac{10}{3}\frac{p}{r}. \tag{12.8}$$

Substituting this expression for the radial pressure gradient into the momentum equation yields

$$\frac{u^2 - a_s^2}{u}\frac{du}{dr} = \frac{2a_s^2}{r} - g_\odot\frac{R_\odot^2}{r^2}, \tag{12.9}$$

where $a_s^2 = 5p/3\rho$ is the local sound speed.

Assuming an isothermal solar corona ($a_s = const.$), we can integrate Eq. (12.9). This is a fairly good approximation for the solar corona. With this assumption we obtain the following solution:

$$\frac{1}{2}u^2 - a_s^2\ln u = 2a_s^2\ln r + g_\odot\frac{R_\odot^2}{r} + C, \tag{12.10}$$

where C is an integration constant. Figure 12.1 shows the mathematically admissible classes of isothermal solutions (depending on the constant, C). These solutions can be divided into five basic categories (denoted by Roman numerals I through V).

Because we are interested in physically meaningful single-valued solutions that describe plasma continuous expansion from the Sun, some of the solution classes can be eliminated right away. Classes I and II describe double-valued solutions, which

[5] Parker, E. N., "The magnetohydrodynamic treatment of the expanding corona," *Astrophys. J.*, **132**, 175, 1960.

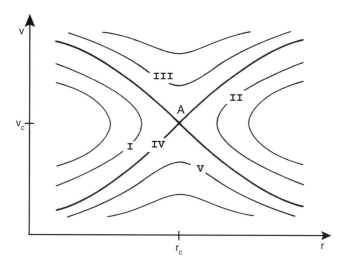

Figure 12.1 Mathematically admissible classes of isothermal solutions of an expanding corona.

are completely unphysical. Class III solutions are supersonic at the base of the corona and therefore cannot meet our physical requirements.

Solutions of type V are physically admissible. They are subsonic everywhere and predict speeds of about 10 km/s at Earth orbit. This *solar breeze* solution was, for a long time, a competing solution with the wind solution. (The idea of a high-speed coronal expansion was so foreign for some older scientists that they never accepted it. As the saying goes: Old ideas never die, scientists do.) The main problem with the solar breeze solution is that it gives finite pressure at infinity (similar to the hydrostatic solution) and this pressure far exceeds the pressure of the interstellar medium. Even when the assumption of an isothermal corona was relaxed, the pressure remained finite at infinity.

The solution favored by Parker was the type IV solution, starting subsonically at the base of the corona and accelerating to supersonic speeds. Since astronomers were unfamiliar with the advances made in fluid dynamics, type IV solutions generated a lot of misconceptions and misunderstandings. The reason was the apparent singularity of Eq. (12.9) at the sonic point, r_c (at the sonic point $u = a_s$).

To understand the underlying physics, let us write Eq. (12.9) the following way:

$$\frac{1}{u}\frac{du}{dr} = \frac{1}{u^2 - a_s^2}\frac{2a_s^2}{r}\left(1 - \frac{\mathcal{U}(r)}{2kT}\right), \tag{12.11}$$

where $\mathcal{U}$ is the gravitational potential of a proton in the solar gravitational field. If we neglect solar gravity, the equation simplifies to

$$\frac{1}{u}\frac{du}{dr} = \frac{1}{u^2 - a_s^2}\frac{2a_s^2}{r}. \tag{12.12}$$

Because this equation has a real singularity at the sonic point, transonic solutions are not allowed. If the flow is subsonic at the solar surface, $du/dr < 0$ for $r > R_\odot$, and the flow will slow down and always remain subsonic. However, in the case of supersonic outflows, $du/dr > 0$ for $r > R_\odot$, and the flow will accelerate and it remains supersonic everywhere.

Elementary fluid dynamics tells us that when one opens a gas reservoir filled with stationary gas to vacuum the gas outflow will be sonic at the surface of the reservoir and it accelerates outward: This outflow is driven by the large pressure gradient force.

The addition of solar gravity introduces a retarding (backward) force reducing the outward expansion of the corona. This retarding force rapidly reduces with increasing distance from the Sun. The effect of the retarding force is that the sonic point moves outward: The outflow starts subsonically and undergoes a sonic transition at a distance where the gravitational potential and the kinetic energy of the random motion are equal (in the case of our ion–electron plasma the condition is $\mathcal{U}(r_\mathrm{c}) = 2kT$). This can be shown by examining the full, time-dependent partial differential equation system describing the spherically symmetric flow. The apparent singularity at $r = r_\mathrm{c}$ is a consequence of the steady-state assumption (mathematically, we switch from a partial differential equation system to a single ordinary differential equation). In other words, when the denominator vanishes in Eq. (12.11), the terms in the parentheses also cancel each other and the value of the fraction remains finite. This solution can easily be recovered from Eq. (12.10) by requiring that $\mathcal{U}(r_\mathrm{c}) = 2kT$ and $u(r_\mathrm{c}) = a_\mathrm{s}$:

$$M^2 - \ln M^2 = 1 + 4\ln\frac{r}{r_\mathrm{c}} + 4\left(\frac{r_\mathrm{c}}{r} - 1\right), \tag{12.13}$$

where $M = u/a_\mathrm{s}$ is the Mach number and

$$r_\mathrm{c} = \frac{3}{5}\frac{mg_\odot R_\odot^2}{2kT}. \tag{12.14}$$

Assuming $T = 2 \times 10^6$ K coronal temperature, we get a sonic point at around $6\ R_\odot$.

The type IV solution given by Eq. (12.13) is the isothermal *solar wind* (see Figure 12.2). This is naturally a greatly simplified situation, because the coronal temperature does not remain constant as it expands. The energetics of the solar wind

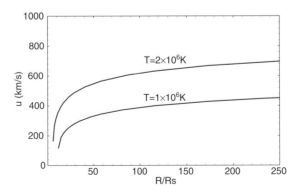

Figure 12.2 Radial expansion speed profiles for an isothermal corona. Earth orbit is at 215 $R_\odot$.

are quite poorly understood (especially near the Sun) and future space missions are needed to understand it. It is well established that the solar wind temperature decreases with radial distance as $r^{-\beta}$, where the polytropic index, β, is less than 1. Such a radial variation still allows for solar wind type solutions. However, some heating mechanism is needed (especially near the critical point) to maintain such a weak radial dependence.

It is important to recognize that the wind solution (type IV) gives zero pressure as $r \to \infty$, even in the case of an isothermal corona. This can be seen from the steady-state continuity equation, which states that $r^2 \rho u = const$. Since the sound speed is constant (the corona is isothermal) and the Mach number logarithmically increases for large r values (see Eq. 12.13), u must also be a slowly increasing function of r. This means that $\rho \to 0$ as $r \to \infty$. Because the pressure is directly proportional to the density, $p \to 0$ as $r \to \infty$.

Table 12.1 summarizes some of the average solar wind properties at the orbit of the Earth. It can be seen that the typical plasma temperature is about a factor of 10 lower than it is in the corona. This implies an "average" polytropic index of about 0.3 to 0.4.

12.3 Interplanetary Magnetic Field (IMF)

12.3.1 Field Lines and Streamlines

The extension of the solar magnetic field into the interplanetary medium can be understood in terms of the magnetic induction equation (Eq. 4.86):

$$\frac{\partial \mathbf{B}}{\partial t} = \nabla \times (\mathbf{u} \times \mathbf{B}). \tag{12.15}$$

(Here we neglected the diffusion term, since the effects of interparticle collisions are negligible in this case.) Next, we express this equation in the coordinate system rotating with the Sun. This can be done with the help of Eq. (8.15), which expresses the time

Table 12.1. *Average solar wind properties at 1 AU.*

Quantity	Value	Units
Electron concentration	7.1	cm^{-3}
Flow speed	450	km/s
Proton temperature	1.2×10^5	K
Electron temperature	1.4×10^5	K
Magnetic field	7.0	nT
Acoustic speed	60	km/s
Alfvén speed	40	km/s
Average collision time	3.5×10^6	s
Travel time to 1 AU	3.5×10^5	s

derivative of an arbitrary vector in the rotating system. Now the flow velocity in the corotating system is

$$\mathbf{u}' = \mathbf{u} - \Omega_{\odot} \times \mathbf{r}. \tag{12.16}$$

The time derivative of the magnetic field vector in the rotating system is the following:

$$\frac{\partial \mathbf{B}'}{\partial t} = \frac{\partial \mathbf{B}}{\partial t} - \Omega_{\odot} \times \mathbf{B}. \tag{12.17}$$

With these expressions the induction equation in the rotating system becomes

$$\frac{\partial \mathbf{B}}{\partial t} + \Omega_{\odot} \times \mathbf{B} = \nabla \times [(\mathbf{u}' + \Omega_{\odot} \times \mathbf{r}) \times \mathbf{B}], \tag{12.18}$$

where we recognized the fact that in nonrelativistic coordinate transformations the magnetic field vector remains unchanged to first-order accuracy, and therefore $\mathbf{B}' = \mathbf{B}$. If we also neglect differential rotation of the Sun ($\Omega_{\odot}$ is constant), we obtain

$$\begin{aligned}
\nabla \times [(\Omega_{\odot} \times \mathbf{r}) \times \mathbf{B}] &= (\Omega_{\odot} \times \mathbf{r})(\nabla \cdot \mathbf{B}) - \mathbf{B}[\nabla \cdot (\Omega_{\odot} \times \mathbf{r})] \\
&\quad + (\mathbf{B} \cdot \mathbf{r})(\Omega_{\odot} \times \mathbf{r}) - [(\Omega_{\odot} \times \mathbf{r}) \cdot \nabla]\mathbf{B} \\
&= (\Omega_{\odot} \times \mathbf{B}) - [(\Omega_{\odot} \times \mathbf{r}) \cdot \nabla]\mathbf{B},
\end{aligned} \tag{12.19}$$

because the magnetic field is always divergenceless and $\nabla \times \mathbf{r} = \mathbf{0}$, $\nabla \mathbf{r} = I$.

Substituting Eq. (12.19) into Eq. (12.18) yields

$$\frac{\partial \mathbf{B}}{\partial t} + [(\Omega_{\odot} \times \mathbf{r}) \cdot \nabla]\mathbf{B} = \nabla \times (\mathbf{u}' \times \mathbf{B}). \tag{12.20}$$

The left-hand side of Eq. (12.20) is the total (convective) derivative of the magnetic field in the corotating system. Therefore we can write

$$\frac{D_{\odot}\mathbf{B}}{D_{\odot}t} = \nabla \times (\mathbf{u}' \times \mathbf{B}), \tag{12.21}$$

where $D_{\odot}/D_{\odot}t$ represents the total time derivative in the system rotating with the Sun.

We consider a situation when the total time derivative of the magnetic field vector vanishes in the corotating frame of reference, $D_{\odot}/D_{\odot}t = 0$. This corresponds to a steady-state axially symmetric physical situation. Using this assumption, we obtain

$$\nabla \times (\mathbf{u}' \times \mathbf{B}) = \mathbf{0}. \tag{12.22}$$

Since the curl of $\mathbf{u}' \times \mathbf{B}$ vanishes, there exists a scalar potential that satisfies the $\nabla \Phi' = \mathbf{u}' \times \mathbf{B}$ condition. In effect, Φ' is the electric potential since the coronal gas is fully ionized, and its electric conductivity is very high. In this case the generalized Ohm's law (Eq. 4.79) implies that $\mathbf{E}' = -\mathbf{u}' \times \mathbf{B}$, and therefore $\mathbf{E}' = -\nabla \Phi'$.

It can easily be shown that both plasma streamlines and magnetic field lines are equipotentials in the rotating frame. Taking the scalar product of $\nabla \Phi'$ with $\mathbf{u}'$ and $\mathbf{B}$

yields

$$\mathbf{u}' \cdot \nabla \Phi' = \frac{\partial \Phi'}{\partial s_u} = \mathbf{u}' \cdot (\mathbf{u}' \times \mathbf{B}) = 0,$$

$$\mathbf{B} \cdot \nabla \Phi' = \frac{\partial \Phi'}{\partial s_B} = \mathbf{B} \cdot (\mathbf{u}' \times \mathbf{B}) = 0, \tag{12.23}$$

where s_u and s_B are the path lengths measured along flow lines and magnetic field lines, respectively. Because the derivatives of the potential vanish along flow lines and magnetic field lines, these are equipotentials.

Next, let us consider a sphere with radius R_s around the Sun. This sphere is at a distance where the solar wind velocity is already significant (near its asymptotic value) and the magnetic field is still close to the solar value ($R_s \approx 10\ R_\odot$). This surface is referred to as the *source surface* of the interplanetary magnetic field. Let us assume for the sake of simplicity that at the source surface both the magnetic field and the flow velocity point radially outward in the corotating frame (in general, this is not a bad assumption). This means that $\nabla \Phi' = \mathbf{u}' \times \mathbf{B} = 0$ everywhere on the source surface. However, all flow lines and field lines are equipotentials, so $\nabla \Phi' = 0$ everywhere in the solar wind (i.e., $\mathbf{E}' = \mathbf{0}$ everywhere). This condition means that

$$\mathbf{u}' \times \mathbf{B} = \mathbf{0} \quad \Longrightarrow \quad \mathbf{u}' \parallel \mathbf{B}. \tag{12.24}$$

This is a fundamentally important result, indicating that under the above assumptions (steady-state axisymmetric solar wind) the magnetic field and plasma flow vectors are always parallel in the frame of reference rotating with the Sun.

12.3.2   Magnetic Field Lines

Equation (12.24) determines the geometry of magnetic field lines in the interplanetary medium. First of all, one must recognize that neither the magnetic field nor the flow velocity will have polar components ($u'_\Theta = 0$ and $B_\Theta = 0$), so we have to derive only the radial and azimuthal components of $\mathbf{u}'$ and $\mathbf{B}$. Since $\mathbf{u}'$ and $\mathbf{B}$ are parallel to each other, the ratio of their components must be the same:

$$\frac{B_\phi}{B_r} = \frac{u'_\phi}{u'_r} = \frac{-(r - R_s)\Omega_\odot \sin \Theta}{u_{sw}}, \tag{12.25}$$

where u_{sw} is the asymptotic solar wind speed (independent of location), Θ is the polar angle of the field line (for a given field line $\Theta = $ const.), and u_{sw} is the plasma speed in the noncorotating system (see Eq. 12.16). Here we assumed that at distances $r \gg R_\odot$ the solar wind is practically radial (in the noncorotating frame). Far from the Sun the velocity components can be written in the following form:

$$u'_r = u_r = \frac{dr}{dt} = u_{sw},$$

$$u'_\phi = r \sin \Theta \frac{d\phi}{dt} = -\Omega_\odot r \sin \Theta, \tag{12.26}$$

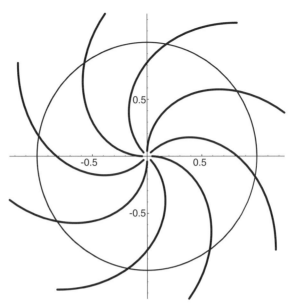

Figure 12.3 The Archimedean spiral shape of the interplanetary magnetic field.

where it is assumed that u_{sw} is the constant radial solar wind velocity outside the source surface. These equations can be solved to give

$$r - R_s = u_{sw}t,$$

$$\phi - \phi_0 = -\Omega_\odot t,$$

(12.27)

where ϕ_0 is the azimuth angle of the "foot" of the field line at the source surface. These two expressions can be combined to obtain the $\phi(r)$ function of a magnetic field line:

$$\phi(r) = \phi_0 - \frac{\Omega_\odot}{u_{sw}}(r - R_s).$$

(12.28)

This curve is the well-known *Archimedean spiral* (sometimes also called the *Parker spiral*). Figure 12.3 shows magnetic field lines in the solar equatorial plane (as seen from the north pole of the Sun) for $R_s = 10\ R_\odot$ and $u_{sw} = 400$ km/s.

The magnitude of the interplanetary magnetic field can be determined in two steps. First, we can express the magnetic field vector with the help of Eq. (12.25):

$$\mathbf{B} = B_r\mathbf{e}_r - B_r\frac{(r - R_s)\Omega_\odot \sin\Theta}{u_{sw}}\mathbf{e}_\phi.$$

(12.29)

In the second step we apply the $\nabla \cdot \mathbf{B} = 0$ Maxwell equation to Eq. (12.29) to obtain

$$\nabla \cdot \mathbf{B} = \frac{1}{r^2}\frac{\partial}{\partial r}(r^2 B_r) + \frac{1}{r\sin\Theta}\frac{\partial B_\phi}{\partial\phi}$$

$$= \frac{1}{r^2}\frac{\partial}{\partial r}(r^2 B_r) - \frac{(r - R_s)\Omega_\odot}{r u_{sw}}\frac{\partial B_r}{\partial\phi} = 0.$$

(12.30)

In the present steady-state axisymmetric model, $\partial B_r/\partial\phi = 0$. This approximation leads to the relation

$$\frac{1}{r^2}\frac{\partial}{\partial r}(r^2 B_r) = 0. \tag{12.31}$$

This can be easily solved to obtain

$$B_r(r) = B_s\left(\frac{R_s}{r}\right)^2, \tag{12.32}$$

because we assumed that the magnetic field is radial at the source surface. Substituting this expression into Eq. (12.29), we get

$$\mathbf{B} = B_s\left(\frac{R_s}{r}\right)^2\mathbf{e}_r - B_s\left(\frac{R_s}{r}\right)^2(r - R_s)\frac{\Omega_\odot\sin\Theta}{u_{sw}}\mathbf{e}_\phi. \tag{12.33}$$

At large distances from the Sun ($r \gg R_s$), this becomes

$$\mathbf{B} = B_s\left(\frac{R_s}{r}\right)^2\mathbf{e}_r - B_s\left(\frac{R_s^2}{r}\right)\frac{\Omega_\odot\sin\Theta}{u_{sw}}\mathbf{e}_\phi. \tag{12.34}$$

Equation (12.34) shows that the radial and azimuthal components of the interplanetary field behave quite differently. The radial component decreases with r^{-2}, whereas the azimuthal component decreases only as r^{-1}. Thus as we go outward in the solar system the magnetic field becomes more and more azimuthal (it "wraps around") in the equatorial plane. At the same time the field behaves quite differently over the solar polar regions. The magnitude of the magnetic field at large distances from the Sun is now

$$B = B_s\left(\frac{R_s}{r}\right)^2\sqrt{1 + r^2\left(\frac{\Omega_\odot\sin\Theta}{u_{sw}}\right)^2}. \tag{12.35}$$

If the magnitude of the magnetic field at the source surface is a few Gauss ($1\,\mathrm{G} = 10^{-4}$ T), then at Earth orbit we get a field strength of 5 to 10 nT, which are typical observed values.

12.4   Coronal Structure and Magnetic Field

A fundamental assumption in our derivation of the solar wind solution was that the corona was spherically symmetric. However, this is a poor approximation in the vicinity of the Sun where most of the acceleration takes place. There are two fundamentally different types of coronal magnetic field structures that eventually result in quite different solar wind and interplanetary magnetic field properties: regions of open magnetic field lines and regions of closed magnetic field lines.

To understand the physical situation, let us consider the corona containing hot plasma and an axisymmetric magnetic field. For the sake of simplicity, we use the single-fluid ideal MHD approximation, and we assume that the plasma temperature is uniformly T everywhere in the corona and that the situation is steady state. In this case the continuity and momentum equations (Equations. 4.80) simply become the following:

$$\nabla \cdot (\rho_{\mathrm{m}} \mathbf{u}) = 0, \tag{12.36}$$

$$\rho_{\mathrm{m}}(\mathbf{u} \cdot \nabla)\mathbf{u} + \nabla p + \rho_{\mathrm{m}} G \frac{M_{\odot}}{r^2}\mathbf{e}_r - \mathbf{j} \times \mathbf{B} = \mathbf{0}. \tag{12.37}$$

The magnetic field and the current density are determined by the steady-state induction equation (Eq. 4.86) and by Ampère's law (Eq. 1.4):

$$\nabla \times (\mathbf{u} \times \mathbf{B}) = 0, \tag{12.38}$$

$$\mathbf{j} = \frac{1}{\mu_0}\nabla \times \mathbf{B}. \tag{12.39}$$

Naturally, these equations are supplemented by the equation of state for perfect gases. Also, the magnetic field must be divergenceless everywhere.

It is assumed that the solution is axisymmetric at all times (all functions depend only on $\mathbf{r}$ and Θ (where Θ is the spherical polar angle). Equations (12.36) through (12.39) completely determine the solution (ρ_{m}, $\mathbf{u}$, $\mathbf{B}$, and $\mathbf{j}$) if we know the boundary conditions at the base of the corona. For instance, one can take a surface dipole field with magnetic field strength of 10^{-4} T (1 G) at the pole with a prescribed density and velocity (where the initial radial velocity, $u_0(\Theta)$, is determined to give the transonic solar wind solution for each flow line).

Equations (12.36) through (12.39) were solved iteratively by Pneuman and Kopp.[6] First, one starts with a dipole distribution and calculates the current system. Next, this current density and magnetic field are used to obtain a new density and flow velocity solution. The new velocity field is substituted into the induction equation to determine a new magnetic field iteration. These steps are successively repeated until the solution converges. In this way we end up with a self-consistent solution of the magnetized plasma flow near the Sun. Such a solution is shown in Figure 12.4.

Dashed lines in Figure 12.4 show the initial dipole magnetic field lines, and solid lines represent the magnetic field configuration of the converged solution. The original dipole configuration gets quite distorted low in the corona, owing to the pressure of the plasma populating the field lines. Magnetic field lines passing through the corona at high latitudes are "drawn outward" by the expanding plasma and become "open" field lines (eventually these open field lines also return to the Sun, but first they extend to very large heliocentric distances).

[6] Pneuman, G. W., and Kopp, R. A., "Gas–magnetic field interactions in the solar corona," *Solar Phys.*, **18**, 258, 1971.

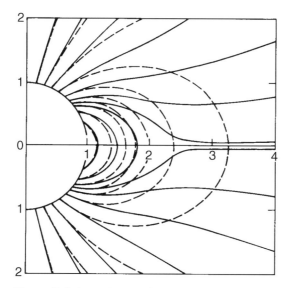

Figure 12.4 Coronal magnetic field line configuration for a dipole field at the base of the corona. The dashed lines are magnetic field lines for a pure dipole field. (From *Pneuman and Kopp.*[6])

Inspection of Figure 12.4 reveals closed field lines, i.e near the dipole equator, but they are distorted outward. These field lines contain coronal plasma in static equilibrium maintained by a balance between the pressure gradient and gravitational forces. Magnetic field lines crossing the base of the corona at higher latitudes are open, with the connected coronal plasma expanding into interplanetary space and forming the solar wind. Open field lines spread in latitude and cover the entire 4π solid angle beyond about three solar radii.

The first open field lines originating from opposite hemispheres come quite close together and extend radially outward. These field lines are of opposite magnetic polarities, and thus the interplanetary magnetic field must change sign rather suddenly within this narrow region. This implies the presence of a thin sheet of very high current density with $\mathbf{j} = \nabla \times \mathbf{B}/\mu_0$ pointing in the azimuthal direction. This current circulates around the dipole axis in the same direction as the original current generating the dipole field. It is this *heliospheric current sheet* that separates fields and plasma flows originating from different hemispheres.

The predicted coronal structures are clearly visible in photographs of the solar corona obtained during solar eclipses. For example, Figure 12.5 shows the magnetic field structure deduced from a photograph taken during the November 1966 eclipse in Peru. Regions of open magnetic field lines contain low density plasma; therefore they appear as dark regions. These regions are also called *coronal holes*.

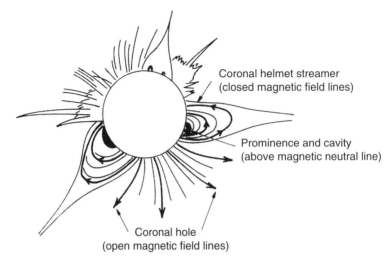

Figure 12.5 Deduced coronal magnetic field structure during the November 1966 solar eclipse. (From *Hundhausen.*[7])

12.5   Solar Wind Stream Structure

In this section we extensively use a recently published summary of the available observational evidence about the stream structure in the solar wind.[8]

Near the ecliptic plane the solar wind tends to be organized into alternating streams of slow and fast streams. In fast streams the solar wind speed typically exceeds 600 km/s, whereas in slow streams the speed is usually less than 350 km/s. These streams tend to be encountered by Earth on several successive solar rotations. High-speed streams are typically unipolar. Sharp, long-lived reversals in magnetic field polarity occur at low speeds close to the leading edges of high-speed streams. The sharp polarity reversals near the leading edges of fast streams correspond to crossings of the heliospheric current sheet.

The long-lived, high-speed streams originate in coronal holes, which are large, nearly unipolar regions in the solar atmosphere. Since these regions are associated with open magnetic field lines, the solar wind expansion is relatively unrestricted by the solar magnetic field. Low-speed flows, in contrast, originate in the outer regions of coronal streamers that straddle regions of magnetic polarity reversals. It is very likely that low-speed solar wind flows are related to complicated open magnetic field line topologies in the coronal streamer belt.

[7] Hundhausen, A. J., "The solar wind," in *Introduction to Space Physics*, M. G. Kivelson and C. T. Russell, eds., p. 91, Cambridge Univ. Press, Cambridge, 1995.

[8] Gosling, J. T., "The solar wind," in *Encyclopedia of the Solar System*, P. Weissman, L.-A. McFadden, and T. Johnson, eds., Academic Press, 1997.

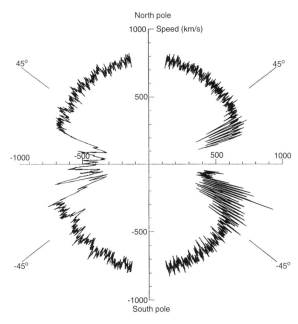

Figure 12.6 Solar wind speed as a function of heliographic latitude as measured by the *Ulysses* spacecraft from 1992 to 1997. (From *Gosling*.[8])

The solar magnetic field undergoes significant changes during the solar cycle. Near solar minimum (including the declining phase of the solar cycle) large, long-lived coronal holes extend from the polar regions to low heliographic latitudes. High-speed solar wind streams originating from these coronal holes are quite commonly observed near the ecliptic plane during this phase of the solar cycle. Near solar maximum, however, the coronal holes are mainly restricted to the polar regions of the Sun, and therefore the associated high-speed solar wind streams cannot be observed near the ecliptic. Figure 12.6 illustrates the variable solar wind at low heliolatitudes and the persistent high-speed flows at high latitudes during the declining phase of the solar cycle and near solar minimum. It shows the solar wind speed as a function of heliospheric latitude (the observations were made by the *Ulysses* spacecraft between 1.4 and 5.4 AU heliocentric distances).

Spatial variations in the solar corona result in the formation of recurring interplanetary structures. Flows of different speed become radially aligned at low heliographic latitudes as the Sun rotates. Thus an azimuthal dependence of solar wind speed results in faster streams "catching up" with slower streams. This convergence of flow near the leading edge of fast streams compresses the plasma and produces a high-pressure region that prevents actual overlap between fast and slow solar wind regions. However, the solar wind cools down as it moves outward, while its velocity keeps slowly increasing. Thus the flow becomes increasingly supersonic with increasing heliocentric distance. The result is that the leading edge of the fast stream keeps steepening and eventually a pair of forward and reverse shocks forms. It is common to refer to the

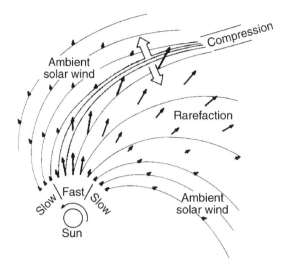

Figure 12.7 Geometry of the interaction between fast
and slow solar wind streams. The plasma is compressed
where the streamlines converge. (From *Pizzo.*[10])

compression at the leading edge of a high-speed stream as an *interaction region.* An
interaction region is illustrated schematically in Figure 12.7. Since these structures
usually live over several solar rotations, the more conventionally used name is *corotating interaction region* or CIR. Fully developed CIRs are routinely observed beyond
Earth's orbit. At 1 AU the interaction between the fast and slow streams usually does
not result in shock formation yet.[9] A pair of shocks are typically formed between
the orbits of Earth and Jupiter. CIRs are associated with recurring geomagnetic disturbances to be discussed later in this book (see Chapter 14). These phenomena are
recurring in the terrestrial frame of reference, since they rotate with the Sun.

It is important to point out that the pattern of a CIR rotates with the Sun, but the
solar wind plasma and associated structures (shocks, magnetic field, etc.) actually
propagate nearly radially outward.

During solar minimum conditions, the solar magnetic field is approximately dipolelike. The orientation of the dipole with respect to the solar rotation axis is quite variable.
Near solar activity minimum the magnetic dipole tends to be nearly aligned with the
rotation axis. The solar dipole is most noticeably tilted relative to the rotation axis
during the declining phase of solar activity. The Sun's magnetic field is not dipolelike
near solar activity maximum.

At times when the solar magnetic axis is substantially tilted with respect to the
rotation axis, the heliospheric current sheet becomes warped into a global structure

---

[9] To zeroth order CIR shocks form because the stream speed greatly exceeds the fast magnetosonic
   speed in the ambient plasma. This condition is typically satisfied beyond Earth orbit. CIRs are
   fully formed by 1 AU, but the bounding shocks are not formed till somewhat larger heliocentric
   distances.
[10] Pizzo, V., "A three-dimensional model of corotating streams in the solar wind. I. Theoretical
   foundations," *J. Geophys. Res.*, **83**, 5563, 1978.

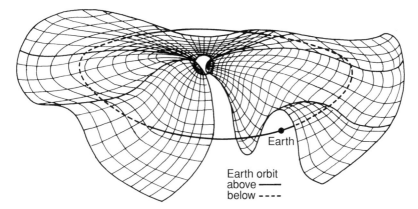

Figure 12.8 Wavy structure of the interplanetary current sheet (the ballerina skirt). [From *Kelly* (*1989*).]

that resembles a ballerina's twirling skirt as illustrated in Figure 12.8. Earth crosses the current sheet at least twice during each solar rotation, sometimes more (if the current sheet is wavy enough). These current sheet crossings are observed at Earth as crossings of *interplanetary magnetic sector boundaries*, when the polarity of the interplanetary magnetic field changes its sign (as observed from Earth).

12.6   Nonrecurring Disturbances in the Solar Wind

In addition to recurrent solar wind variations described above there are nonrecurring transient disturbances. The most interesting type of these phenomena is associated with outward propagating interplanetary shock waves generated by localized energy releases near the Sun (CMEs). The ejected material acts as a piston moving in the ambient solar wind. The slower, "quiet time" solar wind is "snowplowed" by the fast piston of ejected solar material, and eventually a traveling shock is formed ahead of the piston.

The structure of an interplanetary transient shock is shown schematically in Figure 12.9. In the case of major disturbances the structure expands on a broad front. When it arrives at the Earth, its shape is approximately a half circle with radius of 0.5 to 1 AU. Behind the shock a discontinuity is formed between the compressed ambient plasma and the material of the driving plasma. Under some conditions, a reverse shock may also be formed inside the driving gas. The driving plasma of the transient shock is typically cold (indicating a CME origin), and it is sometimes enriched in fully ionized helium.

One of the most interesting and controversial questions of space physics involves the solar origin of these transient interplanetary shocks. For several decades conventional wisdom associated transient interplanetary shocks with solar flares. However, new evidence strongly suggests an alternative picture. According to this growing evidence, transient interplanetary phenomena and related nonrecurring geomagnetic activity are

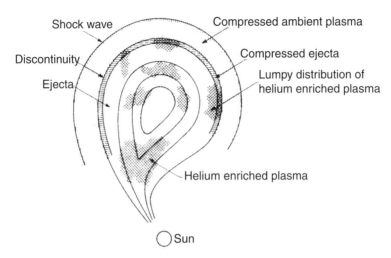

Figure 12.9 Schematic diagram of a possible geometry for a traveling interplanetary shock structure. (From *Bame et al.*[11])

generated by coronal mass ejections rather than by flares. This issue is presently not quite settled and much additional research needs to be done before the origin of transient shocks can be determined.

12.7   The Heliosphere

The *heliosphere* is the region of space influenced by the Sun and its expanding corona, the solar wind. In some respects the heliosphere encompasses the true extent of the solar system.

The idea that a heliosphere might exist first emerged in the 1950s in connection with the propagation of galactic cosmic rays.[12] The basic idea is quite simple. The solar corona is continuously expanding into the interstellar medium. We showed earlier that the pressure of this expanding coronal "bubble" asymptotically approaches zero as the heliocentric distances goes to infinity. However, if there is a finite pressure in the surrounding interstellar medium, the coronal expansion must eventually stop at a point where the solar wind pressure becomes smaller than the interstellar pressure.

The solar system is located in a low-density interstellar cloud believed to be associated with the superbubble seen around the Scorpius–Centaurus Association. This cloud is the environment surrounding the solar system. The solar system is moving through this cloud with a velocity of about 26 km/s. Interstellar charged particles are prevented from entering the heliosphere by plasmaphysical processes (see our discus-

---

[11] Bame, S. J., Asbridge, J. R., Feldman, W. C., Fenimore, E. E., and Gosling, J. T., "Solar wind heavy ions from flare heated coronal plasma," *Solar Phys.*, **62**, 179, 1979.

[12] Davies, L. Jr., "Interplanetary magnetic fields and cosmic rays," *Phys. Rev.*, **100**, 1440, 1955.

Table 12.2. *Estimated properties of the Very Local Interstellar Medium.*

Quantity	Value	Units
V_{sun}	15	km/s
Direction of Sun's galactic motion (apex)	$\ell = 51°, b = 23°$	
V_{cloud}	19	km/s
Direction of cloud's galactic motion	$\ell \approx 155°, b \approx 0°$	
$V_{\text{heliosphere}}$	26	km/s
Direction of Sun's relative motion	$\ell \approx 6°, b \approx 12°$	
Magnitude of magnetic field	0.3–0.5	nT
Direction of magnetic field	$\ell \approx 65°, b \approx 0°$	
H density	0.1	cm^{-3}
He density	0.01	cm^{-3}
Electron density	0.2–0.4	cm^{-3}
Neutral temperature	7,500	K
Electron temperature	7,500	K
Ion temperature	7,500	K

sion below), but interstellar neutrals can penetrate deep into the heliosphere. Over the past several million years, the solar system has traversed a region of interstellar space with very low density and only recently entered the interstellar cloud, the very local interstellar medium (VLISM). The present best estimates of the properties of VLISM are presented in Table 12.2.

In Table 12.2 V_{sun} is the velocity of the solar system with respect to the nearby stars (ℓ and b are galactic longitude and latitude, the galactic center being at $\ell = 0, b = 0$). V_{cloud} is the velocity of the local interstellar cloud, while $V_{\text{heliosphere}}$ is relative velocity of the solar system with respect to the VLISM. Naturally, all data in Table 12.2 are inferred with indirect methods and they are uncertain. Most of the information presented in Table 12.2 comes from a paper by Frisch.[13]

The Sun is a point source of supersonic magnetized plasma that radially expands into a low-pressure external medium. This point source moves with a velocity of ≈ 26 km/s with respect to the very local interstellar medium (VLISM). It is generally assumed that the partially ionized interstellar gas is fairly cold (even though the thermodynamic properties of this region are only poorly known), and consequently this relative motion is most likely supersonic (however, this question is still not fully settled). An immediate consequence of the supersonic motion of the heliosphere in the interstellar plasma is the formation of an upstream bow shock, which decelerates and deflects the interstellar charged particles.

Flow lines of the interstellar plasma do not penetrate into the region dominated by the solar wind flow but flow around a "contact surface," also called the *heliopause*.

---

[13]   Frisch, P. C., "Morphology and ionization of the interstellar cloud surrounding the solar system," *Science*, **265**, 1423, 1994.

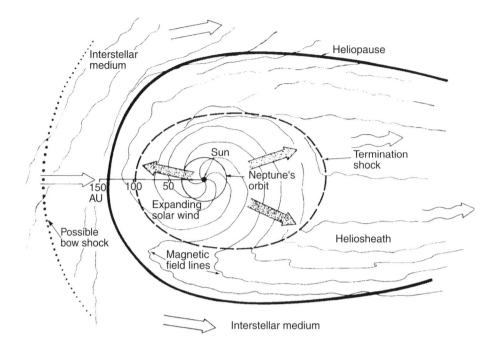

Figure 12.10 Schematic diagram of the heliosphere as seen from above the ecliptic north pole.

At the nose of the heliopause the interstellar gas is nearly stagnating and a density pile-up is expected. The heliopause is considered to be the outer boundary of the heliosphere.

The third large-scale discontinuity is expected to form inside the heliopause. On the upwind (upstream) side of the heliosphere, the initially radially expanding supersonic solar wind must be somehow diverted to the downstream direction; otherwise it would intersect the heliopause. This diversion can only take place in subsonic flow, and therefore the supersonic expansion of the solar wind must be terminated by an "inner shock" or "termination shock." On the downwind (downstream) side, it is expected that a "termination shock" is formed by the expansion of the supersonic solar wind into a low, but finite-pressure external medium (however, the formation of the termination shock on the downwind side is not universally accepted). A schematic representation of this scenario is shown in Figure 12.10.

The characteristic distance from the Sun to the heliopause can be estimated by equating the solar wind ram pressure to the total interstellar pressure (which is the sum of the kinetic, ram, magnetic, and cosmic ray pressures):

$$m_p n_{\text{sw}}(R_{\text{hp}}) u_{\text{sw}}^2 \approx p_{\text{IS}}, \tag{12.40}$$

where R_{hp} is the distance from the Sun to the nose of the dayside heliopause and p_{IS} is the total interstellar pressure. Under steady-state spherically symmetric condition, the continuity equation simplifies to $r^2 n_{sw} u_{sw} = const.$ Therefore one gets the following:

$$m_p \Phi_{sw} \frac{u_{sw}}{R_{hp}^2} \approx p_{IS}, \qquad (12.41)$$

where Φ_{sw} is the solar wind particle flux at 1 AU and R_{hp} is measured in astronomical units. Assuming average solar wind parameters, one obtains $R_{hp} \approx 150$ AU.

Presently, the two Voyager spacecraft are moving toward the heliospheric termination shock, and it is expected that they will reach the termination region of the supersonic solar wind in the next few decades. These spacecraft will be the first man-made objects to leave the region of solar dominance.

12.8   Problems

Problem 12.1 Consider the following idealized model of the solar corona and the solar wind. At a radius of $r_0 = 1.1\ R_s$ the corona is a fully ionized electron–proton plasma that is uniform in latitude and longitude. The plasma at this radial distance has a total density of 10^8 cm^{-3}, a temperature of 3×10^6 K, and a radial magnetic field of 1 G. The plasma pressure at any given point is given by the polytropic relationship $p = p_0(\rho/\rho_0)^\gamma$, where ρ is the mass density, the polytropic index has a value of $\gamma = 1.1$, and p_0 and ρ_0 are the pressure and mass density at the reference point. Assume that the polarity of the magnetic field is positive (outward) in the northern hemisphere and negative (inward) in the southern hemisphere. Assume that the solar rotation is uniform ($\Omega_s = 2.8 \times 10^{-6}$ s^{-1}) and that the pressure of the local interstellar medium is 1.6×10^{-12} dynes/cm^2, and neglect all waves and turbulence.

1. Determine the maximum speed to which the solar wind is accelerated.
2. Determine the density at the critical point.
3. Determine the heliocentric distance of the critical point.
4. Determine the mass loss rate due to the solar wind outflow.
5. Determine the solar wind density at 1 AU.
6. Estimate the heliocentric distance where the solar wind flow speed is equal to the Alfvén speed.
7. Estimate the heliocentric distance of the termination shock.

Problem 12.2 Given an average solar wind, determine how much mass, momentum, and energy is lost by the Sun per unit time. If the solar wind were constant in time, how long would it take to lose all the solar mass at this rate?

Problem 12.3 A comet's ion tail is composed of cometary ions embedded in the solar wind. Consider an ion tail that points $4°$ away from the radial direction. What is the solar wind speed if the azimuthal component of the comet's orbital velocity is 30 km/s?

Problem 12.4 Assume that the solar wind speed is $u_{sw} = 400$ km/s.

1. Find the heliocentric distance where the average interplanetary magnetic field has wrapped itself around once (give the distance in units of AU).

2. What is the angle between the radius vector and the magnetic field vector at this point?

3. What is the magnitude of the magnetic field at this point if the source surface is located at $R_s = 10\,R_\odot$ and if the magnetic field magnitude at the source surface is 10^{-6} T?

4. Determine the number of times the magnetic field has wound around the Sun by a heliocentric distance of 100 AU.

Problem 12.5 In our derivations we assumed that the interplanetary magnetic field lines were equipotentials. This cannot be strictly true owing to the ambipolar electric field term in the generalized Ohm's law caused by the electron pressure gradient. Estimate the electric potential difference along an interplanetary magnetic field line between the solar corona and 1 AU.

Problem 12.6 For large heliocentric distances, the solar wind velocity increases as $\ln r$. Derive the radial dependence of the solar breeze velocity for $r \to \infty$.

Problem 12.7 At its typical velocity of 450 km/s, how long does it take the solar wind to arrive at Mercury, Venus, Earth, Mars, Jupiter, Saturn, Uranus, Neptune, and Pluto?

Chapter 13

Cosmic Rays and Energetic Particles

Understanding the fundamental properties of the energetic particle population in the heliosphere is very important for two reasons:

1. These particles represent considerable hazard for both humans and radiation-sensitive systems in space, because they can penetrate through large amounts of shielding material.
2. They carry information about the large-scale properties of the heliosphere and the galaxy.

High-energy cosmic ray particles carry a large amount of kinetic energy. The deposition of this energy can cause permanent effects in the material through which the cosmic ray particle passes. In the case of biological materials or miniature electronic circuits, these effects can be very serious. In order to provide adequate shielding for radiation-sensitive systems, we need to know the basic properties of the high-energy particle radiation, including its elemental composition, energy spectrum, and temporal variations.

A significant portion of our present knowledge about the global structure of the heliosphere comes from energetic particle observations. These particles travel through space at velocities considerably higher than the characteristic velocities of the local plasma population. Because the propagation of the energetic particles is greatly affected by various physical properties of the medium, energetic particles sample regions of the heliosphere and the galaxy that are currently not accessible to spacecraft.

High-energy particles observed in the heliosphere (outside of planetary magnetospheres) can be divided into three broad categories:

1. galactic cosmic rays coming from outside the solar system,
2. solar energetic particles produced by the Sun, and

3. heliospheric particles accelerated in the heliosphere by processes such as shock acceleration or ion pickup.

This chapter will briefly discuss the main properties of the various energetic particle populations and their relation to the heliosphere.

13.1 Galactic Cosmic Rays

Despite their name, cosmic rays are not a part of the electromagnetic spectrum. Galactic cosmic rays are composed of high-energy nuclei and electrons that propagate through the low-density interstellar medium. The origin of the cosmic radiation remains a matter of scientific debate. Cosmic rays below about 10^{15} eV are believed to originate from galactic sources, whereas very high energy cosmic rays ($> 10^{18}$ eV) are generally assumed to come from extragalactic sources. The flux of galactic cosmic rays (below about 10^{15} eV) is essentially isotropic.

13.1.1 Solar Cycle Modulation of Galactic Cosmic Rays

The bulk of galactic cosmic rays reaching the Earth have energies in the few hundred MeV to few GeV range (see Figure 13.1). The intensity of these galactic cosmic rays exhibits a variation with approximately 11 year periodicity as shown in Figure 13.2. This variation is inversely correlated with the solar activity (as measured by the sunspot number).

Consecutive solar cycles can result in markedly different cosmic ray modulation profiles. During the 1958–64 period, the recovery to maximum cosmic ray intensity was gradual, whereas during 1970–72 the recovery was rapid and was followed by a period of nearly constant cosmic ray intensity. These two kinds of modulation cycles can be found in the long-term cosmic ray modulation data sets, too. In effect, the cosmic ray intensity exhibits a 22-year cycle. This 22-year cycle can be explained in terms of cosmic ray drifts, which will be discussed later in this chapter.

Primary cosmic rays are, by definition, the cosmic radiation incident on the Earth's atmosphere. Cosmic radiation propagating through the atmosphere undergoes nuclear collisions and generates *secondary cosmic rays*. The secondary cosmic rays consist of all known nuclear and subnuclear particles.

The primary cosmic radiation observed at Earth's orbit consists of approximately 83% protons, 13% alpha particles, 1% heavier nuclei, and 3% energetic electrons. This composition is fairly universal from a few hundred MeV to over 10^{14} eV. Above $\sim 10^{14}$ eV the composition is uncertain owing to the absense of direct measurements. The differential energy spectrum of high-energy cosmic rays above ~ 1 GeV/nucleon is proportional to E^{-a}, where E is the kinetic energy per nucleon and a is the spectral index. The value of the spectral index is $a = 2.6$ below about 10^{15} eV and $a \approx 3$ above. Below ~ 1 GeV/nucleon the differential cosmic ray spectrum is not a power law, but

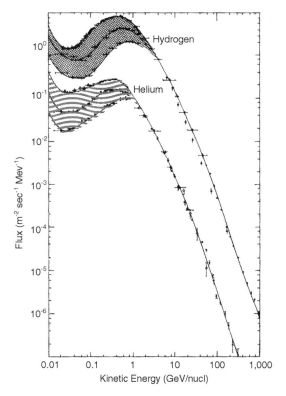

Figure 13.1 Primary cosmic ray differential energy
spectrum. The upper envelope indicates the solar
minimum spectrum while the lower envelope indicates
the spectra for solar maximum conditions. The shaded
areas indicate the range of the solar modulation over a
solar cycle. The hydrogen spectrum has been multiplied
by a factor of 5 for clarity. (From *Webber*.[1])

it exhibits a local maximum at a few hundred MeV/nucleon. Below the maximum the
differential energy spectrum decreases to a few tens of MeV/nucleon. This spectrum
for protons and alpha particles is shown in Figure 13.1. The low-energy cosmic ray
component (below a few GeV/nucleon) changes with time. This change is mainly due
to solar modulation, which will be discussed next.

It is interesting to note that solar modulation effects disappear above a few tens of
GeV/nucleon. This fact is not surprising if one considers the gyroradius of a 1 GeV
proton in the heliospheric magnetic field. Assuming an average field of 0.1 nT in the
outer heliosphere, the corresponding proton gyroradius is about an AU. This means
that the gyroradii of primary cosmic rays with energies of about a few tens of GeV
are comparable to the size of the heliosphere; consequently, heliospheric modulation
is not affecting these particles significantly.

[1]   Webber, W. R., "The energy spectra of the cosmic ray nucleonic and electronic components,"
      Proc. 13th International Cosmic Ray Conference, vol. 5, p. 3568, Denver, 1973.

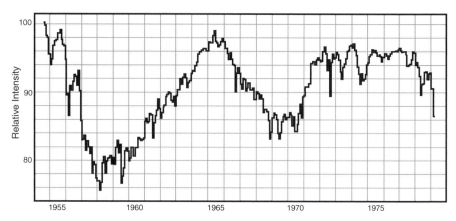

Figure 13.2 Monthly averages of the ∼1 GeV galactic cosmic ray intensity observed by the Mt. Washington neutron monitor from 1954 through 1979. (From *Fisk*.[2])

13.1.2 Diffusion Theory of Cosmic Ray Modulation

The original theory of the modulation of galactic cosmic rays was put forward by Parker (*1963*). Even though this original theory is incomplete (it did not consider the Compton–Getting effect and adiabatic energy loss), it is still important because it introduces the fundamental concepts of energetic particle propagation in the heliosphere. Here we briefly outline the basic elements of this theory. In the next section we will discuss the more complete theoretical description of cosmic ray modulation, which is based on a combination of guiding center drift and diffusion of cosmic ray particles and which is in better agreement with observations.

The basic concept of cosmic ray modulation is quite simple. The solar wind flows nearly radially outward from the Sun and it carries the heliospheric magnetic field. The large-scale structure of the heliospheric magnetic field follows the Parker spiral, but on smaller scales, the magnetic field has many irregularities, such as waves, turbulence, and discontinuities. On the average, the magnetic field irregularities are convected with the solar wind: In a first approximation their velocity is the same as the solar wind velocity.

The charged cosmic ray particles follow helical orbits as they gyrate around the large-scale heliospheric magnetic field. However, magnetic field irregularities result in a continuous series of small pitch-angle changes of these cosmic rays as they propagate in the heliosphere. The result is a diffusion of cosmic ray pitch angles on the magnetic field irregularities, and therefore the cosmic ray distribution function is driven toward isotropy in the frame of reference moving with the irregularities (this frame is practically the solar wind frame of reference). This effect is somewhat moderated by diffusion caused by density gradients.

[2] Fisk, L. A., "Solar modulation of galactic cosmic rays," in *Solar Terrestrial Physics*, R. L. Carovilano and J. M. Forbes, eds., p. 217, D. Reidel, Dordrecht, The Netherlands, 1983.

Since cosmic rays are convected outward with the solar wind plasma by pitch-angle diffusion, this process makes it difficult for the cosmic ray particles to "fight their way" into the inner heliosphere. Not all cosmic ray particles can successfully reach the Earth. Thus the cosmic ray intensity in the inner heliosphere is lower than it is in the interstellar medium.

With the help of this picture, one can obtain the governing equation for the distribution of cosmic rays in the heliosphere. The relevant equation has been discussed earlier in the book in connection with the transport of energetic and superthermal particles. Equation (7.21) describes the transport equation for the momentum space–averaged phase-space distribution function, F_0. This equation is the basis of cosmic ray theory.

For the sake of simplicity, let us neglect the direct effects of the large-scale magnetic field and the term describing momentum-space diffusion. The effect of pitch-angle scattering is taken into account by assuming a nearly isotropic distribution. Further simplification is achieved by assuming steady-state conditions and a spherically symmetric geometry (which is a good first approximation for cosmic rays). In this case Eq. (7.21) becomes

$$u_{\mathrm{sw}}\frac{\partial F_0}{\partial r} - \frac{p}{3}\frac{1}{r^2}\frac{\partial (r^2 u_{\mathrm{sw}})}{\partial r}\frac{\partial F_0}{\partial p} = \frac{1}{r^2}\frac{\partial}{\partial r}\left(r^2 \kappa_r \frac{\partial F_0}{\partial r}\right), \tag{13.1}$$

where u_{sw} is the radial solar wind velocity, κ_r is the radial diffusion coefficient, and $F_0 = F_0(t, r, p)$.

In general, Eq. (13.1) does not have a simple analytic solution. However, one can obtain an approximate solution, which can explain the 11-year modulation of galactic cosmic rays in the heliopshere. First, Eq. (13.1) can be rewritten in the following form:

$$r^2 u_{\mathrm{sw}}\frac{\partial F_0}{\partial r} + \frac{\partial (r^2 u_{\mathrm{sw}})}{\partial r}C F_0 = \frac{\partial}{\partial r}\left(r^2 \kappa_r \frac{\partial F_0}{\partial r}\right), \tag{13.2}$$

where C is the *Compton–Getting coefficient*, defined as

$$C = -\frac{p}{3F_0}\frac{\partial F_0}{\partial p}. \tag{13.3}$$

The Compton–Getting coefficient has been extensively studied in the 1970s. For galactic cosmic rays it is a slowly varying function of r and p. Its value increases from near zero at around 100 MeV to about 1.5 above 1 GeV. One can gain considerable physical insight by approximating the Compton–Getting coefficient with $C = 1$ throughout the modulation energy range. In this approximation Eq. (13.2) becomes

$$\frac{\partial}{\partial r}\left[r^2\left(u_{\mathrm{sw}} F_0 - \kappa_r \frac{\partial F_0}{\partial r}\right)\right] = 0. \tag{13.4}$$

This equation is Parker's well-known convection–diffusion equation (*Parker 1963*):

$$\frac{\partial F_0}{\partial r} = \frac{u_{\mathrm{sw}}}{\kappa_r} F_0. \tag{13.5}$$

The solution of this equation is

$$F_0(r, p) = F_\infty(p) \exp\left(-\int_r^{R_0} dr' \frac{u_{\text{sw}}}{\kappa_r}\right), \tag{13.6}$$

where $F_\infty(p)$ is the isotropic cosmic ray distribution function outside the solar system and R_0 is the radius of the modulation region (cosmic ray modulation boundary).

Equation (13.6) describes the modulation of cosmic rays. It is obvious that at any given location inside the heliosphere the cosmic ray intensity is controlled by the modulation parameter

$$\Phi_{\text{cr}} = \int_r^{R_0} dr' \frac{u_{\text{sw}}}{\kappa_r}, \tag{13.7}$$

which characterizes the efficiency of galactic cosmic rays to reach the inner regions of the heliosphere. Since the modulation parameter varies with the solar cycle, the cosmic ray intensity also exhibits a solar cycle variation.

13.1.3 Modulation Due to Diffusion and Cosmic Ray Drift

In the previous section we outlined the simplest model of cosmic ray modulation. More sophisticated theoretical descriptions also take into account the effects of the large-scale heliospheric magnetic field. The main difference between these models and the original Parker model is that in addition to a changing modulation parameter an interplay between the guiding center drift and diffusion of galactic cosmic rays is thought to cause the observed solar cycle modulation effect. Furthermore, heliospheric transients, such as coronal mass ejections, Forbush decreases, and corotating interaction regions, can also influence the cosmic ray intensity at Earth. Here we outline the basic concepts of the modern modulation theory without going into details of other effects.

The theoretical description of cosmic ray drift[3] is based on the more complete form of Eq. (7.21). Neglecting momentum-space diffusion, this equation can be written as

$$\frac{\partial F_0}{\partial t} + (\mathbf{u} \cdot \nabla)F_0 + (\mathbf{V}_{\text{d}} \cdot \nabla)F_0 - \frac{p}{3}(\nabla \cdot \mathbf{u})\frac{\partial F_0}{\partial p} = \nabla \cdot (\kappa \cdot \nabla F_0), \tag{13.8}$$

where the guiding-center drift velocity of cosmic rays is given by Eq. (7.22). It should be noted that $\mathbf{V}_{\text{d}}$ is divergenceless, so in itself it cannot cause cosmic ray modulation. However, in combination with density gradients in Eq. (13.8) the drift term can result in the observed modulation of the cosmic ray intensity.

---

[3] Jokipii, J. R., Levy, E. H., and Hubbal, W. B., "Effects of particle transport on cosmic rays, I. General properties, application to solar modulation," *Astrophys. J.*, **213**, 861, 1977.

Let us examine the cosmic ray guiding center drift velocity, $\mathbf{V}_d$. For the Archimedean spiral magnetic field lines given by Eq. (12.34) cosmic ray drift velocity can be written as

$$\mathbf{V}_d = \frac{p^2}{3mq} \nabla \times \left[\frac{2H(z) - 1}{B_{\mathrm{S_N}}} \frac{r^2}{R_s^2} \frac{u_{\mathrm{sw}}}{u_{\mathrm{sw}}^2 + (\boldsymbol{\Omega}_\odot \times \mathbf{r})^2} (\mathbf{u}_{\mathrm{sw}} + \boldsymbol{\Omega}_\odot \times \mathbf{r}) \right], \quad (13.9)$$

where $H(z)$ is the Heaviside step function and $B_{\mathrm{S_N}}$ is the source surface magnetic field for the northern hemisphere of the Sun. The factor $2H(z) - 1$ expresses the fact that the source surface field (and consequently the heliospheric magnetic field) changes sign at the heliospheric current sheet, located in the $z = 0$ plane. We also note that in the present model the solar wind velocity is assumed to be constant and radial: $\mathbf{u}_{\mathrm{sw}} = u_{\mathrm{sw}}\mathbf{e}_r$. The curl operator in Eq. (13.9) can be evaluated to obtain

$$\begin{aligned}
\mathbf{V}_d = {} & \frac{2p^2}{3m^2} \frac{A}{R_s\Omega_c} \frac{r^2}{R_s} \delta(z) \frac{u_{\mathrm{sw}}}{u_{\mathrm{sw}}^2 + (\boldsymbol{\Omega}_\odot \times \mathbf{r})^2} \left[\frac{u_{\mathrm{sw}}}{\Omega_\odot r} (\boldsymbol{\Omega}_\odot \times \mathbf{r}) + \Omega_\odot \mathbf{r} \right] \\
& + \frac{2p^2}{3m^2} \frac{A(2H(z)-1)}{R_s\Omega_c} \frac{u_{\mathrm{sw}}}{\left[u_{\mathrm{sw}}^2 + (\boldsymbol{\Omega}_\odot \times \mathbf{r})^2 \right]^2} \left\{ u_{\mathrm{sw}} \frac{r}{R_s} (\mathbf{r} \cdot \boldsymbol{\Omega}_\odot)(\boldsymbol{\Omega}_\odot \times \mathbf{r}) \right. \\
& \left. + \left[u_{\mathrm{sw}}^2 + (\boldsymbol{\Omega}_\odot \times \mathbf{r})^2 \right] (\boldsymbol{\Omega}_\odot \times \mathbf{r}) \times \frac{\mathbf{r}}{R_s} - u_{\mathrm{sw}}^2 \frac{r^2}{R_s} \boldsymbol{\Omega}_\odot \right\}, \quad (13.10)
\end{aligned}$$

where $\boldsymbol{\Omega}_\odot$ is the angular velocity vector of solar rotation, $\Omega_c = |qB_{\mathrm{S_N}}|/m$ is the gyroradius of a cosmic ray particle in the source surface magnetic field, and $A = \mathrm{Sign}(qB_{\mathrm{S_N}})$ characterizes the direction of the source surface magnetic field and the sign of the particle charge. If one uses typical solar wind parameters, the magnitude of the drift velocity for GeV particles is comparable to the solar wind velocity at 1 AU.

For positively charged particles the sign factor A is positive ($A = 1$) when the northern hemisphere interplanetary magnetic field is outward. This was the case for the 1975 sunspot minimum. For the 1965 and 1985 solar minima positive particles had $A = -1$ in the northern hemisphere.

The first term in expression (13.10) describes an azimuthal drift (which is not playing an important role) and a radial drift (outward for $A > 0$) in a narrow region near the current sheet. This term is a consequence of the discountinuous change in the source surface magnetic field at the magnetic equator (B_s changes sign). Note that inside the very narrow current sheet the magnetic field vanishes as it changes sign.

It is very instructive to calculate the field-aligned component of the cosmic ray drift velocity. This can be done by first recognizing that the unit vector along the magnetic field line (outside the current sheet) can be expressed as

$$\mathbf{b} = \frac{\mathbf{B}}{B} = \mathrm{Sign}(B_{\mathrm{S_N}}) \left[2H(z) - 1 \right] \frac{\mathbf{u}_{\mathrm{sw}} + \boldsymbol{\Omega}_\odot \times \mathbf{r}}{\sqrt{u_{\mathrm{sw}}^2 + (\boldsymbol{\Omega}_\odot \times \mathbf{r})^2}}. \quad (13.11)$$

Next, one can obtain the field-aligned component of the drift velocity (outside the current sheet) by taking the scalar product of the **b** and $\mathbf{V}_d$ vectors:

$$\mathbf{V}_{d\parallel} = \text{Sign}(q)\frac{2p^2}{3m^2}\frac{r}{R_s^2\Omega_c}u_{sw}^2(\mathbf{r}\cdot\Omega_\odot)\frac{(\Omega_\odot\times\mathbf{r})^2 - u_{sw}^2}{\left[u_{sw}^2 + (\Omega_\odot\times\mathbf{r})^2\right]^{5/2}}. \tag{13.12}$$

Equation (13.12) tells us quite a bit about the way galactic cosmic rays propagate inside the heliosphere. First of all, near the solar rotation axis (where $\Omega_\odot\times\mathbf{r}\approx\mathbf{0}$) Eq. (13.12) simplifies to the following expression:

$$\mathbf{V}_{d\parallel} = -\text{Sign}(q)\frac{2p^2}{3m^2}\frac{\Omega_\odot}{\Omega_c}\frac{1}{u_{sw}}\frac{r^2}{R_s^2}. \tag{13.13}$$

Thus above the solar pole positively charged galactic cosmic rays drift opposite to the magnetic field direction. Note that the parallel drift velocity is proportional to r^2 and therefore becomes quite large at large heliocentric distances above the solar pole. In plain English, galactic cosmic rays can enter (or leave) the heliosphere very efficiently above the solar poles.

For heliocentric distances $r \gg 1$ AU and away from the poles (at mid- and low heliolatitudes) the corotation velocity ($\Omega_\odot\times\mathbf{r}$) is much larger than the radial velocity. In these regions the field-aligned cosmic drift velocity becomes

$$\mathbf{V}_{d\parallel} = \text{Sign}(q)\frac{2p^2}{3m^2}\frac{u_{sw}^2}{R_s^2\Omega_c\Omega_\odot}\frac{1}{\Omega_\odot r}\frac{\cos\Theta}{\left|\sin^3\Theta\right|}, \tag{13.14}$$

where Θ is the polar angle measured from the solar rotation axis. Equation (13.14) shows that away from the poles the parallel cosmic ray drift velocity decreases with heliocentric distance and therefore it does not play a major role in cosmic ray propagation.

Our results indicate that there are three regions where parallel cosmic ray drift plays an important role in the heliospheric entry (or escape) and propagation of galactic cosmic rays: The current sheet (where the delta function term in Eq. 13.10 plays a role) and the regions above the two solar poles (where the drift velocity increases with heliocentric distance).

Our results mean that during the 1975 solar minimum ($A > 0$ for the northern hemisphere and $A < 0$ for the southern hemisphere) the drift velocity of cosmic ray ions points toward the sun above both poles. During this period, galactic cosmic rays can preferentially move into the heliosphere at the polar regions of the heliosphere. These cosmic rays eventually diffuse out of the heliosphere near the heliospheric current sheet (near the equator) where the drift velocity has a significant outward pointing radial component.

During the 1965 and 1985 solar activity minima, the quantity A was negative for the northern and positive for the southern hemisphere. In this case the cosmic ray drift velocity points outward above both polar regions. Consequently, galactic cosmic ray particles diffuse inward near the heliospheric current sheet and preferentially leave the heliosphere in the two polar regions.

The behavior of cosmic ray drift discussed above explains the differences between even and odd solar activity cycles. During the 1975 solar minimum, the entry of galactic cosmic rays was "helped" by the inward pointing guiding-center drift in the polar regions of the heliosphere, resulting in a fast rise of the cosmic ray intensity at 1 AU, followed by a long flat period. During the 1965 and 1985 activity minima the drift pointed outward, making the heliospheric transport of cosmic rays more difficult in the polar regions. The primary entry mechanism during these periods was inward radial diffusion along the heliospheric current sheet. This diffusive process is quite slow, and it results in a long gradual increase of the cosmic ray intensity at 1 AU.

13.2   Solar Energetic Particles

When the Sun explosively releases a huge amount of energy in the solar atmosphere, it also produces *solar energetic particles*, the so-called *solar cosmic rays*. These particles have energies ranging from a few keV to several GeV. They are primarily protons and electrons, but heavier nuclei can also be found in solar cosmic rays. In this section we will concentrate our attention on the propagation of solar cosmic rays in the interplanetary medium.

Most large and complex solar energetic particle events are accelerated by CME-related shocks in the corona and in the interplanetary space near the Sun. Impulsive solar particle events, in contrast, are accelerated by solar flares. In this section we consider energetic particle propagation in impulsive events.

Following their acceleration in solar flares, energetic particles leave the solar corona and propagate outward in the solar wind. Flare-associated energetic particle events usually begin with an optical flare. There are nearly simultaneous microwave and X-ray bursts followed by longer duration type IV radio emissions. The heavier ($Z \geq 2$) nuclei in the flare-generated solar energetic particle population represent a characteristic sample of the lower corona, whereas the proton to alpha ratio varies greatly from event to event. In general the rise to maximum intensity is about 10 times the radial transit time (v/r). The decay time is typically an order of magnitude longer than the rise time. The observations near 1 AU show that in the early phase of the flare particle event the particles primarily propagate outward along the interplanetary magnetic field line, with very few particles traveling backward (toward the Sun). As time progresses this initial directional anisotropy decays and at later time the particle distribution function approaches isotropy. A typical flare-generated event is shown in Figure 13.3. The plot shows the time evolution of the intensity of > 20 MeV protons.

Modeling flare-generated particle events must account for the above listed typical properties. However, several simplifying assumptions can make the models manageable. First, because the acceleration of energetic particles takes place on a very short time-scale compared to the duration of the event, the acceleration process can be considered impulsive. A second simplification follows from the recognition that in the interplanetary medium the gyroradius of flare-generated energetic particles is small

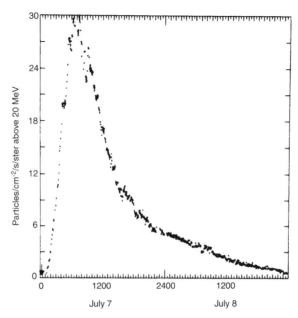

Figure 13.3 A typical flare particle event observed in 1966 by
the *IMP III* spacecraft. (From *Fichtel & McDonald*.[4])

compared to 1 AU (even the most energetic flare particles have gyroradii $<10^6$ km).
This means that in a reasonable approximation flare particles propagate only along
interplanetary field lines, and particle transport across magnetic field lines can be
neglected.

Following their injection into the interplanetary medium the energetic particles
move along interplanetary magnetic field lines. They are also pitch-angle scattered
on magnetic field irregularities moving with the solar wind. Since the distribution of
flare particles is not close to isotropy (there are large directional anisotropies during
the early phase of the flare-generated event), Eq. (7.21) cannot be applied. Instead we
start from the Boltzmann equation and make the following simplifying assumptions:

- The solar wind velocity is small compared to the energetic particle velocity;
  therefore all terms containing **u** can be neglected.

- The energetic particles propagate only along the interplanetary magnetic field
  line (perpendicular transport is neglected).

- The distribution function is isotropic in gyrophase (it does not depend on the
  gyration angle).

- The energy change of flare particles is neglected (only pitch-angle scattering is
  taken into account).

This approximation is basically the same as the one discussed with respect to the
transport of superthermal particles (see Eq. 7.37). Replacing the velocity with p/m in

⁴ Fichtel, C. E., and McDonald, F. B., "Energetic particles from the Sun," *Annu. Rev. Astron.
  Astrophys.*, **5**, 351, 1967.

Eq. (7.37) and neglecting the field-aligned electric field yields the following equation for solar energetic particles: Equation (7.5) becomes

$$\frac{\partial F}{\partial t} + \frac{p\mu}{m}\frac{\partial F}{\partial s} - \frac{p}{m}\frac{1-\mu^2}{2B}\frac{dB}{ds}\frac{\partial F}{\partial \mu} = \frac{F_0 - F}{\tau_c}. \tag{13.15}$$

Here $\mu = \cos\Theta$ is the cosine of the pitch angle, s is distance along the field line, and $F_0(t, s, p, \mu)$ is the gyro phase-averaged distribution function.

In spite of its apparent simplicity Eq. (13.15) is still not easy to solve for an impulsive source. However, much physical insight can be gained by using a highly simplified model of pitch-angle scattering. It is assumed that energetic flare particles can have only two discrete values of pitch angle, either $\Theta = 0$ ($\mu = 1$) or $\Theta = \pi$ ($\mu = -1$). In other words, the perpendicular velocity component of the particles is completely ignored. A scattering flips the pitch angle and the particle reverses direction. For complicated reasons that go beyond the scope of the book, this model seems to describe flare particle events surprisingly well. A big advantage of this simple model is that it not only emphasizes the fundamental underlying physics, but it also leads to closed-form analytic solutions. This model was first suggested in the late 1960s by Fisk and Axford.[5]

Let $F_{\pm}(t, s, p)$ denote the distribution functions of forward and backward propagating particles with momentum p. In this case τ_c is the mean collision time for backscatter. The solid angle–averaged distribution function is simply $F_0 = (F_+ + F_-)/2$. In this case Eq. (13.15) can be written as two coupled equations for $F_{\pm}$:

$$\begin{aligned} \frac{\partial F_+}{\partial t} + \frac{p}{m}\frac{\partial F_+}{\partial s} &= -\frac{F_+ - F_-}{2\tau_c}, \\ \frac{\partial F_-}{\partial t} - \frac{p}{m}\frac{\partial F_-}{\partial s} &= -\frac{F_- - F_+}{2\tau_c}. \end{aligned} \tag{13.16}$$

We note that in this simple case the focusing terms give no contribution.

One can introduce the local particle density and the flux function by

$$\begin{aligned} n &= \frac{B(s)}{B_0}\frac{F_+ + F_-}{2}, \\ S &= \frac{p}{m}\frac{B(s)}{B_0}\frac{F_+ - F_-}{2}, \end{aligned} \tag{13.17}$$

where $B(s)$ and B_0 are the magnetic field magnitude at distance s and at the injection point, respectively. Next we add and subtract the two equations for $F_{\pm}$ to obtain the following for n and S:

$$\begin{aligned} \frac{\partial}{\partial t}\left(\frac{B_0}{B}n\right) + \frac{\partial}{\partial s}\left(\frac{B_0}{B}S\right) &= 0, \\ \frac{\partial}{\partial t}\left(\frac{B_0}{B}S\right) + \frac{p^2}{m^2}\frac{\partial}{\partial s}\left(\frac{B_0}{B}n\right) &= -\frac{1}{\tau_c}\left(\frac{B_0}{B}S\right). \end{aligned} \tag{13.18}$$

[5] Fisk, L. A., and Axford, W. I., "Anisotropies of solar cosmic rays," *Solar Physics*, **7**, 486–498, 1969.

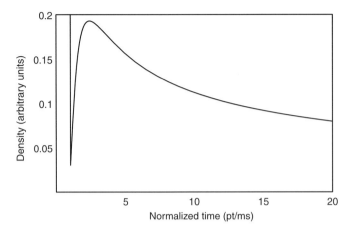

Figure 13.4 Temporal evolution of flare-generated energetic particle density.

One can differentiate the second equation with respect to s and eliminate S using the first equation:

$$\left(\frac{\partial^2}{\partial t^2} + \frac{1}{\tau_c} \frac{\partial}{\partial t} - \frac{p^2}{m^2} \frac{\partial^2}{\partial s^2} \right) \left(\frac{B_0}{B} n \right) = 0. \tag{13.19}$$

Equation (13.19) is the *telegraph equation* for the energetic particle density. This equation is well known in signal propagation theory: It describes the propagation of a finite signal moving in a dispersive medium.

For a well-connected field line (when coronal propagation is negligible) the particles are impulsively injected at $t = 0$ near the Sun ($s = 0$). The solution of Eq. (13.19) then becomes

$$n(t, s, p) = A \frac{m}{p} \frac{B(s)}{B_0} \exp\left(-\frac{m}{2p\tau_c} s \right) \delta\left(t - \frac{m}{p} s \right)$$

$$+ A \frac{1}{2\tau_c} \frac{m}{p} \frac{B(s)}{B_0} \exp\left(-\frac{t}{2\tau_c} \right) \left[I_0(\varphi) + \frac{t}{2\tau_c\varphi} I_1(\varphi) \right] H\left(t - \frac{m}{p} s \right), \tag{13.20}$$

where the normalization constant, A, is proportional to the total number of particles released by the flare with momentum p. Here $I_n(\varphi)$ is the modified Bessel function of the first kind and order n, $H(t)$ is the Heaviside step function, and the dimensionless variable, φ, is defined as

$$\varphi = \frac{\sqrt{t^2 - \frac{m^2}{p^2} s^2}}{2\tau_c}. \tag{13.21}$$

The solution of Eq. (13.20) is shown in Figure 13.4.

We note that the number density contains an initial pulse (i.e., the delta function component), propagating at the particle speed, p/m, and decaying with distance along the field line. The pulse contains those particles that at time t have escaped scattering since the initial release from the flare. Scattered particles make up the rest of the solution. For large time ($t \gg \tau_c$ and $t \gg sm/p$) the solution approaches the asymptotic form

$$n(t, s, p) \approx A \frac{B(s)}{B_0} \frac{1}{\sqrt{\pi \kappa t}} \exp\left(-\frac{s^2}{4\kappa t}\right), \qquad (13.22)$$

where the diffusion coefficient is defined as $\kappa = \tau_c p^2 / m^2$. This diffusive solution can be obtained by neglecting the second-order time derivative in the telegraph equation. At late time this is quite a reasonable solution. Interestingly, the flare particle density profile given by Eq. (13.20) satisfies causality for early times and becomes a diffusive solution for late times.

13.3   Interstellar Pickup Particles and Anomalous Cosmic Rays

13.3.1   Interstellar Pickup Ions

The heliosphere is moving with respect to the very local interstellar medium (VLISM) with a relative velocity of about 26 km/s. Neutral atoms in the interstellar medium penetrate relatively freely into the heliosphere. Since the interstellar neutral gas is relatively cold ($<10^4$ K), both the thermal speed and the bulk speed of the interstellar neutrals are negligible with respect to the solar wind speed (at least inside the termination shock). Hence, in a good approximation the relative speed between the interstellar neutrals and the magnetized solar wind plasma is simply the solar wind speed, u_{sw}.

Interstellar neutral particles can be ionized by photoionization, charge exchange, or electron impact. Inside the termination shock the newly created ions are practically at rest with respect to the supersonic magnetized solar wind. These freshly born ions are accelerated by the motional electric field of the high-speed solar wind flow. The initial ion trajectory is cycloidal, resulting from the superposition of gyration and $\mathbf{E} \times \mathbf{B}$ drift. The velocity-space distribution at this early stage of the pickup process is a ring-beam distribution, as depicted in Figure 13.5, which shows the pickup geometry in velocity space (the coordinate system is at rest with respect to the Sun). It can be seen in Figure 13.5 that newly created ions are initially at rest (point O), and they start gyrating around the magnetic field (the IMF direction is shown by the dashed line). The projection of the ring-beam is denoted by the OO' line. We see from the figure that the gyration speed of the ring is $v_\perp = u_{sw} \sin \alpha$, where α is the angle between the solar wind velocity and magnetic field vectors. It can also be seen that the ring moves with a velocity of $v_\parallel = u_{sw} \cos \alpha$ toward the Sun along the direction of the magnetic field.

The ring-beam distribution of new pickup ions represents an ion beam moving through the solar wind plasma with a velocity of $v_\parallel = u_{sw} \cos \alpha$. This motion generates

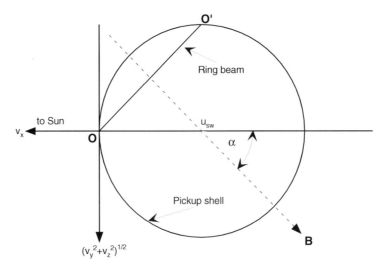

Figure 13.5 Pickup geometry of interstellar particles.

a broad spectrum of magnetohydrodynamic turbulence, which propagates with the Alfvén speed along the magnetic field lines. In the supersonic, superalfvénic solar wind the frequency of the excited waves is much smaller than the cyclotron frequency of the pickup ions. Because the Alfvén speed is much smaller than the solar wind speed, in a first approximation the MHD turbulence moves together with the solar wind plasma. In the frame of reference moving with the turbulence (which is approximately the solar wind frame) the pickup ions interact with the self-generated MHD turbulence without significantly changing their energy. As a result of this process the pitch angles of the pickup ring particles become randomized in the solar wind frame, and they become distributed on the *pickup shell* around the local solar wind velocity. In velocity space the pickup shell is centered at u_{sw} and it has a radius of u_{sw} (see Figure 13.5). The relatively rapid isotropization of the pickup particles on the pickup shell is followed by much slower processes changing the particle energy. Some of these acceleration processes will be discussed later in this chapter.

This model of interstellar ion pickup predicts that a spacecraft (which is nearly at rest in the coordinate system shown in Figure 13.5) observes nonaccelerated pickup ions with velocities between $v = 0$ and $v = 2u_{sw}$. This prediction was clearly confirmed by observations made by the *AMPTE* and *Ulysses* spacecraft. Figure 13.6 shows the velocity distribution of pickup protons as measured by *Ulysses*. At low energies the ambient solar wind dominates the distribution, but above approximately $v = 1.75u_{sw}$ the pickup ion distribution dominates. Inspection of Figure 13.6 reveals that at around $v = 2u_{sw}$ there is a clear break in the distribution, indicating the presence of pickup ions. The high-velocity tail of the distribution is due to accelerated pickup ions, which will be discussed later.

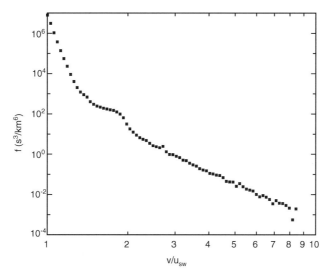

Figure 13.6 Pickup proton velocity distribution as measured by the *Ulysses* spacecraft. (From *Schwadron et al.*[6])

13.3.2   Anomalous Cosmic Rays

In the early 1970s spaceborn cosmic ray detectors revealed the existence of an unexpected low-energy cosmic ray component. Subsequent investigations confirmed the existence of enhanced fluxes of singly ionized hydrogen, helium, nitrogen, oxygen, neon, and carbon with energies from about 1 MeV/nucleon to 30 MeV/nucleon. The observed radial intensity gradient is positive as far as we have measurements (about 60 AU at present), indicating that this cosmic ray component is not of solar origin. These cosmic rays were termed *anomalous cosmic rays*.

The spectrum of anomalous cosmic rays measured during the 1976–77 solar minimum is shown in Figure 13.7. The anomalous cosmic ray component appears between 1 and 30 MeV/nucleon, and it is characterized by large overabundance of helium and oxygen.

The origin of these anomalous cosmic rays remains an area of active research. There is a consensus in the cosmic ray community that the anomalous cosmic rays are accelerated interstellar pickup ions as originally proposed by Fisk, Kozlovsky, and Ramaty.[7]

The open question is how the interstellar pickup ions are accelerated to high energies. At pickup these ions have an energy of about a keV/nucleon in the solar wind frame. This "seed" energy must be increased by about a factor of 10^4 to produce the

---

[6] Schwadron, N. A., Fisk, L. A., and Gloeckler, G., "Statistical acceleration of interstellar pickup ions in corotating interaction regions," *J. Geophys. Res.*, **23**, 2871–2874, 1996.

[7] Fisk, L. A., Kozlovsky, B., and Ramaty, R., "An interpretation of the observed oxygen and nitrogen enhancements in low energy cosmic rays," *Astrophys. J. Lett.*, **190**, L35, 1974.

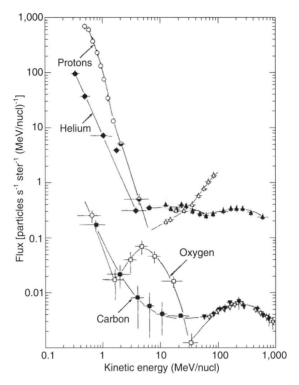

Figure 13.7 The anomalous cosmic ray energy spectrum during the 1976–77 solar minimum. (From *Gloeckler*.[8])

observed anomalous cosmic rays. Several possible mechanisms were proposed over the past twenty years or so, including acceleration at the heliospheric termination shock[9] and acceleration by interplanetary corotating interaction regions. At this point the question of acceleration is still not entirely resolved.

13.4   Energetic Particle Acceleration in the Heliosphere

There are two principal acceleration mechanisms acting in the heliosphere: stochastic acceleration and shock acceleration. In this section we will examine the basic theory of these two acceleration mechanisms.

Stochastic acceleration plays a fundamental role in increasing the velocities of interstellar pickup ions by about a factor of 5 to 10. This preaccelerated pickup ion population is further energized by the forward and reverse shock pairs bounding corotating interaction regions (CIRs). The energization of pickup ions by CIRs in the outer

[8] Gloeckler, G., "Composition of energetic particle populations in interplanetary space," *Rev. Geophys. Space Phys.*, **17**, 569, 1979.

[9] Jokipii, J. R., "Particle acceleration at the termination shock, I. Application to the solar wind and the anomalous component," *J. Geophys. Res.*, **91**, 2929, 1986.

heliosphere (beyond about the orbit of Jupiter) together with later acceleration at the solar wind termination shock is thought to be the main mechanism producing the observed anomalous cosmic ray population.

13.4.1 Stochastic Acceleration

The interplanetary medium is never totally quiet, but it is filled with various types of plasma waves and turbulence. The low-frequency component of magnetohydrodynamic waves play an essential role in the acceleration of pickup ions via the *stochastic acceleration* or *second-order Fermi acceleration* process.

The heliosphere is filled with low-frequency magnetohydrodynamic waves propagating along magnetic field lines (these waves are sometimes called Alvén waves, even though strictly speaking they are not Alfvén waves). These noncompressive waves propagate near the Alfvén speed. In the plasma frame these waves have frequencies near the pickup ion gyrofrequency, and therefore very efficient wave–particle interaction can take place between the waves and pickup ions.

The physical idea behind stochastic acceleration is quite simple. Due to the resonant interaction between low-frequency Alfvén waves (this imprecise expression is often used in the literature) propagating in one direction along the magnetic field line (sunward or antisunward) and the pickup ion population, the particle distribution function is driven toward isotropy in the frame of reference moving with the scattering centers (the particle energy is conserved in the scattering frame). Since we have two groups of scattering centers moving with $\pm V_A$ with respect to the plasma frame, an individual resonant interaction conserves the particle energy in a frame moving with $\pm V_A$ with respect to the plasma. In the plasma frame, however, the particle randomly gains or loses energy, depending on the directions of the particle and the Alfvén wave. This random energy diffusion process leads to stochastic acceleration (and deceleration) of the resonant particles.

This acceleration mechanism can be mathematically demonstrated by considering an appropriate approximation of the transport equation for the momentum space–averaged phase-space distribution function, F_0. We again start from Eq. (7.21). For the sake of simplicity, let us neglect the direct effects of the large-scale magnetic field and the term describing spatial diffusion. The effect of pitch-angle scattering is taken into account by assuming a nearly isotropic distribution. We also add an isotropic source function describing newly ionized particles on a pickup shell. Equation (7.21) then becomes the following:

$$\frac{\partial F_0}{\partial t} + (\mathbf{u} \cdot \nabla) F_0 - \frac{p}{3} (\nabla \cdot \mathbf{u}) \frac{\partial F_0}{\partial p} = \frac{1}{p^2} \frac{\partial}{\partial p} \left(p^2 D_p \frac{\partial F_0}{\partial p} \right) + Q, \qquad (13.23)$$

where $Q(\mathbf{r}, p)$ is an isotropic source function.

Equation (13.23) can be simplified by assuming a spherically symmetric solar wind configuration with a constant radial solar wind velocity, u. In this case Eq. (13.23) becomes

$$u\frac{\partial F_0}{\partial r} = \frac{2u}{3r}p\frac{\partial F_0}{\partial p} + \frac{1}{p^2}\frac{\partial}{\partial p}\left(p^2 D_p \frac{\partial F_0}{\partial p}\right) + Q.$$

(13.24)

Equation (13.23) was solved for a simple, but quite instructive scenario by Fisk.[10] Fisk considered the situation when the momentum dependence of the diffusion coefficient can be written as $D_p = D_0 p^\beta$, where D_0 is a constant, independent of both r and p. He made the additional simplification of assuming that particles are injected at a position r_0 with initial momentum p_0, and he determined the solution at a location, r, where $r \gg r_0$. If β has values between 0 and 2, the solution is

$$F_0(r, p) \propto \frac{1}{r^{(7-2\beta)}}\exp\left(-\frac{1}{3}\frac{7-2\beta}{(2-\beta)^2}\frac{u}{D_0 r}p^{(2-\beta)}\right).$$

(13.25)

Typical values of β are in the range of 0 to 0.5, so solution (13.25) is realistic for stochastic acceleration.

This solution indicates that the momentum (and energy) spectrum of particles energized by stochastic acceleration is an exponential. This is a fairly general result. As we will see next, the energy spectrum due to shock acceleration follows a power law. Therefore, the observed energy spectrum can help us to identify the actual acceleration mechanism.

13.4.2 Shock Acceleration

Shocks occuring in a scattering medium provide the necessary ingredients for systematic acceleration of superthermal charged particles. The shock separates two regions of space with different bulk velocities. We consider shocks where the velocity difference between the upstream and downstream flows is high compared to the Alfvén speed. In this case the scattering turbulence ("scattering centers") can be considered as moving with the bulk flow. Because the scattering centers are assumed to move with the bulk of the plasma, the velocity of the scattering centers is also different in the two regions. In this situation energetic particles can be scattered back to the shock from both regions, and they can be energized by multiple shock crossings. The basic idea is that since the downstream plasma is faster than the plasma upstream of the shock, the energetic particles are, in effect, bouncing between two converging "walls," thus continuously gaining energy (as long as they repeatedly keep crossing the shock). This process is called the *first-order Fermi acceleration* mechanism.

With the help of this picture, one can obtain the governing equation for the shock acceleration of superthermal particles. We again start from Eq. (7.21), which describes the transport equation for the momentum space–averaged phase-space distribution function, F_0.

---

[10] Fisk, L. A., "The acceleration of energetic particles in the interplanetary medium by transit time damping," *J. Geophys. Res.*, **81**, 4633, 1976.

For the sake of simplicity, let us neglect the direct effects of the large-scale magnetic field and the term describing momentum-space diffusion. The effect of pitch-angle scattering is taken into account by assuming a nearly isotropic distribution. The momentum gain due to shock crossings is taken into account by a $Q(x, p)$ source function. Further simplification is achieved by assuming steady-state conditions and a planar shock geometry (with the x direction normal to the shock). In this case Eq. (7.21) becomes the following:

$$\frac{\partial F_0}{\partial t} + u\frac{\partial F_0}{\partial x} - \frac{p}{3}\frac{\partial u}{\partial x}\frac{\partial F_0}{\partial p} = \frac{\partial}{\partial x}\left(\kappa\frac{\partial F_0}{\partial x}\right) + Q, \tag{13.26}$$

where u is the bulk speed of the scattering medium perpendicular to the shock, $\kappa(x, p)$ is the diffusion coefficient, and $Q(x, p)$ represents particle sources and sinks as well as nonadiabatic energy changes. Equation (13.26) can also be written as

$$\frac{\partial F_0}{\partial t} + \frac{\partial}{\partial x}\left(uF_0 - \kappa\frac{\partial F_0}{\partial x}\right) - \frac{\partial u}{\partial x}\frac{1}{p^2}\frac{\partial}{\partial p}\left(\frac{p^3}{3}F_0\right) = Q. \tag{13.27}$$

We consider the situation in which a shock wave is situated at $x = 0$. The flow velocity in the downstream and upstream directions is u_1 and u_2, respectively, with $u_1 > u_2$. We are looking for steady-state solutions and assume that $F_0 \to F_{0_1}(p)$ as $x \to -\infty$ and that F_0 remains finite as $x \to +\infty$ ($F_0 \to F_{0_2}(p)$). Also, we assume that there is a source at $x = 0$ corresponding to nonadiabatic particle acceleration at the shock, $Q(x, p) = Q_0(p)\delta(x)$. Then the derivative of the flow velocity becomes a delta function and Eq. (13.27) can be written as

$$\frac{\partial}{\partial x}\left(uF_0 - \kappa\frac{\partial F_0}{\partial x}\right) = \left[Q_0 - \frac{u_1 - u_2}{3}\frac{1}{p^2}\frac{\partial}{\partial p}(p^3F_0)\right]\delta(x). \tag{13.28}$$

Because of the delta function in Eq. (13.28), the solution for $x < 0$ will differ from that for $x > 0$. For the upstream region ($x < 0$) Eq. (13.28) can be integrated from $x' = -\infty$ to $x' = x$ to give

$$\frac{\partial F_0(x, p)}{\partial x} = \frac{u_1}{\kappa(x, p)}[F_0(x, p) - F_{0_1}]. \tag{13.29}$$

This equation can be integrated for the $x < 0$ region, provided that

$$\lim_{x\to-\infty}\int_0^x dx'\frac{u_1}{\kappa(x', p)} = 0. \tag{13.30}$$

If condition (13.30) is satisfied, the solution of Eq. (13.30) for $x < 0$ values is the following:

$$F_0(x, p) = F_{0_1} + C_0\exp\left(\int_0^x dx'\frac{u_1}{\kappa(x', p)}\right), \tag{13.31}$$

where C_0 is an integration constant (its value will be determined later).

For the downstream region ($x > 0$), the delta function makes a contribution. Let us integrate Eq. (13.28) from $x' = -\infty$ to $x' = x > 0$:

$$\kappa \frac{\partial F_0}{\partial x} = u_2 F_0 - u_1 F_{0_1} + \frac{u_1 - u_2}{3} \frac{1}{p^2} \frac{\partial}{\partial p} \left(p^3 F_{0_2} \right) - Q_0. \tag{13.32}$$

For $x > 0$ values, this equation yields unbounded (infinite) solutions, except when $F_0 = const$. This means that for $x > 0$ the solution is $F_0(x, p) = F_{0_2}(p)$. Furthermore, the following relation between F_{0_1} and F_{0_2} must be satisfied:

$$u_2 F_{0_2} - u_1 F_{0_1} + \frac{u_1 - u_2}{3} \frac{1}{p^2} \frac{\partial}{\partial p} \left(p^3 F_{0_2} \right) - Q_0 = 0. \tag{13.33}$$

Note that Eq. (13.33) is a differential equation for the downstream momentum spectrum:

$$\frac{\partial}{\partial p} \left[p^{\left(\frac{3u_1}{u_1 - u_2} \right)} F_{0_2} \right] = \left(\frac{3u_1}{u_1 - u_2} \right) p^2 \left(F_{0_1} + \frac{Q_0}{u_1} \right). \tag{13.34}$$

It is reasonable to assume that there are no accelerated energetic particles far upstream of the shock. Thus $F_{0_1} = 0$. One can also assume that the "injection" (acceleration) at the shock adds an average momentum to the accelerated particles, p_0. In this approximation one can write

$$Q_0(p) = \hat{Q}_0 p_0^{\left(\frac{3u_1}{u_1 - u_2} \right)} \delta(p - p_0), \tag{13.35}$$

and the solution of Eq. (13.34) is

$$F_{0_2}(p) = \frac{3}{u_1 - u_2} \hat{Q}_0 \left(\frac{p}{p_0} \right)^{-\left(\frac{3u_1}{u_1 - u_2} \right)} H(p - p_0), \tag{13.36}$$

where H is again the Heaviside step function. For strong shocks, $u_1/u_2 = 4$, and therefore one obtains an accelerated momentum spectrum of

$$F_{0_2}(p) \propto \left(\frac{p}{p_0} \right)^{-4}. \tag{13.37}$$

This is a very important and surprisingly general result: The momentum spectrum of particles accelerated by strong shocks is proportional to p^{-4}.

Finally, we can determine the constant, C_0, that appears in Eq. (13.31). The value of C_0 can be obtained by requiring that the distribution function remain continuous at $x = 0$. This means that

$$C_0 = F_{0_2} - F_{0_1}. \tag{13.38}$$

In summary, the distribution function of shock accelerated particles is

$$
F_0(x, p) = \begin{cases} \dfrac{3}{u_1 - u_2} \hat{Q}_0 \left(\dfrac{p}{p_0} \right)^{-\left(\frac{3u_1}{u_1 - u_2} \right)} \exp \left(\int\limits_0^x dx' \dfrac{mu_1}{\kappa(x',p)} \right), & x < 0, \\[4mm] \dfrac{3}{u_1 - u_2} \hat{Q}_0 \left(\dfrac{p}{p_0} \right)^{-\left(\frac{3u_1}{u_1 - u_2} \right)}, & x > 0. \end{cases} \tag{13.39}
$$

It is interesting to note that in planar type infinite geometries losses at free escape boundaries, cooling, or some other loss processes result in exponential cutoffs. The power law type spectrum, consequently, remains valid in the immediate vicinity of such boundaries, irrespective of the details of the solution.

13.5   Problems

Problem 13.1 Assume that the spectrum of galactic cosmic rays is proportional to $T^{-\beta}$, where $T = E - mc^2$ is the relativistic kinetic energy and $\beta > 0$ is the spectral index. Calculate the Compton–Getting coefficient for this distribution. Express the result with the help of the

$$
\alpha = \frac{T + 2mc^2}{T + mc^2}
$$

parameter.

Problem 13.2 Calculate the maximum energy (in keV) of potassium and sodium ions picked up near the Moon if the solar wind velocity is 450 km/s. What is the ratio of the gyroradii of these pickup ions to the radius of the Moon if the magnetic field strength is 5 nT?

Problem 13.3 Newly born interstellar pickup ions are picked up at the solar wind velocity. As a result of this process the source term in the Boltzmann equation due to ion pickup is

$$
\frac{\delta F}{\delta t} = \frac{n_0}{\tau_i} \delta^3 (\mathbf{v} - \mathbf{u}_{sw}),
$$

where τ_i is the ionization lifetime of neutrals, $\mathbf{u}_{sw}$ is the local solar wind velocity, and n_0 is the density of interstellar neutral particles.

1. What is the rate of change of the plasma pressure due to particle pickup?
2. Assume that the ambient solar wind density is 10 cm^{-3} and that the thermal velocity is $v_{th} = u_{sw}/10$. How many pickup ions are needed per unit volume to double the ambient solar wind pressure?

(*Hint*: Start from Eq. 4.7.)

Chapter 14

The Terrestrial Magnetosphere

The terrestrial magnetosphere comprises the region of space where the properties of naturally occurring ionized gases are controlled by the presence of Earth's magnetic field. This very broad definition means that the terrestrial magnetosphere extends from the bottom of the ionosphere to more than ten Earth radii (R_e) in the sunward direction and to several hundred R_e in the antisunward direction.

The magnetosphere is formed as a result of the interaction of the supersonic, superalfvénic, magnetized solar wind with the intrinsic magnetic field of the Earth. To understand this interaction, we first briefly discuss the main characteristics of the intrinsic terrestrial magnetic field and then turn our attention to the interaction between this intrinsic field and the solar wind.

14.1   The Intrinsic Magnetic Field

A couple of hundred years ago Gauss showed that the magnetic field at the surface of the Earth can be described as the gradient of a scalar potential. In general the near Earth magnetic field can be expressed as

$$\mathbf{B} = -\nabla \Phi_{\text{tot}} = -\nabla (\Phi_{\text{int}} + \Phi_{\text{ext}}), \tag{14.1}$$

where Φ_{int} and Φ_{ext} represent scalar potentials due to intrinsic and external sources, and Φ_{tot} is the sum of these two potentials (describing the total geomagnetic field). In general, planetary magnetic potentials are expressed as an infinite series using associated Legendre polynomials:

$$\Phi_{\text{int}} = R_e \sum_{n=1}^{\infty} \sum_{m=0}^{n} \left(\frac{r}{R_e} \right)^{-(n+1)} P_n^m (\cos \theta) \left[g_n^m \cos(m\phi) + h_n^m \sin(m\phi) \right] \tag{14.2}$$

and

$$\Phi_{\text{ext}} = R_{\text{e}} \sum_{n=1}^{\infty} \sum_{m=0}^{n} \left(\frac{r}{R_{\text{e}}}\right)^n P_n^m(\cos\theta) \left[G_n^m \cos(m\phi) + H_n^m \sin(m\phi)\right], \quad (14.3)$$

where θ and ϕ are geographic colatitude and east longitude, respectively. The coefficients g_n^m, G_n^m, h_n^m, and H_n^m are empirical coefficients determined with the help of ground-based and spaceborne measurements. Finally, the functions $P_n^m(\cos\theta)$ are associated Legendre polynomials with Schmidt normalization:

$$P_n^m(\cos\theta) = \begin{cases} P_n(\cos\theta), & m = 0, \\ \sqrt{\frac{2(n-m)!}{(n+m)!}} (1 - \cos^2\theta)^{m/2} \frac{d^m P_n(\cos\theta)}{d(\cos\theta)^m}, & m > 0, \end{cases} \quad (14.4)$$

where $P_n(\cos\theta)$ is the Legendre polynomial.

It is clear that the $n = 1$ term in the series describes the magnetic dipole field, $n = 2$ is the magnetic quadrupole, and so on. In the case of Earth the intrinsic magnetic field is dominated by the dipole term. The coefficients of expansion exhibit significant secular (long-term) variations and new sets of coefficients are published regularly. The intrinsic dipole field parameters for 1990 are summarized in Table 14.1.

The axis of the dipole differs from the axis of terrestrial rotation: The tilt angle is approximately $10.8°$. The relationship between magnetic coordinates (magnetic latitude, λ_{M}, and magnetic longitude, ϕ_{M}) and geographic coordinates is the following:

$$\sin\lambda_{\text{M}} = \sin\lambda \sin\lambda_{\text{N}} + \cos\lambda \cos\lambda_{\text{N}} \cos(\phi - \phi_{\text{N}}), \quad (14.5)$$

$$\sin\phi_{\text{M}} = \frac{\cos\lambda \sin(\phi - \phi_{\text{N}})}{\cos\lambda_{\text{M}}}, \quad (14.6)$$

where $\lambda = \pi/2 - \theta$, and the subscript N refers to the coordinates of the North pole.

In addition to the magnetic dipole tilt, the Earth's rotation axis is also inclined $23.5°$ to the ecliptic plane. As a consequence of the Earth's daily rotation and its orbiting around the Sun, the angle between the Sun–Earth line and the terrestrial magnetic dipole varies between about $56°$ and $90°$. The variation of this angle has important consequences for the configuration of the terrestrial magnetosphere (such as daily,

Table 14.1. *Terrestrial dipole parameters in 1990.*

Quantity	Symbol	Value
Dipole moment	$\frac{\mu_0}{4\pi} M_{\text{e}}$	7.84×10^{15} T m^3
		30.4μT R_{e}^3
Tilt of dipole axis		$10.8°$
Geographic latitude of magnetic north pole	λ_{N}	$79.2°$
Geographic longitude of magnetic north pole	ϕ_{N}	$289°$E

semiannual, and annual variations), since the large-scale morphology is controlled to a large extent by the interaction of the magnetized solar wind with the terrestrial magnetic field.

The dipole approximation is more or less valid up to a few Earth radii. Beyond this distance the terrestrial magnetic field significantly deviates from the dipole field because of the interaction with the magnetized solar wind. The magnetic field of a dipole was discussed earlier in this book (see Eqs. 1.75–1.80). In general, closed geomagnetic field lines are characterized by the L parameter or by the *invariant latitude*. As discussed before, the L parameter expresses the planetocentric distance (measured in planetary radii) of a dipolar magnetic field line at the dipole equator. The invariant latitude is the magnetic latitude where a field line reaches the surface of the Earth. The invariant latitude is related to the L parameter by

$$\Lambda = \arccos \sqrt{\frac{1}{L}}. \tag{14.7}$$

14.2   Interaction of the Solar Wind with the Terrestrial Magnetic Field

14.2.1   The Chapman–Ferraro Model

By the 1930s it had been recognized that, at least occasionally, high-speed, quasineutral streams of solar corpuscular radiation may interact with the dipole magnetic field of the Earth. These low-density plasma streams were thought to contain an equal number of positively and negatively charged particles. Collisional effects were thought to play only a negligible role in this low-density plasma, and therefore the streams were expected to be highly conducting. In 1930 (before the birth of modern plasma physics) Chapman and Ferraro proposed that as a highly conducting solar plasma cloud approaches the terrestrial dipole field, it can be considered as a moving conducting surface. Since the magnetic field cannot penetrate into this conducting surface, a current system is generated at the surface, which, in effect, cancels the dipole field. In other words, the Earth will see an advancing "mirror dipole" generated by currents flowing at the highly conducting surface of the approaching stream.[1] The interaction with the solar stream "compresses" the Earth's magnetic field. After the solar stream passes the Earth, the terrestrial magnetic field again expands and the original configuration is restored. This interaction is shown schematically in Figure 14.1.

In the Chapman–Ferraro model the sunward boundary of the compressed geomagnetic cavity is located halfway between the geomagnetic dipole and its mirror image. However, the real situation is much more complicated, since certain conservation laws must be obeyed at all times (such as the conservation of mass, momentum, and energy).

---

[1] Chapman, S., and Ferraro, C. A., "A new theory of magnetic storms," *Nature*, **126**, 129, 1930.

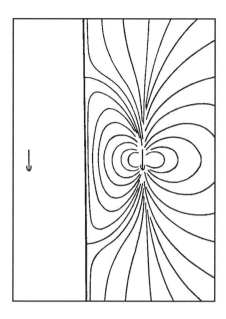

Figure 14.1 Interaction of a dipole field with a moving conducting wall.

The three-dimensional interaction of the magnetized solar wind with the Earth (and with other planets) is quite complex, involving a number of complicated plasmaphysical phenomena. However, the basic physics is quite simple: An extended magnetized obstacle is placed into a supersonic, superalfvénic plasma flow. As a result of the interaction between the high-speed solar wind plasma flow and the body, various magnetohydrodynamic discontinuities and current systems are formed, which result in extended regions where the magnetic field and plasma flow patterns are controlled by the interaction.

14.2.2   The Bow Shock and the Magnetopause

Next, we apply our results about shocks and discontinuities to the flow of the solar wind around the magnetic field of the Earth. The terrestrial magnetic dipole can be envisioned as a stationary dipole placed into a high-speed magnetized plasma flow. Because the flow around the Earth is supersonic and superalfvénic, we can expect the development of a standing shock wave ahead of the Earth. Such a shock is necessary to slow down the solar wind so that the flow can be diverted around the dipole. Since the solar wind is much faster than the Alfvén speed, the intermediate shock solution is not relevant, and we need to look at compressive shock solutions.

In the solar wind both the sonic and Alfvénic Mach numbers are large (typically $M = u_{sw}/a \geq 10$ and $M_A = u_{sw}/V_A \geq 10$). Therefore the shock jump conditions can be greatly simplified. First, the effects of the upstream thermal pressure and magnetic field can be neglected altogether. Also, the downstream pressure is much larger than the tangential component of the magnetic pressure. In this case we obtain the following jump conditions (see Eqs. 6.30–6.33):

$$\Phi_{\mathrm{sw_n}}(u_{\mathrm{n2}} - u_{\mathrm{sw_n}}) + p_2 = 0, \tag{14.8}$$

$$u_{\mathrm{n2}} B_{\mathrm{t2}} - u_{\mathrm{sw_n}} B_{\mathrm{sw_t}} = 0, \tag{14.9}$$

$$u_{\mathrm{t2}} - u_{\mathrm{sw_t}} - \frac{B_{\mathrm{sw_n}}}{\Phi_{\mathrm{sw_n}} \mu_0}(B_{\mathrm{t2}} - B_{\mathrm{sw_t}}) = 0, \tag{14.10}$$

$$\frac{1}{2}\Phi_{\mathrm{sw_n}}\left(u_{\mathrm{n2}}^2 - u_{\mathrm{sw_n}}^2\right) + \frac{\gamma}{\gamma - 1}p_2 u_{\mathrm{n2}} = 0, \tag{14.11}$$

where $\Phi_{\mathrm{sw_n}} = \rho_{\mathrm{sw}} u_{\mathrm{sw_n}} = \rho_{m2} u_{\mathrm{n2}}$ is the solar wind mass flux normal to the shock. The subscript 2 refers to conditions downstream from the shock; the subscripts n and t denote the normal and tangential components, respectively.

Equations (14.8) and (14.11) can be combined to obtain the velocity jump across the shock:

$$u_{\mathrm{n2}} = \frac{\gamma - 1}{\gamma + 1}u_{\mathrm{sw_n}} = \frac{1}{4}u_{\mathrm{sw_n}}. \tag{14.12}$$

The other jumps can be readily obtained:

$$\rho_2 = \frac{\gamma + 1}{\gamma - 1}\rho_{\mathrm{sw}} = 4\rho_{\mathrm{sw}}, \tag{14.13}$$

$$p_2 = \frac{2}{\gamma + 1}\rho_{\mathrm{sw}} u_{\mathrm{sw_n}}^2 = \frac{3}{4}\rho_{\mathrm{sw}} u_{\mathrm{sw_n}}^2, \tag{14.14}$$

$$B_{\mathrm{t2}} = \frac{\gamma + 1}{\gamma - 1}B_{\mathrm{sw_t}} = 4B_{\mathrm{sw_t}}, \tag{14.15}$$

$$u_{\mathrm{t2}} - u_{\mathrm{sw_t}} = \frac{2}{\gamma - 1}\frac{B_{\mathrm{sw_n}} B_{\mathrm{sw_t}}}{\mu_0 \rho_{\mathrm{sw}} u_{\mathrm{sw}}} = 3\frac{B_{\mathrm{sw_n}} B_{\mathrm{sw_t}}}{\mu_0 \rho_{\mathrm{sw}} u_{\mathrm{sw}}}, \tag{14.16}$$

where $\gamma = 5/3$ was used for the specific heat ratio of the solar wind. Equation (14.15) shows that the tangential component of the magnetic field increases across the shock; consequently, the terrestrial shock is a fast shock. This shock, which is called a *bow shock*, separates the free streaming solar wind from the region where the presence of the Earth's magnetic field significantly modifies the space environment.

Equations (14.13) through (14.16) describe the jump conditions at the bow shock, but they do not define its location. The location of the bow shock is basically determined by the shape and size of the magnetospheric "obstacle," the magnetopause. In a first approximation the dayside magnetopause can be considered as a tangential discontinuity that separates the shocked solar wind from the region dominated by the terrestrial magnetic dipole field (see Figure 14.2). The region between the bow shock and the magnetospheric obstacle (the magnetopause) is called the *magnetosheath*.

At a tangential discontinuity (which is a good first approximation for the dayside magnetopause) no plasma transport across the discontinuity occurs and the normal component of the magnetic field is zero. At the same time the tangential magnetic field

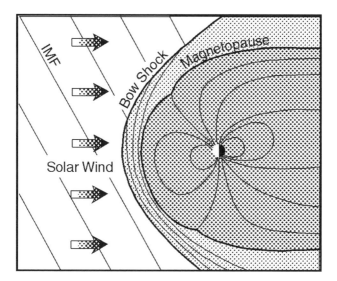

Figure 14.2 Schematic representation of the bow shock and the magnetopause.

component, the density, and the pressure can all be discontinuous. The total pressure (thermal plus magnetic), however, is balanced at the magnetopause as indicated by Eq. (6.19).

For the sake of simplicity, we consider the nose region of the magnetosphere. The calculation can also be carried out for other regions, but the mathematics is somewhat more complicated.

Inside the magnetopause, the total pressure is dominated by the compressed terrestrial dipole field and the thermal pressure can be neglected. Outside the magnetopause, the total pressure is due to a combination of the thermal and magnetic pressures of the shocked solar wind (since the flow is tangential to the surface). In a first approximation one can neglect the contribution from the tangential magnetic field component (B_{sw_t} is small). In this case Eq. (14.14) gives the thermal pressure outside the magnetopause. The pressure balance equation now becomes

$$\rho_{sw} u_{sw_n}^2 = \mathcal{F} \frac{B_e^2}{\mu_0} \left(\frac{R_e}{R_{MP}} \right)^6, \tag{14.17}$$

where $\mathcal{F}$ is a magnetic field compression factor (its value is about 2), B_e is the equatorial magnetic field at the surface of the Earth, and R_{MP} is the standoff distance of the magnetopause (the distance from the center of the Earth to the nose of the magnetopause). This expression can be easily evaluated. For average solar wind conditions one gets $R_{MP} \approx 10\, R_e$.

The ratio $(R_{BS} - R_{MP})/R_{MP}$ has been found empirically to be 1.1 times ρ_2/ρ_{sw} (where R_{BS} is the subsolar distance of the bow shock from the center of the planet). This finding is also consistent with the prediction of the hypersonic approximation (which is applicable here, since both the sonic and Alfvénic Mach numbers are much larger than unity). In the case of a strong shock the density jump is 4 (see Eq. 14.13),

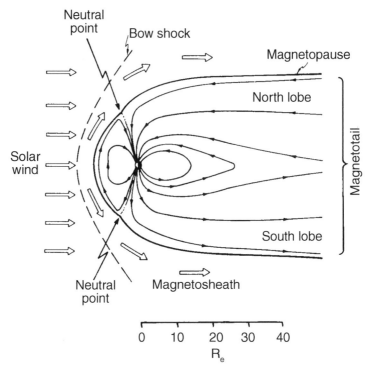

Figure 14.3 Noon–midnight meridional cross section of the simplest model of the magnetosphere. (From *Vasyliunas.*[2])

and therefore one gets

$$R_{BS} \approx 1.275 \, R_{MP} \approx 13 \, R_e. \tag{14.18}$$

The magnetosheath is filled with shocked solar wind plasma, which flows around the magnetopause. This plasma is much denser, hotter, and slower than the ambient solar wind flow. The magnetic field lines are draped around the magnetopause and the level of magnetic field turbulence is quite a bit higher than it is in the solar wind.

14.2.3 The Magnetospheric Cavity

The magnetospheric cavity is formed because the fast moving, magnetized solar wind cannot penetrate the magnetopause. It is a fairly complicated three-dimensional object with a blunt nose region and an extended tail. In the absence of viscous tangential stresses the forces acting normal to the magnetopause balance each other. In this case the magnetopause would enclose a teardrop-shaped closed cavity.

Figure 14.3 shows a schematic representation of the magnetosphere in the noon–midnight meridian. In contrast to the compressed configuration of the dayside magnetosphere, the magnetic field lines on the nightside are highly stretched along

[2] Vasyliunas, V. M., "Large-scale morphology of the magnetosphere," in *Solar Terrestrial Physics*, R. L. Carovilano and J. M. Forbes, eds., p. 243, D. Reidel, Dordrecht, The Netherlands, 1983.

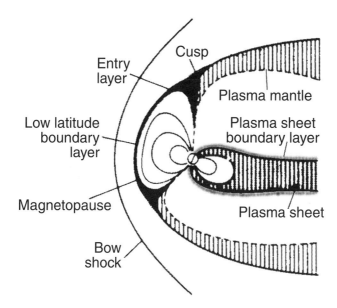

Figure 14.4 Schematic representation of the principal boundary regions in the magnetosphere.

the solar wind flow direction, and they form the *geomagnetic tail* or *magnetotail*. The reversal of the magnetic field between the northern and southern *geomagnetic lobes* implies the existence of a *tail current sheet* (often called the *neutral sheet*) with electric current flowing from dawn to dusk across the tail.

At the interface between the day and night sides of the magnetosphere lie the *polar cusps*. The field lines at the center of the cusps topologically correspond to the polar field lines. In the Chapman–Ferraro model these field lines intersect the magnetopause at the neutral point (where $B = 0$). In reality, the polar cusps are not single lines but finite-sized regions where the magnetosheath plasma extends down to the atmosphere.

There are several regions just inside the magnetopause that are topologically related to the magnetospheric cavity but are populated by magnetosheath-type plasma. These regions are called *magnetospheric boundary layers*. These boundary layers are schematically shown in Figure 14.4.

At the cusps the magnetosheath plasma extends deep into the magnetosphere (actually into the dense regions of the atmosphere). This region is referred to as the *polar cusp* or the *magnetospheric cleft*. Its outermost parts (near the magnetopause) are called the *entry layer*.

The *low latitude boundary layer* (LLBL) is a very thin region just inside the magnetopause (it can be half an R_e thick). The plasma flow in this region resembles that in the magnetosheath. The presence of such a plasma population indicates that the magnetopause is not a perfectly conducting shielding layer that can not be penetrated by plasma particles. However, the question of how and where the magnetosheath plasma crosses the magnetopause is still controversial. The LLBL is also thought to extend down into the dayside ionosphere. For southward IMF conditions the *high latitude boundary layer* (HLBL) or *entry layer* is located on the dayside poleward from LLBL.

The *plasma mantle* is located in a region of open magnetic field lines. Plasma from the magnetosheath expands into this region as the field lines are convected downward along the geomagnetic tail. The density in the plasma mantle decreases from the magnetosheath value, but the flow is primarily parallel to the magnetosheath flow with a small inward component (toward the neutral sheet).

The *plasma sheet* contains heated (compressed) plasma extending along the center of the magnetotail. Geomagnetic field lines within the plasma sheet are highly stretched, and the plasma there is hot and dense. Just outside the plasma sheet is the *plasma sheet boundary layer*, which contains somewhat lower density and less energetic plasma. The plasma sheet boundary layer extends earthward deep into the ionosphere.

Finally, between the plasma sheet and the plasma mantle lie the *magnetotail lobes*, characterized by very low plasma densities and by relatively steady and strong magnetic fields. The magnetic lobes are dominated by magnetic pressure.

14.3   Magnetospheric Current Systems

Currents are produced by moving electrically charged particles. In turn, current systems generate magnetic fields. In the magnetosphere there are five principal current systems that play an important role in the formation of the magnetosphere:

- the magnetopause current,
- the tail current,
- the ring current,
- field-aligned (Birkeland) currents, and
- ionospheric currents.

In this section we will examine the first four of these currents, their generation mechanisms, and their interrelations. The ionospheric currents are discussed separately.

14.3.1   The Magnetopause Current

The magnetopause current (also called the *Chapman–Ferraro current*) is the current system flowing around the magnetopause. This current system generates a magnetic field that "prevents" the terrestrial dipole field from penetrating into the solar wind.

Let us start from the single-fluid MHD equations (Eqs. 4.79) and neglect the effects of gravity. The steady-state continuity and momentum equations can now be written

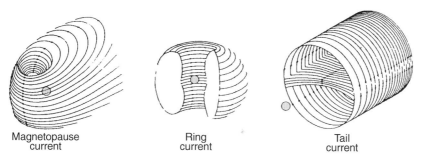

Magnetopause   Ring   Tail
  current   current   current

Figure 14.5 Schematic representation of the magnetospheric current systems. (From *Olsen.*[3])

in the following form:

$$\nabla \cdot (\rho_m \mathbf{u}) = 0,$$

(14.19)

$$(\rho_m \mathbf{u} \cdot \nabla)\mathbf{u} + \nabla p - \mathbf{j} \times \mathbf{B} = \mathbf{0}.$$

With the help of the continuity equation, the momentum equation can be written as

$$\nabla(\rho_m u^2 + p) = \mathbf{j} \times \mathbf{B}.$$

(14.20)

In the simplest model of the magnetopause current system one neglects the magnetic field outside the magnetopause and the contribution of particles inside it. In other words, the magnetopause current system separates the shocked magnetosheath plasma from the "empty" magnetic dipole field inside. In this model the current density is given by

$$\mathbf{j}_{MP} = \frac{\mathbf{B}_{dp}}{B_{dp}^2} \times \nabla(\rho_m u^2 + p),$$

(14.21)

where $\mathbf{B}_{dp}$ is the magnetospheric field at the magnetopause. The total pressure (ram pressure plus thermal pressure) in the magnetosheath is approximately equal to the free streaming solar wind kinetic pressure. Therefore one can estimate the magnitude of the magnetopause current as

$$j_{MP} \approx \frac{1}{B_{dp}} \frac{\rho_{m\,sw} u_{sw}^2}{\Delta},$$

(14.22)

where Δ is the thickness of the magnetopause. Substituting typical values ($B_{dp} \approx 3 \times 10^{-8}$ T, $\Delta \approx 500$ km, $n_{sw} \approx 8$ cm^{-3}, $u_{sw} \approx 400$ km/s), we get $j_{MP} \approx 10^{-7}$ A/m^2. A schematic representation of the magnetopause current system is shown in Figure 14.5.

[3] Olsen, P. W., "The geomagnetic field and its extension into space," *Adv. Space Res.*, **2**, 13, 1982.

14.3.2 The Tail Current

The geomagnetic tail is formed by the highly stretched ("open") magnetic field lines originating from the polar ionosphere (called the *polar cap*). The magnetic flux leaving the polar cap is the radial component of the magnetic field integrated over the polar cap area (see Eqs. 1.90 and 1.91):

$$\Phi_{PC}(R_e, \lambda_{PC}) = 2\pi r^2 \int_{\lambda_{PC}}^{\pi/2} B_r(R_e, \lambda') \cos\lambda' d\lambda'$$

$$= -2\pi B_e R_e^2 \cos^2\lambda_{PC}, \tag{14.23}$$

where λ_{PC} is the magnetic latitude of the equatorward edge of the polar cap and B_e is the surface magnetic field strength at the terrestrial equator.

The flux in a tail lobe is

$$\Phi_T = \frac{1}{2}\pi R_T^2 B_T \tag{14.24}$$

where B_T is the magnetic field magnitude in the tail, and we assumed the cross section of the tail lobe to be a semicircle with radius R_T.

The magnetic flux emerging from a polar cap must be conserved in the corresponding lobe of the geomagnetic tail (i.e., $\Phi_{PC}(R_e, \lambda_{PC}) = \Phi_T$). Equating these fluxes yields the following equation for the tail radius:

$$R_T = \sqrt{\frac{4B_e}{B_T}} \cos\lambda_{PC} R_e. \tag{14.25}$$

Using typical values ($\lambda_{PC} \approx 15°$, $B_e \approx 3 \times 10^{-5}$ T, and $B_T \approx 2 \times 10^{-8}$ T), we get $R_T \approx 20 R_e$.

The tail magnetic field is naturally generated by a current system. The relation between the current and the field is given by the MHD approximation of Ampère's law:

$$\nabla \times \mathbf{B} = \mu_0 \mathbf{j}. \tag{14.26}$$

Ampère's law also can be written in integral form:

$$\oint_C \mathbf{B} \cdot d\mathbf{s} = \mu_0 \int_S \mathbf{j} \cdot d\mathbf{S}, \tag{14.27}$$

where C is a closed curve bounding the surface S.

One can calculate the current density in the tail current sheet by taking the closed integration path in the noon–midnight meridional plane in a way that it encloses the current sheet. The magnetic field changes sign across the current sheet, and therefore Eq. (14.27) now yields

$$2B_T = \mu_0 I, \tag{14.28}$$

where I is the total sheet current density (current per unit tail length). Using $B_T = 20$ nT, we obtain, $I = 30$ mA/m or $I = 2 \times 10^5$ A/R_e. This huge current density can have major effects on the ionosphere. Figure 14.5 shows a schematic representation of the magnetotail current.

14.3.3   The Ring Current

The ring current is primarily composed of geomagnetically trapped 10 to 200 keV particles (H^+ and O^+) bouncing along closed geomagnetic field lines. It is typically located between 3 and 6 R_e. The essence of the ring current is the *gradient–curvature drift*, which was extensively discussed earlier in this book (see Eqs. 1.73 and 1.111). We again note that the direction of the gradient–curvature drift is opposite for electrons and ions, thus creating a net current.

Let us examine the ring current for a mathematically simple situation. Let us assume that all ring current particles are equatorially trapped (their equatorial pitch angle is $90°$) on an "average" field line characterized by L. In this case the gradient–curvature drift becomes the following (see Eq. 1.111):

$$\mathbf{v}_{GC} = -\frac{3}{2}\frac{mv^2}{q}\frac{L^2}{B_e R_e}\mathbf{e}_\phi. \tag{14.29}$$

In this approximation all particles are located at the equator at a planetocentric distance of $L R_e$. If we denote the total number of ring current particles of type "t" by N_t, the total current, I_ϕ, can be expressed as

$$I_\phi = -3\frac{L^2}{B_e R_e}\sum_{t=e,i} N_t \frac{m_t v_t^2}{2}. \tag{14.30}$$

However, we want to express the total current carried by the ring current in terms of the total energy of ring current particles, E_{RC}. In the present approximation

$$E_{RC} = 2\pi L R_e \left(\sum_{t=e,i} N_t \frac{m_t v_t^2}{2}\right). \tag{14.31}$$

Substituting Eq. (14.31) into expression (14.30) yields the following:

$$I_\phi = -\frac{3L E_{RC}}{2\pi B_e R_e^2}. \tag{14.32}$$

It is well-known in electricity and magnetism that Ampère's law can be integrated to express the magnetic field with the help of the current density:

$$\mathbf{B}(\mathbf{r}) = \frac{\mu_0}{4\pi}\int d^3 r' \frac{\mathbf{j}(\mathbf{r'}) \times (\mathbf{r} - \mathbf{r'})}{|\mathbf{r} - \mathbf{r'}|^3}. \tag{14.33}$$

This is the *Biot–Savart equation*. For the magnetic field at the center of a circular loop of radius r the Biot–Savart equation yields

$$\mathbf{B} = \frac{\mu_0 I}{2r}\mathbf{e}_z, \tag{14.34}$$

where the z axis is perpendicular to the plane of the loop.

In the case of the ring current the magnetic field perturbation created by the ring current at the center of the Earth can be expressed as

$$\Delta \mathbf{B} = -\frac{3\mu_0 E_{RC}}{4\pi B_e R_e^3}\mathbf{e}_z, \tag{14.35}$$

where $\mathbf{e}_z$ is pointing along the magnetic axis.

It is important to note that the magnetic field perturbation is independent of L. The sign of the magnetic field perturbation is negative, indicating that the ring current "weakens" the geomagnetic field. The magnetic field perturbation is a very good measure of the total energy of the ring current particles. One of the most important geomagnetic indices, the D_{st} index, is, in effect, the magnetic perturbation caused by the ring current (but measured at the surface of the Earth and not at the center). The D_{st} index is used to characterize *magnetic storms*, which are usually accompanied by strong variations of the ring current. Figure 14.5 shows a schematic representation of the ring current.

The drift path of a ring current particle is not necessarily a closed orbit. The reason is that there are electric fields in the magnetosphere. These electric fields are generated by the moving solar wind and by the rotation of the Earth. The magnetospheric electric fields will be discussed later (Section 14.4.4). Here we only note that near the equatorial plane the magnetospheric electric field typically points from dawn to dusk.

In the inner magnetosphere the magnetic field is reasonably well represented by the terrestrial dipole field. Therefore the $\mathbf{E} \times \mathbf{B}$ guiding center drift in the equatorial plane can be expressed as

$$v_E = \left|\frac{\mathbf{E} \times \mathbf{B}_d}{B_d^2}\right|_{eq} = \frac{E}{B_d} = L^3 \frac{E}{B_e}, \tag{14.36}$$

where the subscript eq refers to the equatorial plane, $\mathbf{B}_d$ is the dipole field, and B_e is the magnetic field strength at the terrestrial equator. In the present case the electric field vector is directed from dawn to dusk. For equatorially bouncing electrons and ions, this electric drift is sunward (the direction of v_E is independent of the particle charge and hence this drift does not produce electric current), and it is proportional to L^3. However, because the gradient–curvature drift for these particles is proportional to L^2 (see Eq. 14.29), at larger geocentric distances the electric drift typically dominates over the magnetic gradient–curvature drift. The resulting ion drift orbits are schematically shown in Figure 14.6. Close to Earth the magnetic drift dominates and the drift paths

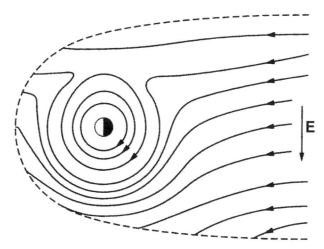

Figure 14.6 Drift paths for ring current and partial ring current ions in the equatorial plane.

are closed trajectories. This is the region of the *ring current*. Further out in the inner magnetosphere the sunward drifting particles (due to the magnetospheric electric field) are deflected around the Earth by the magnetic drift. These particles form the *partial ring current*.

A fraction of the ring current is carried by relativistic electrons and very energetic (MeV) ions. These particles represent *penetrating radiation* that can penetrate deep into dense materials and thus cause radiation damage to spacecraft systems or to humans. These *radiation belt* or *Van Allen belt* particles contribute relatively little to the ring current, but they are important for safety considerations.

14.3.4 Field-Aligned Currents

In 1908 Kristian Birkeland of Norway concluded from his studies of auroral phenomena that auroral zone current systems are closed by vertical electrical currents extending beyond the ionosphere. However, this idea was rejected by the leading authority of space science in the 1930s, Sydney Chapman, and consequently Birkeland's idea was not taken seriously until the late 1960s when a combination of satellite and ground-based observations provided convincing evidence for the existence of field-aligned currents. Today we call these field-aligned currents *Birkeland currents*.

We discussed field-aligned currents to some extent in connection with ionospheric conductivities and currents. There we showed that the current system closes in the polar ionosphere through Hall and Pedersen currents. Thus there must be two separate field-aligned currents in the auroral ionosphere: one downward and the other upward. These field-aligned currents are often called *Region 1 currents*, which flow outward from the ionosphere, and *Region 2 currents*, which flow into the ionosphere. Region 1

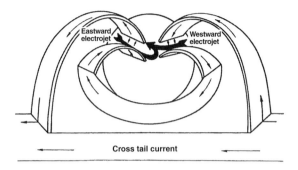

Figure 14.7 Connection between ionospheric and magnetospheric currents. (From *Swift*.[4])

is a pair of currents at higher latitudes, whereas Region 2 is a pair of currents at lower latitudes (see Figure 14.7).

An interesting feature of the field-aligned currents can be obtained from the conservation of charge (Eq. 4.59). Under a steady-state condition this equation simply states that the current density is divergenceless: $\nabla \cdot \mathbf{j} = 0$. But since we know that field-aligned currents are parallel to the magnetic field, we can write

$$\mathbf{j}_{\parallel} = \alpha \mathbf{B},\qquad(14.37)$$

and therefore

$$\nabla \cdot \mathbf{j}_{\parallel} = \nabla \cdot (\alpha \mathbf{B}) = (\mathbf{B} \cdot \nabla)\alpha = B\frac{\partial \alpha}{\partial s} = 0.\qquad(14.38)$$

This means that the quantity α is conserved along a magnetic field line.

The field-aligned currents close through the ionosphere, but how are they generated in the magnetosphere? Observations indicate that the ring current is not uniform, and parts of the ring current close through field-aligned currents into the ionosphere. This situation is shown Figure 14.7.

Finally, Figure 14.8 shows a schematic representation of our present understanding of the magnetospheric boundaries and current systems.

14.4 Plasma Convection in the Magnetosphere

14.4.1 The Axford–Hines and the Dungey Models

The coupling between the solar wind and the magnetosphere produces plasma convection inside the magnetosphere. The existence of such a systematic convection pattern

---

[4] Swift, D. W., "Auroral mechanisms and morphology," *Rev. Geophys. Space Phys.*, **17**, 681, 1979.

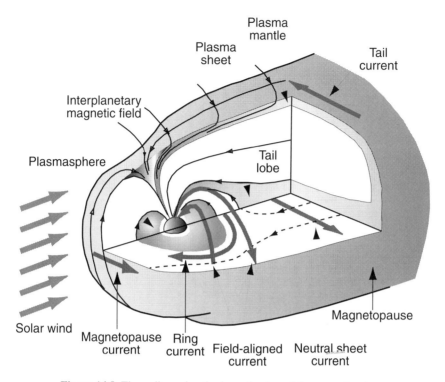

Plasma
mantle

Plasma
sheet

Tail
current

Interplanetary
magnetic field

Plasmasphere

Tail
lobe

Solar wind

Magnetopause
current

Ring
current

Field-aligned
current

Neutral sheet
current

Magnetopause

Figure 14.8 Three-dimensional schematic view of the magnetosphere.

was originally suggested by Axford and Hines[5] and by Dungey.[6] Both of these models were based on observed ionospheric convection patterns.

Axford and Hines introduced the concept of viscous interaction between the solar wind and the magnetosphere. This model is shown schematically in Figure 14.9. Plasma in the outside boundary layer of the magnetosphere and in the magnetically connected high-latitude ionosphere flows in the antisunward direction as if "dragged" by frictional interaction with the shocked solar wind in the magnetosheath. In the inner magnetosphere and in the magnetically connected lower-latitude polar ionosphere the flow is sunward, because the circulation pattern must be closed in this *closed magnetosphere* model. The Axford–Hines model has several very attractive features, but it leaves the fundamental issue unanswered: What is the nature of the frictional interaction between the magnetosheath plasma and the plasma inside the magnetopause?

In many respects the Dungey model is quite similar to the Axford–Hines model. However, Dungey's model attributed the momentum transfer from the shocked solar wind to the magnetosphere to magnetic reconnection (and not to viscous coupling).

[5] Axford, W. I., and Hines, C. O, "A unifying theory of high latitude geophysical phenomena and magnetic storms," *Can. J. Phys.*, **39**, 1433, 1961.
[6] Dungey, J. W., "Interplanetary magnetic field and the auroral zones," *Phys. Rev. Lett.*, **6**, 47, 1961.

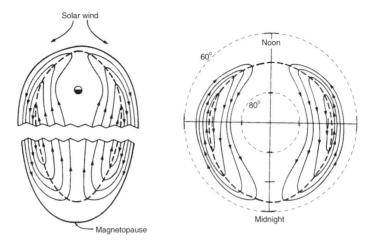

Figure 14.9 Streamlines in the Axford—Hines convection model. The left panel shows the magnetospheric equatorial plane and the right panel shows convection in the polar cap ionosphere. Streamlines near the magnetopause are magnetically connected ionospheric streamlines near the magnetic pole. (From *Hill.*[7])

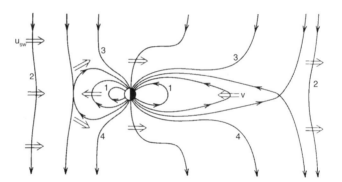

Figure 14.10 The Dungey magnetosphere model in the noon–midnight meridian. [From *Lyons & Williams (1984)*].

The Dungey model in the noon–midnight meridian for southward IMF is shown in Figure 14.10 (southward IMF is the only configuration that results in an essentially two-dimensional geometry). The model predicts two *magnetic reconnection* points, where the interplanetary and geomagnetic field lines "merge." It is this magnetic reconnection that makes it possible for the solar wind magnetic field lines (and consequently the motional electric field) to penetrate into the magnetosphere and drive plasma convection. A very important factor in magnetospheric dynamics is the *reconnection rate*, which remains a subject of intensive research.

[7] Hill, T. W., "Solar wind magnetosphere coupling," in *Solar Terrestrial Physics*, R. L. Carovilano and J. M. Forbes, eds., p. 261, D. Reidel, Dordrecht, The Netherlands, 1983.

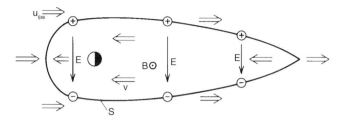

Figure 14.11 Velocity field in the equatorial plane. The surface S is
charged and it separates the regions of sunward and antisunward
convection. [From *Lyons & Williams* (*1984.*)]

Figure 14.11 shows the overall convective flow in the magnetosphere. The figure shows the equatorial plane with a magnetic field that is the superposition of the
terrestrial intrinsic field and the IMF. The surface S, which is positively charged on the
dawn side and negatively charged on the dusk side, separates the regions of sunward
and antisunward convections. If the solar wind does not penetrate into the region of
closed magnetic field lines, the surface S separates the regions of open and closed
field lines. The presence of the dawn to dusk electric field significantly modifies the
convection pattern and introduces a dawn–dusk anisotropy into the flow.

When a southward pointing interplanetary magnetic field line (denoted by 2 in
Figure 14.10) reaches the subsolar magnetopause, it will merge with a closed terrestrial
magnetic field line (denoted by 1 in Figure 14.10). The merged field lines split into
two open field lines (marked by 3 and 4 in Figure 14.10). These field lines have
one end connected to the Earth and the other end open into the solar wind. The solar
wind convects the open end of these field lines downwind. The other end of the field
line is "dragged" antisunward in the polar cap ionosphere by this convection. At
the nightside end of the magnetosphere the two open ends of field lines 3 and 4
meet and reconnect. This process leaves a closed but stretched terrestrial field line
in the magnetotail and a somewhat stretched interplanetary field line downstream of
the magnetosphere (marked by 2). Because of magnetic tension, this field line will
eventually relax to a regular solar wind field line.

14.4.2   Magnetic Diffusion

The motion of the bulk plasma population and the time evolution of magnetic field
lines are coupled by the induction equation (Eq. 4.85):

$$\frac{\partial \mathbf{B}}{\partial t} = \nabla \times (\mathbf{u} \times \mathbf{B}) + \eta_m \nabla^2 \mathbf{B}, \tag{14.39}$$

where $\eta_m = 1/\bar{\sigma}_0 \mu_0$ is the magnetic viscosity. It should be noted that $\bar{\sigma}_0$ is not
necessarily the collisional conductivity, but it can also be the result of complicated
plasmaphysical processes that lead to anomalous (collisionless) conductivity.

In nearly stagnating plasmas the first term in Eq. (14.39) can be neglected and,
consequently, the time evolution of the magnetic field is simply determined by magnetic

diffusion:

$$\frac{\partial \mathbf{B}}{\partial t} = \eta_{\rm m} \nabla^2 \mathbf{B}. \tag{14.40}$$

Equation (14.40) indicates that the magnetic field tends to diffuse away in a stagnating plasma. In a one-dimensional case (when all quantities vary only in the x direction) Eq. (14.40) can be simplified even further. For the sake of simplicity, we assume that in a first approximation $\eta_{\rm m}$ is constant and that the magnetic field is in the z direction (the slight variation of the magnetic field in the y direction required by the $\nabla \cdot \mathbf{B} = 0$ condition can be neglected). In this case $\mathbf{B} = B(x, t)\, \mathbf{e}_z$ and Eq. (14.40) becomes

$$\frac{\partial B}{\partial t} = \eta_{\rm m} \frac{\partial^2 B}{\partial x^2}. \tag{14.41}$$

Let us assume that at $t = 0$ the magnetic field can be described by a slowly varying function with a characteristic scale length of L_B,

$$B(x, 0) = B_0 \operatorname{erf}\left(-\frac{x}{2L_B}\right), \tag{14.42}$$

where erf is the error function (see Appendix D). With this initial condition, we get the following solution for Eq. (14.41):

$$B(x, t) = B_0 \operatorname{erf}\left(\frac{x}{2\sqrt{\eta_{\rm m} t + L_B^2}}\right). \tag{14.43}$$

We see that as time goes on, the initial magnetic field "diffuses" away with a characteristic time-scale of

$$\tau_{\rm d} = \frac{L_B^2}{\eta_{\rm m}}. \tag{14.44}$$

Thus for very small magnetic viscosity values this characteristic time-scale can be very large. In the extreme situation, when $\eta_{\rm m} \to 0$, the diffusion time becomes infinite and the magnetic field becomes "frozen" into the plasma.

Note that the magnetic flux going through a small surface element in the (x, y) plane, $dA = dx\,dy$, is

$$d\Phi_{\rm m}(x, t) = B(x, t)\, dx\, dy. \tag{14.45}$$

Next, we integrate the magnetic flux for a surface area that is symmetric around $x = 0$:

$$\int_{-x_0}^{x_0} \int_0^{y_0} B(x, t)\, dx\, dy = 0. \tag{14.46}$$

Equation (14.46) means that the magnetic flux encircled by a closed loop remains constant with time, even though the magnetic field changes. One can visualize this situation as magnetic field lines frozen into the plasma and actually moving with the plasma as it moves.

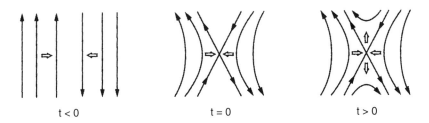

Figure 14.12 Schematic representation of magnetic field reconnection.

The relative importance of the convective and diffusive terms in the induction equation (Eq. 14.39) is characterized by the *magnetic Reynolds number, R_m,* defined as

$$R_m = \frac{|\nabla \times (\mathbf{u} \times \mathbf{B})|}{|\eta_m \nabla^2 \mathbf{B}|} \approx \mu_0 \bar{\sigma}_0 L_B u. \tag{14.47}$$

The magnetic Reynolds number is a very useful quantity for determining the importance of diffusion in a plasma. For high magnetic Reynolds number flows, magnetic diffusion can be completely neglected. In this case the plasma flow simply "carries" the magnetic field. This is the "frozen in" limit. However, in flows with $R_m \approx 1$ the effects of diffusion are very important. In these flows the magnetic field is not frozen into the plasma, and the plasma can readily flow across magnetic field lines.

14.4.3   Magnetic Reconnection

Magnetic reconnection or *magnetic merging* is a process in which several existing magnetic field lines are cut in the region of reconnection and the free ends are reconnected to other field lines. This process locally changes the magnetic topology.

To illustrate the fundamental physics of magnetic reconnection, let us consider the following very simple scenario. Two semi-infinite slabs of plasma with antiparallel frozen-in magnetic field lines (in the $\pm z$ direction) are moving toward each other very slowly (see Figure 14.12). Similar magnetic topologies occur around thin current sheets at the magnetopause and in the tail.

When the two slowly moving slabs of plasma reach each other, a nearly stagnating plasma topology is created with discontinuous magnetic field at the interface ($x = 0$). Assuming that the antiparallel magnetic field in the two slabs is $\pm B_0$ in the z direction, the time evolution of the magnetic field will be described by Eq. (14.41). The initial condition is now a discontinuous jump from $B_z = B_0$ to $B_z = -B_0$ (see Figure 14.13); therefore $L_B = 0$. In this case the solution becomes

$$B_z(x, t) = -B_0 \operatorname{erf}\left(\frac{x}{2\sqrt{\eta_m t}}\right). \tag{14.48}$$

At $t = 0$, Eq. (14.48) describes the discontinuous solution, which at $x = 0$ jumps from $B_z = B_0$ to $B_z = -B_0$, while at later times the discontinuity "erodes." This can be seen in Figure 14.13, which shows the variation of the z component of the magnetic

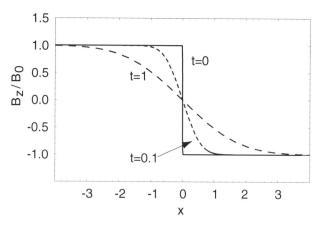

Figure 14.13 Schematic representation of the time evolution of
merging magnetic field lines. Distance and time are normalized to
x/x_0 and $\eta_m t/x_0^2$, where x_0 is an arbitrary length.

field. The curves represent $\eta_m t/x_0^2 = 0$, 0.1, and 1 values, respectively (where x_0 is
an arbitrary length). We see that as time goes on, the magnetic field "dissipates" near
the tangential discontinuity. It is obvious that the thickness of the transition region
(the current sheet) increases with time as $\sqrt{4\eta_m t}$.

Note that even though the magnetic flux remains conserved (see Eq. 14.46), the
magnetic energy density, $B^2/2\mu_0$, decreases with time. With the help of the MHD
form of Ampère's law and Eq. (14.40), we see that

$$\frac{1}{2\mu_0}\frac{\partial B^2}{\partial t} = -\mathbf{E} \cdot \mathbf{j}, \tag{14.49}$$

where the electric field, $\mathbf{E}$, is nonzero in the current layer separating the two plasma
slabs. Equation (14.49) means that the magnetic energy is ohmically converted into
heat (note that in the present simple case $\mu_0 j = \partial B/\partial x$).

Equation (14.48) describes magnetic reconnection (or merging). Right after the
oppositely oriented magnetic field lines come into close contact with each other in
a region where the magnetic Reynolds number is near unity, an X-type magnetic
configuration is formed, and the field lines "reconnect." The magnetic field magnitude
is zero near the center of the *X-point* (also called the *neutral point*). At later times
the plasma gets transported toward the neutral point, while the merged field lines
are transported away from the reconnection point (see Figure 14.12). The merged
field lines will be populated with a mixed plasma originating from the two sides of
the reconnection configuration. As long as magnetic flux is transported toward the
neutral point the reconnection process continues. The *reconnection rate* is primarily
determined by the rate of magnetic flux transport toward the reconnection region.

Next, we examine the energy balance in the reconnection process. We showed
that without a continuous supply of magnetic energy the magnetic field slowly

dissipates in the reconnection region. Nearly steady-state (or long lasting) reconnection requires that magnetic energy be continuously transported into the region of reconnection.

Let us explore the energy balance of magnetic reconnection using our very simple model of the dayside reconnection at the magnetopause. Let us assume that magnetic energy is resupplied into the region of reconnection. Consequently, the magnetic field profile in the x direction is given by Eq. (14.48) with a characteristic current layer thickness of $L = \sqrt{4\eta_m \tau_m}$, where τ_m is the time-scale of magnetic energy dissipation within the reconnection layer.

Without reconnection, the line integral of the magnetic energy density from $-x_0$ to x_0 is $x_0 B_0^2 \mu_0$. Then the average magnetic energy between $-x_0$ and x_0 is $\bar{\varepsilon}_{m0} = B_0^2/2\mu_0$. In the case of a reconnected field geometry (when the magnetic field is given by Eq. 14.48) the average magnetic energy between $-x_0$ and x_0 is the following:

$$
\bar{\varepsilon}_m = \frac{1}{2x_0} \int\limits_{-x_0}^{x_0} dx \frac{B^2}{2\mu_0} = \frac{B_0^2}{2\mu_0} \operatorname{erf}^2\left(\frac{x_0}{L}\right)
$$
$$
+ \frac{B_0^2}{\mu_0} \frac{L}{x_0} \frac{1}{\sqrt{\pi}} \left[\operatorname{erf}\left(\frac{x_0}{L}\right) \exp\left(-\frac{x_0^2}{L^2}\right) - \frac{1}{\sqrt{2}} \operatorname{erf}\left(\frac{\sqrt{2}x_0}{L}\right) \right]. \qquad (14.50)
$$

We average for a region that is thicker than the current layer. If $x_0/L > 3$ (which is a very reasonable requirement), the averaged magnetic energy density can be approximated by

$$
\bar{\varepsilon}_m = \frac{B_0^2}{2\mu_0} - \frac{B_0^2}{\mu_0} \frac{1}{\sqrt{2\pi}} \frac{L}{x_0}. \qquad (14.51)
$$

Finally, the change (from the initial state) of the average magnetic energy density between $-x_0$ and x_0 is

$$
\Delta\bar{\varepsilon}_m = \bar{\varepsilon}_{m0} - \bar{\varepsilon}_m = \frac{B_0^2}{\mu_0} \frac{1}{\sqrt{2\pi}} \frac{L}{x_0}. \qquad (14.52)
$$

If the surface area of the magnetic reconnection region (in the plane of the tangential discontinuity) is $\mathcal{A}$, then the total energy converted from magnetic energy to plasma kinetic energy is

$$
\Delta E_m = \Delta\bar{\varepsilon}_m 2x_0 \mathcal{A} = 2 \frac{B_0^2}{\mu_0} \frac{\mathcal{A}L}{\sqrt{2\pi}}. \qquad (14.53)
$$

Under steady-state conditions, this energy is converted at the same rate as magnetic energy is transported into the reconnection region through area $\mathcal{A}$. If the normal component of the upstream plasma velocity is u, then the characteristic time of magnetic energy transport into the reconnection region is $\tau_m = L/u$. The total energy conversion rate in the reconnection region is then

$$\Delta W_{\mathrm{m}} = \frac{\Delta E_{\mathrm{m}}}{\tau_{\mathrm{m}}} = \frac{B_0^2}{2\mu_0} \frac{\mathcal{A}u}{\sqrt{2\pi}} = \frac{1}{2} \rho \bar{V}_A^2 \frac{\mathcal{A}u}{\sqrt{2\pi}}, \tag{14.54}$$

where $\bar{V}_A$ is the average Alfvén speed in the reconnection region ($\bar{V}_A^2 = B_0^2/\mu_0\rho$).

Equation (14.54) reveals a very interesting result: In steady-state reconnection the energy converted from magnetic energy to plasma kinetic energy is controlled by the average Alvén speed; it does not depend directly on the details of the reconnection process.

14.4.4 Convection Electric Field

The ultimate driver of magnetospheric convection is the momentum of the solar wind. Owing to the very large conductivity of the plasma there is no electric field in the plasma frame, but an observer on Earth detects an electric field of

$$\mathbf{E}_{\mathrm{c}} = -\mathbf{u}_{\mathrm{c}} \times \mathbf{B}, \tag{14.55}$$

where $\mathbf{u}_{\mathrm{c}}$ is the velocity of the magnetic flux tube. This is the *convection electric field*. For southward IMF, the convection electric field points from dawn to dusk.

If the solar wind cannot penetrate the magnetopause, the convection electric field is also excluded from the magnetosphere. Magnetic reconnection is the process that enables interplanetary magnetic field lines to connect to terrestrial field lines and thus makes it possible for the convection electric field to penetrate into the magnetosphere.

Under steady-state conditions, the collisionless induction equation becomes simply

$$\nabla \times (\mathbf{u}_{\mathrm{c}} \times \mathbf{B}) = \mathbf{0}, \tag{14.56}$$

which means that $\nabla \times \mathbf{E}_{\mathrm{c}} = \mathbf{0}$ everywhere in the magnetosphere, and consequently, the electric field can be expressed in terms of a scalar potential, Φ_{c}, as $\mathbf{E}_{\mathrm{c}} = -\nabla\Phi_{\mathrm{c}}$. It can easily be seen that the magnetic field lines and the plasma flow lines are equipotentials, because

$$\mathbf{u} \cdot \nabla\Phi_{\mathrm{c}} = \frac{\partial\Phi_{\mathrm{c}}}{\partial s_u} = \mathbf{u}_{\mathrm{c}} \cdot (\mathbf{u}_{\mathrm{c}} \times \mathbf{B}) = 0,$$
$$\mathbf{B} \cdot \nabla\Phi_{\mathrm{c}} = \frac{\partial\Phi_{\mathrm{c}}}{\partial s_B} = \mathbf{B} \cdot (\mathbf{u}_{\mathrm{c}} \times \mathbf{B}) = 0. \tag{14.57}$$

In a first approximation the convection electric field, $\mathbf{E}_{\mathrm{c}}$, is constant in the equatorial plane of the magnetosphere. Therefore the convection potential can be approximated as

$$\Phi_{\mathrm{c}} = -E_{\mathrm{c}} R_{\mathrm{e}} L \cos\psi, \tag{14.58}$$

where ψ is the azimuth angle. Thus the convection potential in the equatorial plane linearly increases with the L parameter.

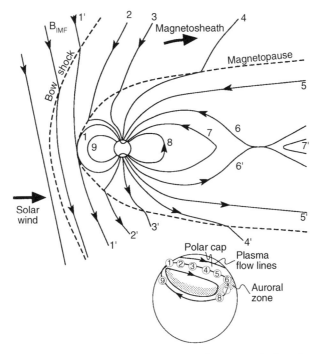

Figure 14.14 Schematic representation of magnetospheric and ionospheric convections driven by reconnection. (From *Hughes*.[8])

The convection electric field in the solar wind is about 20 mV/m (assuming $u_{sw} = 400$ km/s and $B_{sw} = 5$ nT), corresponding to about a 640 kV potential drop across the geomagnetic tail (assuming a tail diameter of 50 R_e). The observed electric potential drop accross the region of open field lines in the ionosphere is about ten times smaller. This implies that about 10% of the interplanetary magnetic flux that impacts the geometric cross section of the magnetosphere reconnects with the geomagnetic field.

For the sake of simplicity, let us consider the case of southward IMF. This situation is schematically shown in Figure 14.10, and the motion of magnetic field lines has already been discussed. Here we have another look at the plasma convection associated with the motion of the reconnected magnetic field lines.

Figure 14.14 also illustrates how magnetic field lines reconnected at the dayside magnetopause are convected tailward by the solar wind. The numbered field lines show the transport of a field line as it is convected downwind from the dayside magnetopause (field line 1–1') to the tail (field line 5–5'). Field lines 6 and 6' reconnect in the tail forming a newly closed geomagnetic field line and a new interplanetary field line. The new interplanetary field line (denoted by 7') is highly stretched, and it contains plasma of terrestrial origin. This field line forms part of the geomagnetic tail. It is convected downwind and ultimately returns to the ambient solar wind.

[8] Hughes, J., "The magnetopause, magnetotail, and magnetic reconnection," in *Introduction to Space Physics*, M. G. Kivelson and C. T. Russell, eds., p. 227, Cambridge University Press, Cambridge, 1995.

The newly closed geomagnetic field line (denoted by 7) is also stretched, and it relaxes by flowing earthward. Near the Earth, the field line eventually moves around the planet and returns to the dayside. The inset in Figure 14.14 shows the ionospheric footprints of the numbered field lines in the northern high-latitude ionosphere and the corresponding ionospheric convection. The resulting flow pattern is antisunward convection in the polar cap and a return flow at subauroral latitudes.

Near the Earth, there is an additional electric field that generates plasma convection. This electric field, due to the rotation of the Earth, is called the *corotation electric field*. It is caused by the motion of the magnetic field lines and the plasma associated with it.

The electric field measured by an observer in the nonrotating frame of reference is given by

$$\mathbf{E}_{cr} = -(\mathbf{\Omega}_r \times \mathbf{r}) \times \mathbf{B} = -(\mathbf{B} \cdot \mathbf{\Omega}_r)\mathbf{r} + (\mathbf{B} \cdot \mathbf{r})\mathbf{\Omega}_r, \tag{14.59}$$

where $\Omega_r = 7.292 \times 10^{-5}$ s^{-1} is the angular velocity of the Earth. In the equatorial plane of the terrestrial dipole field this expression simplifies to

$$\mathbf{E}_{cr} = -\Omega_r R_e \frac{B_e}{L^2} \mathbf{e}_r, \tag{14.60}$$

where $\mathbf{e}_r$ is the radial unit vector. This corresponds to the following corotation potential in the equatorial plane:

$$\Phi_{cr} = -\Omega_r R_e^2 \frac{B_e}{L} = -\frac{92 \, \text{kV}}{L}. \tag{14.61}$$

The corotation potential is rotationally symmetric in the equatorial plane: The equipotential lines form concentric circles and the potential value linearly decreases with L.

Because the corotation potential decreases radially, its effect on the plasma decreases with increasing geocentric distance. At the same time the convection potential increases with L (see Eq. 14.58). The sum of the corotation and convection potentials determines the drift path of the plasma particles. This total potential field is shown in Figure 14.15.

Inspection of Figure 14.15 reveals two topologically different regions of equipotentials. Near the Earth, the equipotentials are closed contours (elongated toward dusk), whereas at larger distances the equipotentials are open curves. The region of closed equipotentials is the plasmasphere, which corotates with the Earth. Since the convection trajectories are also equipotentials (see Eq. 14.57), the plasmasphere also corotates with the planet. The last closed equipotential (or convection path), called *the plasmapause*, separates the regions of corotating and noncorotating plasma populations. The plasma density drops over an order of magnitude at the plamapause,[9] since the plasmapause, in effect, separates plasmas of ionospheric and magnetospheric origin.

---

[9] Gringauz, K. I., "The structure of the ionized gas envelope of Earth from direct measurements in the USSR of local charged particle concentrations," *Planet. Space Sci.*, **11**, 281, 1963.

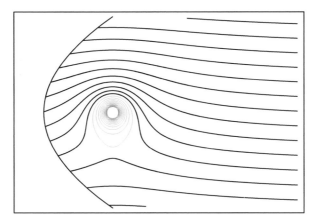

Figure 14.15 Equipotential contours of the combined
convection and corotation potentials in the equatorial plane.

14.5   High-Latitude Electrodynamics

The convecting magnetospheric flux tubes directly connect to the high-latitude ionosphere. The ionospheric footprints of these flux tubes move in the collisional ionosphere, where the electrical conductivity is finite. The resulting convection (and potential) pattern is quite interesting, and it mirrors the magnetospheric transport of plasma
and magnetic flux.

14.5.1   Polar Cap Convection for Southward IMF

In the polar regions of the ionosphere (poleward from the auroral oval) the geomagnetic
field lines are directly connected to the solar wind magnetic field. Because the magnetic
field lines are electric equipotentials (see Eq. 14.57), the motional electric field maps
down from the solar wind to the ionosphere and results in plasma convection.

In the frame of reference moving with the Earth the solar wind electric field is the
following:

$$\mathbf{E}_{sw} = -\mathbf{u}_{sw} \times \mathbf{B}_{sw}. \tag{14.62}$$

If we assume that the interplanetary magnetic field is southward (this is a relatively
well understood configuration from the point of view of solar wind–magnetosphere
interaction), the solar wind electric field points from dawn to dusk (see Figure 14.16).
The electric field maps down to the ionosphere along the connected field lines, where
a dawn-to-dusk directed electric field, $\mathbf{E}_I$, is generated.

The ionospheric electric field, $\mathbf{E}_I$, is much larger than $\mathbf{E}_{sw}$, because the magnetic
field lines converge as they connect to the polar ionosphere: The potential difference
between the field lines remains the same, but the ionospheric electric field (potential

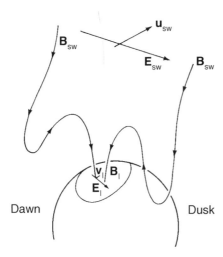

Figure 14.16 Schematic representation of the solar wind electric field (in the terrestrial frame). [From *Kelley* (*1989*).]

Dawn

Dusk

gradient) is about 50 times larger than in the solar wind. The magnetic field is about a factor of 10^4 stronger in the ionosphere than in the solar wind. The combination of the geomagnetic field and the ionospheric electric field drives the plasma motion in the ionospheric F region with a velocity of

$$\mathbf{u}_I = \frac{\mathbf{E}_I \times \mathbf{B}_I}{B_I^2}, \tag{14.63}$$

where $\mathbf{B}_I$ is the ionospheric magnetic field. The magnitude of this plasma flow velocity is about $50/10^4 = 5 \times 10^{-3}$ times the solar wind speed (about 1 km/s). It can also be shown that the direction of convection in the polar ionosphere is antisunward.

14.5.2 Ionospheric Convection Velocities

Any theoretical description of ionospheric plasma convection must explain how the convection is driven. To answer this question let us consider the momentum equation of an ion species in the dense region of the ionosphere (below the F_2 peak). For the sake of mathematical simplicity, let us make the following (quite justified) simplifying assumptions:

1. the plasma is weakly ionized (ion–neutral collisions dominate) and nearly isothermal, and the major ions are singly charged;
2. the neutral atmosphere is stationary and the effects of gravity can be neglected;
3. steady-state conditions prevail; and
4. the convection speed is small compared to the local sound speed (second-order terms in the flow velocity can be neglected).

In this approximation the momentum equation of an ion species (Eq. 4.35) becomes

$$kT_i \nabla n_i - en_i(\mathbf{E} + \mathbf{u}_i \times \mathbf{B}) = -m_i n_i \bar{v}_{in} \mathbf{u}_i, \tag{14.64}$$

where the subscripts i and n refer to ions and neutrals, respectively.

Equation (14.64) can also be written in the form of

$$\mathbf{A} \cdot \mathbf{u}_i = -\frac{kT_i}{m_i \bar{v}_{in}} \frac{\nabla n_i}{n_i} + \frac{e}{m_i \bar{v}_{in}} \mathbf{E}, \tag{14.65}$$

where the elements of tensor $\mathbf{A}$ can be expressed as

$$\mathbf{A}_{jk} = \delta_{jk} - \frac{\Omega_{ci}}{\bar{v}_{in}} \epsilon_{jkl} b_l, \tag{14.66}$$

where Ω_{ci} is the ion gyrofrequency and $\mathbf{b}$ is the unit vector along the magnetic field line. The tensor $\mathbf{A}$ can be inverted to obtain

$$(A^{-1})_{jk} = \frac{\bar{v}_{in}^2}{\bar{v}_{in}^2 + \Omega_{ci}^2} \left[\delta_{jk} + \left(\frac{\Omega_{ci}}{\bar{v}_{in}} \right)^2 b_j b_k + \frac{\Omega_{ci}}{\bar{v}_{in}} \epsilon_{jkl} b_l \right]. \tag{14.67}$$

One can now express the ion velocity with the help of Eqs. (14.65) and (14.67):

$$\mathbf{u}_i = \frac{\bar{v}_{in}}{\bar{v}_{in}^2 + \Omega_{ci}^2} \frac{e}{m_i} \left[\mathbf{E} + \frac{\Omega_{ci}}{\bar{v}_{in}} \mathbf{E} \times \mathbf{b} \right]$$

$$- \frac{\bar{v}_{in}}{\bar{v}_{in}^2 + \Omega_{ci}^2} \frac{kT_i}{m_i} \left[\frac{\nabla n_i}{n_i} + \left(\frac{\Omega_{ci}}{\bar{v}_{in}} \right)^2 \frac{1}{n_i} \frac{dn_i}{ds} \mathbf{b} + \frac{\Omega_{ci}}{\bar{v}_{in}} \frac{\nabla n_i \times \mathbf{b}}{n_i} \right], \tag{14.68}$$

where s is the distance along the magnetic field line. When deriving Eq. (14.68), the field-aligned electric field component was neglected (this is justified since the field lines are assumed to be equipotentials).

Equation (14.68) helps us to understand plasma convection in various regions of the ionosphere. Let us apply it first to the gyration-dominated region (> 160 km) of the polar cap ionosphere, where $\Omega_{ci}/\bar{v}_{in} \gg 1$. In this region the electric field, $\mathbf{E}_{\mathrm{I}}$, is from dawn to dusk, and it is proportional to the motional electric field in the solar wind (with an amplification of ≈ 50). This electric field affects the ions to a much greater extent than does the pressure gradient force. This means that the ion velocity in the polar cap is

$$\mathbf{u}_i = \frac{\Omega_{ci}}{\bar{v}_{in}^2 + \Omega_{ci}^2} \frac{e}{m_i} \mathbf{E}_{\mathrm{I}} \times \mathbf{b} \approx \frac{\mathbf{E}_{\mathrm{I}} \times \mathbf{B}}{B^2} = \mathbf{u}_{\mathrm{I}}. \tag{14.69}$$

It is clear that $\mathbf{u}_{\mathrm{I}}$ points in the antisunward direction. Hence in the polar cap the ionospheric convection is antisunward everywhere.

Magnetic field lines originating from the auroral zone connect to magnetospheric regions with antisunward plasma convection. Thus the ionospheric electric field in the auroral zone points opposite to the electric field in the polar cap (dusk to dawn). According to Eq. (14.69), this electric field results in sunward convection in the auroral zone. This convection pattern is shown schematically in Figure 14.17.

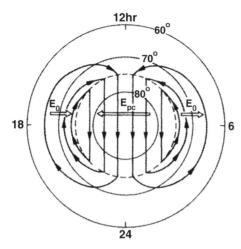

Figure 14.17 Schematic representation of plasma convection in the northern hemisphere polar cap and auroral zone. [From *Kelley (1989)*.]

14.6   Magnetic Activity and Substorms

14.6.1 S_q Current and the Equatorial Electrojet

The best understood type of geomagnetic activity is the *quiet time diurnal variation*, S_q. On quiet days midlatitude magnetic observatories record systematic variations in each magnetic field component. The daily variation is caused by an ionospheric current system, which is fixed with respect to the Sun. Two cells of current circulate around focal points located near local noon and $30°$ magnetic latitudes north and south of the equator. Currents in both cells flow from dawn to dusk at the equator. This diurnal variation is primarily attributed to a dynamo created by Hall currents in the ionosphere. These currents are driven by winds brought about by solar heating and, to some extent, solar and lunar tides. Figure 14.18 shows a global view of the average S_q current system.

At the equator, the southern and northern S_q currents form an extended jetlike dawn-to-dusk current, the *equatorial electrojet*. This electrojet is stronger than the sum of the two S_q currents. The combination of the special magnetic field geometry at the equator and the nearly perpendicular incidence of the solar radiation (which enhances ionization) results in an enhanced special conductivity, which leads to significant amplification of the equatorial current.

Near the equator, the magnetic field is horizontal and it points from south to north. Let us choose this direction to be our x axis: $\mathbf{B} = B\mathbf{e}_x$. Neglecting field-aligned currents, the current density can be written as the sum of the Pedersen and Hall currents (see Eq. 10.93):

$$\mathbf{j} = \bar{\sigma}_P \, \mathbf{E}_\perp - \bar{\sigma}_H \, (\mathbf{E} \times \mathbf{b})$$
$$= \bar{\sigma}_P \, E_0\mathbf{e}_y - \bar{\sigma}_H \, (E_0\mathbf{e}_y \times \mathbf{e}_x) = \bar{\sigma}_P \, E_0\mathbf{e}_y + \bar{\sigma}_H \, E_0\mathbf{e}_z, \qquad (14.70)$$

because the S_q Pedersen current is eastward in the y direction. The z axis points

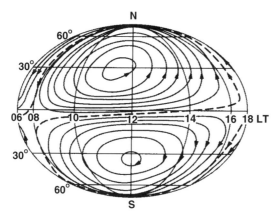

Figure 14.18 Global view of the average S_q current
system. [From *Baumjohann & Treuumann (1996).*]

downward. In Eq. (14.70) the quantity E_0 is the electric field associated with the S_q
current system.

It follows from Eq. (14.70) that the S_q electric field drives a secondary Hall current
downward in the z direction. This Hall current will build up a space charge in the
ionosphere with an excess negative charge density at the top of the ionosphere and
a positive charge density at the bottom. This space charge generates a polarization
electric field, $\mathbf{E}_p = -E_1\mathbf{e}_z$, which under equilibrium conditions drives a secondary
Pedersen current neutralizing the downward pointing Hall current:

$$\bar{\sigma}_H E_0 = \bar{\sigma}_P E_1. \tag{14.71}$$

In addition, the polarization electric field will drive a secondary Hall current in the y
direction:

$$\mathbf{j}_p = -\bar{\sigma}_H (\mathbf{E}_p \times \mathbf{b}) = \bar{\sigma}_H E_1 (\mathbf{e}_z \times \mathbf{e}_x)$$
$$= \bar{\sigma}_H E_1 \mathbf{e}_y = \frac{\bar{\sigma}_H^2}{\bar{\sigma}_P} E_0 \mathbf{e}_y. \tag{14.72}$$

The total current is the sum of the S_q and the polarization currents:

$$\mathbf{j}_{tot} = \left(\bar{\sigma}_P + \frac{\bar{\sigma}_H^2}{\bar{\sigma}_P} \right) E_0 \mathbf{e}_y = \bar{\sigma}_C E_0 \mathbf{e}_y, \tag{14.73}$$

where $\bar{\sigma}_C$ is the *Cowling conductivity*:

$$\bar{\sigma}_C = \bar{\sigma}_P + \frac{\bar{\sigma}_H^2}{\bar{\sigma}_P}. \tag{14.74}$$

Typically the Hall conductivity in the equatorial ionosphere is three to four times
larger than the Pedersen conductivity. This implies that the Cowling conductivity is
an order of magnitude larger than the Pedersen conductivity, which explains the strong
amplification of the equatorial electrojet.

14.6.2 Magnetic Storms

Magnetospheric storms are large, prolonged disturbances of the magnetosphere caused by variations in the solar wind. Many magnetic storms follow solar flares or coronal mass ejections, which produce large nonrecurring interplanetary disturbances that interact with the magnetosphere. The impulse from the interplanetary disturbance impulsively compresses the magnetosphere. This sudden compression rapidly increases the magnetopause current, which is observed as a sudden increase in the horizontal component of the surface magnetic field. This *sudden commencement* can be seen in midlatitude magnetograms. It typically has a rise time of a few minutes. The rise time corresponds to the propagation time of MHD waves from the magnetopause to the observation point. When not followed by later storm phases, this phenomenon is called a *sudden impulse*. The initial compressive phase of the magnetic storm typically lasts two to eight hours.

Most magnetic storms are related to long periods (several hours) when the interplanetary magnetic field has a significant southward component. This is the most favorable configuration for magnetic reconnection at the dayside magnetopause. In contrast, purely northward IMF only allows minimal dayside reconnection, and, consequently, it minimizes the transfer of magnetic flux into the geomagnetic tail. Extended periods (several hours) of southward IMF lead to the *main phase* of magnetic storms.

The increased dayside reconnection increases the penetration of the solar wind motional electric field into the magnetosphere, and, consequently, it increases magnetospheric convection. The enhanced duskward electric field increases the number of particles injected into the ring current. Stronger electric fields not only lead to more energetic ring current ions but also result in an earthward expansion of the ring current region. Heavy ionospheric particles are also accelerated outward and a part of this population is eventually accommodated into the ring current.

It was discussed earlier in this chapter that the ring current causes decreases in the horizontal component of the surface magnetic field. These decreases, measured by the D_{st} index, can reach hundreds of nanotesla during strong magnetic storms. As long as the injection of new particles continues, the ring current will grow and it will asymptotically approach a saturation value, when particle sources and sinks balance each other. The time period during which the ring current increases and the horizontal magnetic component decreases is called the *main phase* of the magnetic storm. The main phase typically lasts from a few hours to about a day.

As soon as the southward component of the IMF weakens or disappears, the ring current starts to decay and the horizontal component of the surface magnetic field gradually returns to its normal value. This is the *recovery phase* of the magnetic storm. The recovery phase occurs in several steps. As the southward IMF component weakens, the reconnection rate decreases. This reduced reconnection rate results in decreasing electric fields, which in turn decrease the injection of new particles into the ring current and move the convection boundary outward. The ionosphere starts to fill the depleted flux tubes within this expanded boundary with cold ionospheric plasma.

Thus in the inner region of the expanded ring current the cold ionospheric plasma overlaps the energetic ring current ion population. The interaction between the two plasma populations increases the ring current loss rate due to growing plasma waves (which pitch-angle scatter the ring current particles) and via direct charge exchange. These processes result in a gradual decay of the ring current, which is the recovery phase of the magnetic storm. The recovery phase typically takes several days.

14.6.3   Substorms

Large-amplitude transient magnetic field fluctuations observed during the night in high-latitude regions (such as Scandinavia) have attracted the attention of scientists for most of the twentieth century. The auroral signatures of these transient magnetic perturbations were termed *auroral substorms* by Akasofu,[10] and the magnetic signatures themselves became known as *polar magnetic substorms*. The complete phenomenon, which includes magnetospheric, auroral, and ionospheric disturbances, became known as a *magnetospheric substorm*.

A *magnetospheric substorm* is a time period of enhanced energy input into the magnetosphere from the solar wind and its subsequent dissipation in the magnetosphere–ionosphere system. It is termed substorm because the main phase of large magnetic storms often appear to be the superposition of many smaller storms, each of which contributes to the growth of the ring current. The intermittent and impulsive nature of substorms account for the noisy nature of the main phase of magnetic storms.

Isolated substorms are generated by relatively brief (about one hour) periods of southward IMF. Many of the details of the complicated plasmaphysical processes leading to the development and evolution of magnetospheric substorms are still hotly debated (for a summary of the present models, we refer to a set of review articles recently published in the *Journal of Geophysical Research*.[11])

Individual substorms usually follow periods of northward IMF configuration. During such periods there is very little magnetic reconnection going on near the dayside magnetopause. When the IMF turns negative, the reconnection rate increases considerably at the front of the magnetopause. This leads to a subsequent increase in the magnetic flux transferred to the geomagnetic tail. The increased reconnection also increases the magnetic energy converted to plasma kinetic energy, and the plasma flow in the magnetospheric boundary layer also increases. This process of energy storage in the tail is referred to as the *growth phase* of the magnetospheric substorm. The growth phase usually lasts for about an hour. After that time period too much magnetic flux and thus magnetic energy has been accumulated in the tail. The tail becomes unstable and releases its surplus energy. This is the time of *substorm onset* and marks the beginning of the *substorm expansion phase*.

The substorm expansion phase typically lasts about thirty to sixty minutes. During this period of time, auroral activity in the auroral ovals is greatly enhanced. During

[10] Akasofu, S.-I., "The development of the auroral substorm," *Planet. Space Sci.*, **12**, 273, 1964.
[11] *J. Geophys. Res.*, **101**, A6, 12,937–13,113, June 1996.

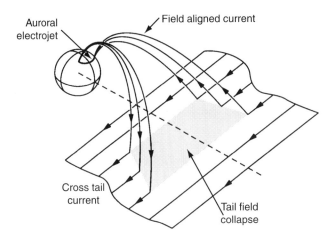

Figure 14.19 Schematic representation of the generation of the substorm electrojet by near-Earth current disruption. (From *McPherron et al.*[12])

the growth phase, auroral arcs can be observed in this region. At substorm onset these arcs suddenly brighten, and they may even fill the whole sky. The dramatic increase in auroral brightness is usually accompanied by large-amplitude magnetic disturbances in the auroral oval.

In addition to the increased auroral activity the ionospheric current flow is also greatly enhanced during the substorm expansion phase. The current increases in two ways. First, because of enhanced convection the *auroral electrojet* increases. This current starts to increase during the growth phase of the substorm. The second way the current flow increases is through the *substorm electrojet* or *substorm current wedge*, which is related to the unloading of magnetic energy in the tail. The current in the substorm electrojet flows westward in the midnight sector. This current is coupled to the cross-tail currents via field-aligned currents, and it is closely associated with the reconfiguration of the magnetotail.

The formation of the substorm current wedge is a consequence of the reconnection process taking place in the near-Earth magnetotail. During the growth phase, the magnetic flux in the tail lobes increases and the cross-tail current sheet (which separates the two lobes) becomes thinner and thinner due to the increased pressure from the lobes. When some portion of the current sheet reaches an appropriate threshold, reconnection begins spontaneously near the center of the current sheet. This reconnection disrupts the cross-tail current and part of the current is diverted toward the ionosphere along magnetic field lines connecting the reconnection region to the polar region of Earth. The closure of disrupted current in the ionosphere is the substorm current wedge. This process is shown in Figure 14.19.

[12] McPherron, R. L., Russell, C. T., and Aubry, M., "Satellite studies of magnetospheric substorms on August 15, 1968, 9. Phenomenological model for substorms," *J. Geophys. Res.*, **78**, 3131, 1973.

About an hour after the substorm onset, the ionospheric current flow and the bright aurora start to decrease and the *substorm recovery phase* begins. This phase lasts a couple of hours and ends when the magnetosphere returns to a "quiet" state. However, during extended periods of significant southward IMF, the recovery of one substorm may overlap with the growth phase of the next substorm.

14.6.4   Geomagnetic Activity Indices

Magnetic indices are widely used to characterize the dynamic state of various aspects of the magnetosphere–ionosphere system. Here we briefly describe some of the most commonly used magnetic indices.

Geomagnetic indices are based on ground magnetograms that record the magnetic field vector as a function of time. The magnetic field vector is fully determined by three components, but different observatories measure somewhat different independent quantities to characterize the magnetic field vector. In standard terminology the magnitude of the geomagnetic field vector is called *total intensity* and it is denoted by F; the magnitude H of the horizontal component of the magnetic field vector is called *horizontal intensity*; and the vertical magnetic field component, Z, is called *vertical intensity*. The northward and eastward components of the horizontal field are denoted by X and Y, respectively. X, Y, and Z comprise the Cartesian components of the magnetic field vector (we note that the Z component points downward). The angle, D, between north and the horizontal magnetic field component is called *declination*, while the angle, I, between the horizontal component and the total magnetic field vector is called *inclination*. The quantities F, H, X, Y, Z, D, and I are called *magnetic elements*.

General Activity Indices   The range, R, of a magnetic quantity for a given time interval is defined as the difference between the highest and lowest values after subtracting the regular daily variation.

The *K index* is a measure of the irregular short-term variations of standard magnetograms and characterizes the general level of disturbance at a given observatory. The K index is defined for each 3-hour interval on the basis of the largest value of the 3-hour ranges in X, Y, D, or H. The K value for a given value of R is found from a lookup table in which the location of the station is taken into account. Because the K index reflects primarily auroral zone activity, stations nearer to the auroral zone have higher sensitivity. K values are integers ranging from 0 through 9 with a roughly logarithmic scaling.

The K_p *index* (the subscript p stands for planetary) is probably the most widely used geomagnetic index. Originally it was intended to characterize the worldwide geomagnetic activity level; however, it is most sensitive to auroral zone activity and quite insensitive to some other types of disturbance. The K_p index is based on K indices from about a dozen stations. The individual station indices are first standardized through tables that translate the integer (0–9) K value into a fractional value with quantized units of $1/3$ (28 possible values, 0, $1/3$, $2/3$, ..., 9). The K_p index is then defined as the arithmetic average of these fractional values.

Auroral Electrojet Indices   The auroral electrojet indices were defined to obtain a measure of the strength of the auroral electrojet that is relatively uncontaminated by ring current effects. These indices are calculated from corrected H component plots measured by a worldwide chain of auroral zone magnetometers. Monthly mean values are first subtracted from each station's H-component time series, and the resulting curves are plotted in a single plot. In the next step an upper and a lower envelope are determined that represent the highest and lowest corrected H values measured by any of the participating observatories. At any instant the *AU index* (auroral upper index) is defined as the maximum positive disturbance recorded by any station in the chain. Similarly, the *AL index* (auroral lower index) is defined as the lowest negative disturbance recorded by the magnetometer chain. The *AE index* and the *AO index* are now defined as $AE = AU - AL$ and $AO = (AU + AL)/2$.

Ring Current Index   The hourly D_{st} *index* is a measure of the strength of the current. It is the average around the world of the adjusted residuals of the H components measured by low magnetic latitude observatories. Adjustment is made for the quiet day levels and for the station's magnetic latitude. In the manner in which it is derived, the D_{st} index is similar to the auroral AE index.

14.7   Problems

Problem 14.1 A magnetic field line crosses the magnetic equator at 4 R_e. Assuming that the Earth's field is dipolar, where does this magnetic field line intersect the surface of the Earth?

Problem 14.2 The polar cap of Jupiter can be approximated by a circular cap of extent $10°$ from the dipole axis.

1. Calculate the magnetic flux content of the Jovian tail.
2. Calculate the cross-sectional radius of the tail assuming that the magnetic field strength in the tail is 10^{-9} T.

Problem 14.3 The total magnetic energy contained in volume $\mathcal{V}$ is given by

$$E_{\mathcal{V}} = \int_{\mathcal{V}} dV \frac{B^2}{2\mu_0}. \tag{14.75}$$

1. Calculate the magnetic energy of Earth's dipole field contained outside the surface of Earth.
2. Calculate the the total magnetic energy for a magnetosphere including both the IMF and the dipole field. What is the energy difference between the southward and northward IMF configurations?

Problem 14.4 What is the subsolar distance to Mercury's magnetopause in planetary radii for a solar wind velocity of 400 km/s and a density of 20 cm^{-3}? What is the magnitude of the magnetic field at the subsolar magnetopause?

Problem 14.5 During magnetic storms, the ring current can reach a million amperes. Assume that the ring current is located at $L = 5$ and that the average energy of the ring current particles is 10 keV. Calculate the total energy carried by ring current particles.

Problem 14.6 During the growth phase of a geomagnetic substorm, the polar cap is observed to expand. Assume that the polar cap expands from an initial radius of $15°$ to $20°$ and that the radius of the tail remains fixed at $x_0 = -10\ R_e$ at a value of $R_T = 18\ R_e$. For $x < x_0$, calculate the tail radius, the lobe field strength, and the plasma sheet pressure as a function of x.

Problem 14.7 It has been suggested that a substorm onset might be triggered when the current sheet thickness becomes less than the gyroradius of a thermal proton in the current sheet. During the growth phase, increasing lobe magnetic field strength compresses the plasma sheet. Assume that the plasma sheet compresses adiabatically and that the initial current sheet thickness is 1 R_e, the initial ion temperature is 4 keV, and that in the current sheet $B_z = 2$ nT at all times. By what fraction must the lobe magnetic field increase to reach this critical triggering threshold? Is this answer reasonable?

Problem 14.8 The cross-tail current is primarily carried by plasma sheet protons, which have a density of $n = 0.3$ cm^{-3}. If the current sheet is 1 R_e thick, and the lobe magnetic field strength is 20 nT, calculate the average ion velocity required for the ions to carry the current. Compare this velocity to the average ion thermal speed if the average ion energy is 4 keV.

Appendix A

Physical Constants

Table A.1. *Universal constants*

Quantity	Symbol	Value	Units (SI)
Avogadro number	N_A	6.0221×10^{23}	mol^{-1}
Boltzmann constant	k	1.3807×10^{-23}	J K^{-1}
Universal gas constant	$R = N_A k$	8.3145	$\text{J K}^{-1} \text{mol}^{-1}$
Elementary charge	e	1.6022×10^{-19}	C
Faraday constant	$F = N_A e$	$96,485$	C mol^{-1}
Electron mass	m_e	9.1094×10^{-31}	kg
Proton mass	m_p	1.6726×10^{-27}	kg
Proton/electron mass ratio	$\frac{m_p}{m_e}$	$1,836.1$	
Electron charge to mass ratio	$\frac{e}{m_e}$	1.7588×10^{11}	C kg^{-1}
Atomic mass unit (amu)	m_u	1.6605×10^{-27}	kg
Gravitational constant	G	6.6726×10^{-11}	$\text{m}^3 \text{ s}^{-2} \text{ kg}^{-1}$
Planck constant	h	6.6261×10^{-34}	J s
	$\hbar = \frac{h}{2\pi}$	1.0546×10^{-34}	J s
Speed of light in vacuum	c	2.9979×10^8	m s^{-1}
Permeability of free space	μ_0	$4\pi \times 10^{-7}$	H m^{-1}
Permittivity of free space	$\varepsilon_0 = \frac{1}{\mu_0 c^2}$	8.8542×10^{-12}	F m^{-1}
Bohr radius	$a_0 = \frac{\varepsilon_0 h^2}{\pi m_e e^2}$	5.2918×10^{-11}	m
Bohr cross section	πa_0^2	8.7974×10^{-21}	m^2
Rydberg constant	$R_\infty = \frac{m_e e^4}{8\varepsilon_0^2 c h^3}$	1.0974×10^7	m^{-1}
Stefan-Boltzmann constant	σ_{SB}	5.6705×10^{-8}	$\text{W m}^{-2} \text{ K}^{-4}$
Classical electron radius	$r_e = \frac{e^2}{4\pi \varepsilon_0 m_e c^2}$	2.8179×10^{-15}	m

Table A.2. *Conversion factors*

Quantity	Symbol	Value	Units (SI)
1 day		$86,400$	s
1 year		3.1557×10^7	s
Speed of 1 eV electron	$\sqrt{\frac{2e}{m_e}}$	5.9319×10^5	m s^{-1}
Speed of 1 eV proton	$\sqrt{\frac{2e}{m_p}}$	1.3841×10^4	m s^{-1}
Wavelength of a 1 eV photon	$\lambda_0 = \frac{hc}{e}$	1.2398×10^{-6}	m
Wavenumber of a 1 eV photon	$k_0 = \frac{1}{\lambda_0}$	8.0655×10^5	m^{-1}
Frequency of a 1 eV photon	$\nu_0 = \frac{e}{h}$	2.4180×10^{14}	Hz
Energy of a 1 eV photon	$h\nu_0$	1.6022×10^{-19}	J
Energy associated with 1 K	$\frac{k}{e}$	8.6174×10^{-5}	eV
Temperature associated with 1 eV	$\frac{e}{k}$	$11,604$	K
calorie	cal	4.1868	J
1 torr	torr	1.3332×10^2	Pa

Table A.3. *Solar physics parameters*

Quantity	Symbol	Value	Units (SI)
Solar mass	$M_\odot$	1.99×10^{30}	kg
Solar radius	$R_\odot$	6.96×10^8	m
Mean density	$\rho_\odot$	1.41×10^3	kg m^{-3}
Surface gravitational acceleration	$g_\odot$	274	m s^{-2}
Escape velocity at surface	v_∞	618	km s^{-1}
Effective blackbody temperature	T_{eff}	5,770	K
Luminosity	$L_\odot$	3.83×10^{26}	W
Radiative flux density	$\mathcal{F}_\odot$	6.28×10^7	W m^{-2}
Solar constant (intensity at 1 AU)	$f_\odot$	1.36×10^3	W m^{-2}
Typical equatorial magnetic field strength	B_{eq}	10^{-4}	T
Magnetic field strength in sunspots	B_{max}	0.25–0.35	T
Equatorial rotation period	$T_\odot$	26	days
Equatorial angular velocity	$\Omega_\odot$	2.80×10^{-6}	s^{-1}
Inclination of equator to ecliptic		7	deg

Table A.4. *Heliospheric parameters at 1 AU*

Quantity		Symbol	Value	Units (SI)
Astronomical Unit		AU	1.496×10^{11}	m
Equatorial solar wind	mean density	n	8.7	cm^{-3}
	mean speed	u	470	$km\ s^{-1}$
	particle flux	nu	3.8×10^8	$cm^{-2}\ s^{-1}$
	α particle abundance	A_α	4.7	%
	proton temperature	T_p	1.2×10^5	K
	electron temperature	T_e	1.4×10^5	K
	α temperature	T_α	5.8×10^5	K
	mean magnetic field	B	6.2	nT
Polar solar wind	mean density	n	2.5	cm^{-3}
	mean speed	u	770	$km\ s^{-1}$
	mean magnetic field	B	2.5	nT
	particle flux	nu	1.9×10^8	$cm^{-2}\ s^{-1}$

Table A.5. *Terrestrial parameters*

Quantity	Symbol	Value	Units (SI)
Equatorial radius	R_e	6,378	km
Surface gravitational acceleration	g	9.8067	$m\ s^{-1}$
Angular velocity	Ω_r	7.292×10^{-5}	s^{-1}
Mass of Earth	M_e	5.9736×10^{24}	kg
Magnetic dipole moment	$\frac{\mu_0}{4\pi} M_d$	7.84×10^{15}	$T\ m^3$

Table A.6. *Planetary parameters*

Planet	Mean solar distance (AU)	Period of solar revolution	Rotation period	Orbital Eccentricity	Mass (Earth = 1)	Equatorial radius (km)	Surface gravity (Earth = 1)	Magnetic moment (Earth = 1)
Mercury	0.387	87.97^d	58.65^d	0.2056	0.0553	2,440	0.378	6.1×10^{-4}
Venus	0.723	224.701^d	243.02^d (retrograde)	0.0068	0.8151	6,052	0.907	$< 10^{-4}$
Earth	1.000	365.256^d	23.9345^h	0.0167	1.0000	6,378	1.000	1
Mars	1.524	686.98^d	24.6229^h	0.0934	0.1075	3,393	0.377	$< 2 \times 10^{-4}$
Jupiter	5.203	11.862^y	9.9250^h	0.0485	317.83	71,492	2.364	2.0×10^4
Saturn	9.539	29.457^y	10.500^h	0.0557	95.162	60,268	0.916	580
Uranus	19.182	84.011^y	17.24^h (retrograde)	0.0472	14.536	25,559	0.889	49
Neptune	30.057	164.79^y	16.11^h	0.0086	17.147	24,766	1.125	28
Pluto	39.53	247.68^y	122.46^h	0.2482	0.0021	1,137	0.068	?

Table A.7. *Atmospheric parameters*

Quantity	Symbol	Value	Units (SI)
Standard temperature	T_0	273.15	K
Standard pressure	p_0	1.0133×10^5	Pa
Heat capacity at constant volume (at the surface)	C_v	712	J/kg
Heat capacity at constant pressure (at the surface)	C_p	996	J/kg
Molar volume at STP[1]	$V_0 = \frac{RT_0}{p_0}$	2.2414×10^{-2}	m^3
Density of air at STP	ρ_0	1.293	$kg\ m^{-3}$
Molecular mass of dry air	m_{air}	28.966	amu
Column density of dry air	$\mathcal{N}$	2.15×10^{29}	molecules m^{-2}
Adiabatic lapse rate	$\Gamma_{ad} = \frac{g}{C_p}$	9.8	$K\ km^{-1}$

Note: [1]STP: Standard Temperature and Pressure.

Table A.8. *Thermospheric composition*

Height z (km)	[N_2] (m^{-3})	[O] (m^{-3})	[O_2] (m^{-3})	[He] (m^{-3})	[H] (m^{-3})
86	1.13×10^{20}	8.6×10^{16}	3.0×10^{19}	7.6×10^{14}	—
90	5.5×10^{19}	2.4×10^{17}	1.5×10^{19}	4.0×10^{14}	—
100	9.2×10^{18}	4.3×10^{17}	2.2×10^{18}	1.1×10^{14}	—
150	3.1×10^{16}	1.8×10^{16}	2.8×10^{15}	2.1×10^{13}	3.7×10^{11}
200	2.9×10^{15}	4.1×10^{15}	1.9×10^{14}	1.3×10^{13}	1.6×10^{11}
250	4.8×10^{14}	1.4×10^{15}	2.5×10^{13}	9.7×10^{12}	1.2×10^{11}
300	9.6×10^{13}	5.4×10^{14}	3.9×10^{12}	7.6×10^{12}	1.0×10^{11}
400	4.7×10^{12}	9.6×10^{13}	1.3×10^{11}	4.9×10^{12}	9.0×10^{10}
500	2.6×10^{11}	1.8×10^{13}	4.6×10^{9}	3.2×10^{12}	8.0×10^{10}
750	2.7×10^{8}	3.7×10^{11}	1.8×10^{6}	1.2×10^{12}	6.2×10^{10}
1,000	4.6×10^{5}	9.6×10^{9}	1.3×10^{3}	4.9×10^{11}	5.0×10^{10}

Source: Chamberlain & Hunten (1987).

Table A.9. *A model atmosphere*

Height z (km)	Temperature T (K)	Pressure p (mb)	Number density n (m^{-3})	Mean molec. weight (g/mole)	Pressure scale height (km)
0	288	1.013×10^3	2.547×10^{25}	28.96	8.434
5	256	5.405×10^2	1.531×10^{25}	28.96	7.496
10	223	2.650×10^2	8.598×10^{24}	28.96	6.555
15	217	1.211×10^2	4.049×10^{24}	28.96	6.372
20	217	5.529×10^1	1.849×10^{24}	28.96	6.382
25	222	2.549×10^1	8.334×10^{23}	28.96	6.536
30	227	1.197×10^1	3.828×10^{23}	28.96	6.693
35	237	5.746×10^0	1.760×10^{23}	28.96	7.000
40	250	2.871×10^0	8.308×10^{22}	28.96	7.421
45	264	1.491×10^0	4.088×10^{22}	28.96	7.842
50	271	7.978×10^{-1}	2.135×10^{22}	28.96	8.047
60	247	2.196×10^{-1}	6.439×10^{21}	28.96	7.368
70	220	5.221×10^{-2}	1.722×10^{21}	28.96	6.570
80	198	1.052×10^{-2}	3.838×10^{20}	28.96	5.962
86	187	3.734×10^{-3}	1.447×10^{20}	28.95	5.621
90	187	1.836×10^{-3}	7.120×10^{19}	28.91	5.640
100	195	3.201×10^{-4}	1.190×10^{19}	28.40	6.010
110	240	7.104×10^{-5}	2.140×10^{18}	27.27	7.720
120	360	2.538×10^{-5}	5.110×10^{17}	26.20	12.09
130	469	1.250×10^{-5}	1.930×10^{17}	25.44	16.29
140	560	5.403×10^{-6}	9.320×10^{16}	24.75	20.03
150	634	4.542×10^{-6}	5.190×10^{16}	24.10	23.38
160	696	3.040×10^{-6}	3.160×10^{16}	23.49	26.41
180	790	1.527×10^{-6}	1.400×10^{16}	22.34	31.70
200	855	8.474×10^{-7}	7.189×10^{15}	21.30	36.18
220	899	5.015×10^{-7}	4.049×10^{15}	20.37	40.04
240	930	3.106×10^{-7}	2.429×10^{15}	19.56	43.41
260	951	1.989×10^{-7}	1.529×10^{15}	18.85	46.35
280	966	1.308×10^{-7}	9.818×10^{14}	18.24	48.93
300	976	8.770×10^{-8}	6.518×10^{14}	17.73	51.19
350	990	3.450×10^{-8}	2.528×10^{14}	16.64	55.83
400	996	1.452×10^{-8}	1.068×10^{14}	15.98	59.68
450	998	6.248×10^{-9}	4.687×10^{13}	15.25	63.64
500	999	3.024×10^{-9}	2.197×10^{13}	14.33	68.79
750	1000	2.260×10^{-10}	1.646×10^{12}	6.58	161.1
1000	1000	7.514×10^{-11}	5.445×10^{11}	3.94	288.2

Source: Chamberlain & Hunten (1987).

Appendix B

Vector and Tensor Identities and Operators

B.1 Vector and Tensor Identities

Below are some vector and tensor identities resulting in scalar or vector quantities. In these relations f represents a scalar function, $f = f(r)$, while $\mathbf{A}(r)$, $\mathbf{B}(r)$, $\mathbf{C}(r)$, and $\mathbf{D}(r)$ are vector quantities. The symbol ∇ represents the configuration-space differential operator.

$$\mathbf{A} \cdot (\mathbf{B} \times \mathbf{C}) = (\mathbf{A} \times \mathbf{B}) \cdot \mathbf{C} = \mathbf{B} \cdot (\mathbf{C} \times \mathbf{A})$$

$$= (\mathbf{B} \times \mathbf{C}) \cdot \mathbf{A} = \mathbf{C} \cdot (\mathbf{A} \times \mathbf{B}) = (\mathbf{C} \times \mathbf{A}) \cdot \mathbf{B}, \tag{B.1}$$

$$\mathbf{A} \times (\mathbf{B} \times \mathbf{C}) = \mathbf{C} \times (\mathbf{B} \times \mathbf{A}) = (\mathbf{A} \cdot \mathbf{C})\mathbf{B} - (\mathbf{A} \cdot \mathbf{B})\mathbf{C}, \tag{B.2}$$

$$(\mathbf{A} \times \mathbf{B}) \cdot (\mathbf{C} \times \mathbf{D}) = (\mathbf{A} \cdot \mathbf{C})(\mathbf{B} \cdot \mathbf{D}) - (\mathbf{A} \cdot \mathbf{D})(\mathbf{B} \cdot \mathbf{C}), \tag{B.3}$$

$$(\mathbf{A} \times \mathbf{B}) \times (\mathbf{C} \times \mathbf{D}) = [(\mathbf{A} \times \mathbf{B}) \cdot \mathbf{D}]\mathbf{C} - [(\mathbf{A} \times \mathbf{B}) \cdot \mathbf{C}]\mathbf{D}, \tag{B.4}$$

$$\nabla \cdot (f\mathbf{A}) = f\nabla \cdot \mathbf{A} + \mathbf{A} \cdot \nabla f, \tag{B.5}$$

$$\nabla \times (f\mathbf{A}) = f\nabla \times \mathbf{A} + \nabla f \times \mathbf{A}, \tag{B.6}$$

$$\nabla \cdot (\mathbf{A} \times \mathbf{B}) = \mathbf{B} \cdot (\nabla \times \mathbf{A}) - \mathbf{A} \cdot (\nabla \times \mathbf{B}), \tag{B.7}$$

$$\nabla \times (\mathbf{A} \times \mathbf{B}) = \mathbf{A}(\nabla \cdot \mathbf{B}) - \mathbf{B}(\nabla \cdot \mathbf{A}) + (\mathbf{B} \cdot \nabla)\mathbf{A} - (\mathbf{A} \cdot \nabla)\mathbf{B}, \tag{B.8}$$

$$\mathbf{A} \times (\nabla \times \mathbf{B}) = (\nabla\mathbf{B}) \cdot \mathbf{A} - (\mathbf{A} \cdot \nabla)\mathbf{B}, \tag{B.9}$$

$$\nabla(\mathbf{A} \cdot \mathbf{B}) = \mathbf{A} \times (\nabla \times \mathbf{B}) + \mathbf{B} \times (\nabla \times \mathbf{A}) + (\mathbf{A} \cdot \nabla)\mathbf{B} + (\mathbf{B} \cdot \nabla)\mathbf{A}, \tag{B.10}$$

$$\nabla^2\mathbf{A} = \nabla(\nabla \cdot \mathbf{A}) - \nabla \times (\nabla \times \mathbf{A}), \tag{B.11}$$

$$\nabla \times \nabla f = \mathbf{0}, \tag{B.12}$$

$$\nabla \cdot (\nabla \times \mathbf{B}) = 0, \tag{B.13}$$

$$\nabla \cdot (\mathbf{A}\mathbf{B}) = (\nabla \cdot \mathbf{A})\mathbf{B} + (\mathbf{A} \cdot \nabla)\mathbf{B}. \tag{B.14}$$

B.2   Differential Operators in Curvilinear Coordinates

B.2.1   Spherical Coordinates

In spherical coordinates the vector $\mathbf{A}$ is given by three independent coordinates, A_r, A_Θ, and A_ϕ. The vector itself is expressed as $\mathbf{A} = A_r \mathbf{e}_r + A_\Theta \mathbf{e}_\Theta + A_\phi \mathbf{e}_\phi$, where $\mathbf{e}_r$, $\mathbf{e}_\Theta$, and $\mathbf{e}_\phi$ are orthonormal unit vectors in the r, Θ, and ϕ directions (here Θ is the polar angle or colatitude). In these relations f represents a scalar function, $f = f(r)$, while $\mathbf{A}(r)$ and $\mathbf{B}(r)$ are vector quantities. The symbol $\mathbf{T}$ refers to a tensor with two indices, T_{ij}. Finally, ∇ represents the configuration space differential operator.

Divergence of a vector:

$$\nabla \cdot \mathbf{A} = \frac{1}{r^2} \frac{\partial (r^2 A_r)}{\partial r} + \frac{1}{r \sin \Theta} \frac{\partial (\sin \Theta A_\Theta)}{\partial \Theta} + \frac{1}{r \sin \Theta} \frac{\partial A_\phi}{\partial \phi}. \tag{B.15}$$

Gradient of a scalar quantity:

$$\nabla f = \frac{\partial f}{\partial r} \mathbf{e}_r + \frac{1}{r} \frac{\partial f}{\partial \Theta} \mathbf{e}_\Theta + \frac{1}{r \sin \Theta} \frac{\partial f}{\partial \phi} \mathbf{e}_\phi. \tag{B.16}$$

Curl of a vector:

$$\nabla \times \mathbf{A} = \left[\frac{1}{r \sin \Theta} \frac{\partial (\sin \Theta A_\phi)}{\partial \Theta} - \frac{1}{r \sin \Theta} \frac{\partial A_\Theta}{\partial \phi} \right] \mathbf{e}_r$$

$$+ \left[\frac{1}{r \sin \Theta} \frac{\partial A_r}{\partial \phi} - \frac{1}{r} \frac{\partial (r A_\phi)}{\partial r} \right] \mathbf{e}_\Theta + \left[\frac{1}{r} \frac{\partial (r A_\Theta)}{\partial r} - \frac{1}{r} \frac{\partial A_r}{\partial \Theta} \right] \mathbf{e}_\phi. \tag{B.17}$$

Laplacian of a scalar:

$$\nabla^2 f = \frac{1}{r^2} \frac{\partial}{\partial r} \left(r^2 \frac{\partial f}{\partial r} \right) + \frac{1}{r^2 \sin \Theta} \frac{\partial}{\partial \Theta} \left(\sin \Theta \frac{\partial f}{\partial \Theta} \right) + \frac{1}{r^2 \sin^2 \Theta} \frac{\partial^2 f}{\partial \phi^2}. \tag{B.18}$$

Laplacian of a vector:

$$\nabla^2 \mathbf{A} = \left[\nabla^2 A_r - \frac{2 A_r}{r^2} - \frac{2}{r^2} \frac{\partial A_\Theta}{\partial \Theta} - \frac{2 \cot \Theta A_\Theta}{r^2} - \frac{2}{r^2 \sin \Theta} \frac{\partial A_\phi}{\partial \phi} \right] \mathbf{e}_r$$

$$+ \left[\nabla^2 A_\Theta + \frac{2}{r^2} \frac{\partial A_r}{\partial \Theta} - \frac{A_\Theta}{r^2 \sin^2 \Theta} - \frac{2 \cos \Theta}{r^2 \sin^2 \Theta} \frac{\partial A_\phi}{\partial \phi} \right] \mathbf{e}_\Theta$$

$$+ \left[\nabla^2 A_\phi - \frac{A_\phi}{r^2 \sin^2 \Theta} + \frac{2}{r^2 \sin \Theta} \frac{\partial A_r}{\partial \phi} - \frac{2 \cos \Theta}{r^2 \sin^2 \Theta} \frac{\partial A_\Theta}{\partial \phi} \right] \mathbf{e}_\phi. \tag{B.19}$$

Components of $(\mathbf{A} \cdot \nabla)\mathbf{B}$:

$$(\mathbf{A} \cdot \nabla)\mathbf{B} = \left[A_r \frac{\partial B_r}{\partial r} + \frac{A_\Theta}{r} \frac{\partial B_r}{\partial \Theta} + \frac{A_\phi}{r \sin \Theta} \frac{\partial B_r}{\partial \phi} - \frac{A_\Theta B_\Theta + A_\phi B_\phi}{r} \right] \mathbf{e}_r$$

$$+ \left[A_r \frac{\partial B_\Theta}{\partial r} + \frac{A_\Theta}{r} \frac{\partial B_\Theta}{\partial \Theta} + \frac{A_\phi}{r \sin \Theta} \frac{\partial B_\Theta}{\partial \phi} + \frac{A_\Theta B_r - \cot \Theta A_\phi B_\phi}{r} \right] \mathbf{e}_\Theta$$

$$+ \left[A_r \frac{\partial B_\phi}{\partial r} + \frac{A_\Theta}{r} \frac{\partial B_\phi}{\partial \Theta} + \frac{A_\phi}{r \sin \Theta} \frac{\partial B_\phi}{\partial \phi} + \frac{A_\phi B_r + \cot \Theta A_\phi B_\Theta}{r} \right] \mathbf{e}_\phi .$$

$$(B.20)$$

Divergence of a tensor:

$$\nabla \cdot \mathbf{T} = \left[\frac{1}{r^2} \frac{\partial (r^2 T_{rr})}{\partial r} + \frac{1}{r \sin \Theta} \frac{\partial (\sin \Theta\, T_{\Theta r})}{\partial \Theta} \right.$$

$$\left. + \frac{1}{r \sin \Theta} \frac{\partial T_{\phi r}}{\partial \phi} - \frac{T_{\Theta\Theta} + T_{\phi\phi}}{r} \right] \mathbf{e}_r$$

$$+ \left[\frac{1}{r^2} \frac{\partial (r^2 T_{r\Theta})}{\partial r} + \frac{1}{r \sin \Theta} \frac{\partial (\sin \Theta\, T_{\Theta\Theta})}{\partial \Theta} \right.$$

$$\left. + \frac{1}{r \sin \Theta} \frac{\partial T_{\phi\Theta}}{\partial \phi} + \frac{T_{\Theta r} - \cot \Theta\, T_{\phi\phi}}{r} \right] \mathbf{e}_\Theta$$

$$+ \left[\frac{1}{r^2} \frac{\partial (r^2 T_{r\phi})}{\partial r} + \frac{1}{r \sin \Theta} \frac{\partial (\sin \Theta\, T_{\Theta\phi})}{\partial \Theta} \right.$$

$$\left. + \frac{1}{r \sin \Theta} \frac{\partial T_{\phi\phi}}{\partial \phi} + \frac{T_{\phi r} + \cot \Theta\, T_{\phi\Theta}}{r} \right] \mathbf{e}_\phi . \qquad (B.21)$$

B.2.2   Cylindrical Coordinates

In cylindrical coordinates the vector $\mathbf{A}$ is given by three independent coordinates, A_r, A_ϕ, and A_z. The vector itself is expressed as $\mathbf{A} = A_r \mathbf{e}_r + A_\phi \mathbf{e}_\phi + A_z \mathbf{e}_z$, where $\mathbf{e}_r$, $\mathbf{e}_\phi$, and $\mathbf{e}_z$ are orthonormal unit vectors in the r, ϕ, and z directions. In these relations f represents a scalar function, $f = f(r)$, while $\mathbf{A}(r)$ and $\mathbf{B}(r)$ are vector quantities. The symbol $\mathbf{T}$ refers to a tensor with two indices, T_{ij}. Finally, ∇ represents the configuration space differential operator.

Divergence of a vector:

$$\nabla \cdot \mathbf{A} = \frac{1}{r} \frac{\partial (r A_r)}{\partial r} + \frac{1}{r} \frac{\partial A_\phi}{\partial \phi} + \frac{\partial A_z}{\partial z} . \qquad (B.22)$$

Gradient of a scalar quantity:

$$\nabla f = \frac{\partial f}{\partial r} \mathbf{e}_r + \frac{1}{r} \frac{\partial f}{\partial \phi} \mathbf{e}_\phi + \frac{\partial f}{\partial z} \mathbf{e}_z . \qquad (B.23)$$

Curl of a vector:

$$\nabla \times \mathbf{A} = \left[\frac{1}{r}\frac{\partial A_z}{\partial \phi} - \frac{\partial A_\phi}{\partial z}\right]\mathbf{e}_r + \left[\frac{\partial A_r}{\partial z} - \frac{\partial A_z}{\partial r}\right]\mathbf{e}_\phi + \left[\frac{1}{r}\frac{\partial (r A_\phi)}{\partial r} - \frac{1}{r}\frac{\partial A_r}{\partial \phi}\right]\mathbf{e}_z.$$

$$\text{(B.24)}$$

Laplacian of a scalar:

$$\nabla^2 f = \frac{1}{r}\frac{\partial}{\partial r}\left(r\frac{\partial f}{\partial r}\right) + \frac{1}{r^2}\frac{\partial^2 f}{\partial \phi^2} + \frac{\partial^2 f}{\partial z^2}. \qquad \text{(B.25)}$$

Laplacian of a vector:

$$\nabla^2 \mathbf{A} = \left[\nabla^2 A_r - \frac{2}{r^2}\frac{\partial A_\phi}{\partial \phi} - \frac{A_r}{r^2}\right]\mathbf{e}_r + \left[\nabla^2 A_\phi + \frac{2}{r^2}\frac{\partial A_r}{\partial \phi} - \frac{A_\phi}{r^2}\right]\mathbf{e}_\phi + \nabla^2 A_z \mathbf{e}_z.$$

$$\text{(B.26)}$$

Components of $(\mathbf{A} \cdot \nabla)\mathbf{B}$:

$$\begin{aligned}
(\mathbf{A} \cdot \nabla)\mathbf{B} = {} & \left[A_r\frac{\partial B_r}{\partial r} + \frac{A_\phi}{r}\frac{\partial B_r}{\partial \phi} + A_z\frac{\partial B_r}{\partial z} - \frac{A_\phi B_\phi}{r}\right]\mathbf{e}_r \\
& + \left[A_r\frac{\partial B_\phi}{\partial r} + \frac{A_\phi}{r}\frac{\partial B_\phi}{\partial \phi} + A_z\frac{\partial B_\phi}{\partial z} + \frac{A_\phi B_r}{r}\right]\mathbf{e}_\phi \\
& + \left[A_r\frac{\partial B_z}{\partial r} + \frac{A_\phi}{r}\frac{\partial B_z}{\partial \phi} + A_z\frac{\partial B_z}{\partial z}\right]\mathbf{e}_z.
\end{aligned} \qquad \text{(B.27)}$$

Divergence of a tensor:

$$\begin{aligned}
\nabla \cdot T = {} & \left[\frac{1}{r}\frac{\partial (r T_{rr})}{\partial r} + \frac{1}{r}\frac{\partial T_{\phi r}}{\partial \phi} + \frac{\partial T_{zr}}{\partial z} - \frac{T_{\phi\phi}}{r}\right]\mathbf{e}_r \\
& + \left[\frac{1}{r}\frac{\partial (r T_{r\phi})}{\partial r} + \frac{1}{r}\frac{\partial T_{\phi\phi}}{\partial \phi} + \frac{\partial T_{z\phi}}{\partial z} + \frac{T_{\phi r}}{r}\right]\mathbf{e}_\phi \\
& + \left[\frac{1}{r}\frac{\partial (r T_{rz})}{\partial r} + \frac{1}{r}\frac{\partial T_{\phi z}}{\partial \phi} + \frac{\partial T_{zz}}{\partial z}\right]\mathbf{e}_z.
\end{aligned} \qquad \text{(B.28)}$$

Appendix C

Some Important Integrals

A definite integral that frequently appears in kinetic theory is

$$L_n(\beta) = \int_0^\infty dt\, t^n\, e^{-\beta t^2}, \qquad \beta > 0, \quad n = 0, 1, 2, 3, \ldots. \tag{C.1}$$

This integral can be evaluated for even and odd values of n:

$$L_{2n}(\beta) = \int_0^\infty dt\, t^{2n}\, e^{-\beta t^2} = \frac{(2n-1)(2n-3)\cdots 1}{2^{n+1}\beta^{(n+12)}}\sqrt{\pi},$$

$$\beta > 0, \quad n = 0, 1, 2, 3, \ldots, \tag{C.2}$$

$$L_{2n+1}(\beta) = \int_0^\infty dt\, t^{2n+1}\, e^{-\beta t^2} = \frac{n!}{2\beta^{(n+1)}}, \qquad \beta > 0, \quad n = 0, 1, 2, 3, \ldots. \tag{C.3}$$

For specific values of n, one obtains

$$L_0(\beta) = \frac{1}{2}\sqrt{\frac{\pi}{\beta}}, \qquad L_1(\beta) = \frac{1}{2\beta},$$

$$L_2(\beta) = \frac{\sqrt{\pi}}{4\beta^{3/2}}, \qquad L_3(\beta) = \frac{1}{2\beta^2},$$

$$L_4(\beta) = \frac{3\sqrt{\pi}}{8\beta^{5/2}}, \qquad L_5(\beta) = \frac{1}{\beta^3}, \tag{C.4}$$

$$L_6(\beta) = \frac{15\sqrt{\pi}}{16\beta^{7/2}}, \qquad L_7(\beta) = \frac{3}{\beta^4},$$

$$L_8(\beta) = \frac{105\sqrt{\pi}}{32\beta^{9/2}}, \qquad L_9(\beta) = \frac{12}{\beta^5}.$$

One can also obtain the following recursive formula for the evaluation of the definite integral, L_n:

$$L_{n+2}(\beta) = -\frac{dL_n(\beta)}{d\beta}. \tag{C.5}$$

The $L_n(\beta)$ function can also be expressed in terms of the gamma function:

$$L_n(\beta) = \frac{\Gamma\left(\frac{n+1}{2}\right)}{2\,\beta^{\frac{n+1}{2}}}. \tag{C.6}$$

Some frequently used integrals involving the error function (see Section 8.5.3) include the following (all these integrals are only defined for $\beta > 0$ values):

$$\int_0^x dt\, e^{-\beta t^2} = \frac{1}{2}\sqrt{\frac{\pi}{\beta}}\, \mathrm{erf}(\sqrt{\beta}\, x), \tag{C.7}$$

$$\int_0^x dt\, t\, e^{-\beta t^2} = \frac{1}{2\beta}\left(1 - e^{-\beta x^2}\right), \tag{C.8}$$

$$\int_0^x dt\, t^2\, e^{-\beta t^2} = \frac{1}{2\beta}\left[\frac{1}{2}\sqrt{\frac{\pi}{\beta}}\, \mathrm{erf}(\sqrt{\beta}\, x) - x\, e^{-\beta x^2}\right], \tag{C.9}$$

$$\int_0^x dt\, t^{n+2}\, e^{-\beta t^2} = -\frac{d}{d\beta}\int_0^x dt\, t^n\, e^{-\beta t^2}. \tag{C.10}$$

Appendix D

Some Useful Special Functions

D.1 The Dirac Delta Function

The Dirac delta function, $\delta(x)$, is a very convenient mathematical concept (strictly speaking it is not a function) that singles out a particular value of the variable x. The delta function is characterized by the following basic properties:

$$\delta(x - x_0) = \begin{cases} 0, & \text{for } x \neq x_0, \\ \infty, & \text{for } x = x_0, \end{cases} \tag{D.1}$$

but in such a way that the integral of the delta function satisfies the relation

$$\int_{x_0-\varepsilon}^{x_0+\varepsilon} dx\, \delta(x - x_0) = 1 \tag{D.2}$$

for any $\varepsilon > 0$ value. This means that the delta function has a very sharp peak at $x = x_0$, but in such a way that the area under the peak is unity. It follows from the above definition that for any smooth function, $f(x)$,

$$\int_a^b dx\, f(x)\, \delta(x - x_0) = \begin{cases} f(x_0), & \text{if } a < x_0 < b, \\ 0, & \text{otherwise.} \end{cases} \tag{D.3}$$

The derivative of the delta function, $\delta'(x)$, can be interpreted the following way:

$$\int_a^b dx\, f(x)\delta'(x - x_0) = \int_a^b dx\, \frac{d}{dx}[f(x)\delta(x - x_0)] - \int_a^b dx\, f'(x)\, \delta(x - x_0)$$

$$= [f(x)\delta(x - x_0)]_{x=a}^{x=b} - f'(x_0) = -f'(x_0) \tag{D.4}$$

for $a < x_0 < b$.

Other fundamental properties of the delta function include:

$$\delta(-x) = \delta(x), \tag{D.5}$$

$$\delta'(-x) = -\delta'(x), \tag{D.6}$$

$$x\,\delta(x) = 0, \tag{D.7}$$

$$x\,\delta'(x) = -\delta(x), \tag{D.8}$$

$$\delta(a\,x) = \frac{1}{a}\delta(x), \tag{D.9}$$

$$x\,\delta(x - y) = y\,\delta(x - y). \tag{D.10}$$

The following are various examples of analytical representations of the Dirac δ function (where the positive parameter, ϵ, is taken in the limit $\epsilon \to 0$):

$$\delta(x) = \begin{cases} \lim\limits_{\varepsilon \to 0} \frac{1}{\varepsilon}, & \text{for } -\frac{\varepsilon}{2} < x < \frac{\varepsilon}{2}, \\ 0, & \text{otherwise}, \end{cases} \tag{D.11}$$

$$\delta(x) = \lim_{\varepsilon \to 0} \frac{1}{\pi} \frac{\varepsilon}{\varepsilon^2 + x^2}, \tag{D.12}$$

$$\delta(x) = \lim_{\varepsilon \to 0} \frac{1}{\varepsilon \sqrt{2\pi}} \exp\left(-\frac{x^2}{2\varepsilon^2}\right), \tag{D.13}$$

$$\delta(x) = \frac{1}{2\pi} \int_{-\infty}^{\infty} dk\, e^{ikx}. \tag{D.14}$$

D.2   The Heaviside Step Function

The Heaviside step function (named after Oliver Heaviside, who first introduced it), $H(x)$, is defined as

$$H(x) = \begin{cases} 0, & \text{for } x < 0, \\ 1, & \text{for } x > 0. \end{cases} \tag{D.15}$$

The function $H(x)$ is piecewise continuous in $-\infty < x < \infty$ with a single jump discontinuity at $x = 0$, where $H(x)$ is undefined. The Heaviside step function sometimes is also called the *unit-step function*, because it represents the turn-on action of an ideal switch. The derivative of the Heaviside step function is the delta function:

$$H'(x) = \delta(x). \tag{D.16}$$

The symmetry relation of the Heaviside step function is the following:

$$H(-x) = 1 - H(x). \tag{D.17}$$

D.3   **The Error Function**

The error function, erf(x), is defined by the relation

$$\text{erf}(x) = \frac{2}{\sqrt{\pi}} \int_0^x dt \, e^{-t^2}. \tag{D.18}$$

This integral can be evaluated numerically and is tabulated in Table D.1.

The error function is a monotonically increasing function of x. Obviously erf(0) = 0, and it can be shown that

$$\lim_{x \to \infty} \text{erf}(x) = 1. \tag{D.19}$$

The behavior of the error function is illustrated in Figure D.1. It follows from its definition that the error function satisfies the following symmetry relation:

$$\text{erf}(-x) = -\text{erf}(x). \tag{D.20}$$

The derivative of the error function is an exponential:

$$\frac{d}{dx} \text{erf}(x) = \frac{2}{\sqrt{\pi}} e^{-x^2}. \tag{D.21}$$

The general series expansion of the error function can be written in the following form:

$$\text{erf}(x) = \frac{2e^{-x^2}}{\sqrt{\pi}} \sum_{k=0}^{\infty} \frac{2^k}{1 \cdot 3 \cdot \ldots \cdot (2k+1)} x^{2k+1}. \tag{D.22}$$

Table D.1. *Tabulated values of the error function.*

x	erf (x)	x	erf (x)	x	erf (x)
0.00	0.000000	0.85	0.770668	1.70	0.983790
0.05	0.056372	0.90	0.796908	1.75	0.986672
0.10	0.112463	0.95	0.820891	1.80	0.989091
0.15	0.167996	0.00	0.842701	1.85	0.991111
0.20	0.222703	1.05	0.862436	1.90	0.992790
0.25	0.276326	1.10	0.880205	1.95	0.994179
0.30	0.328627	1.15	0.896124	2.00	0.995322
0.35	0.379382	1.20	0.910314	2.05	0.996258
0.40	0.428392	1.25	0.922900	2.10	0.997021
0.45	0.475482	1.30	0.934008	2.15	0.997639
0.50	0.520500	1.35	0.943762	2.20	0.998137
0.55	0.563323	1.40	0.952285	2.25	0.998537
0.60	0.603856	1.45	0.959695	2.30	0.998857
0.65	0.642029	1.50	0.966105	2.35	0.999111
0.70	0.677801	1.55	0.971623	2.40	0.999311
0.75	0.711156	1.60	0.976348	2.45	0.999469
0.80	0.742101	1.65	0.980376	2.50	0.999593

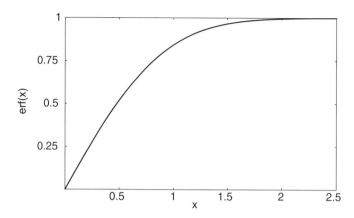

Figure D.1 The error function.

For very large argument values ($x \gg 1$), another, much simpler expansion series can be found:

$$\text{erf}(x) = 1 - \frac{e^{-x^2}}{x\sqrt{\pi}}\left(1 - \frac{1}{2x^2} - \cdots\right), \quad x \gg 1. \tag{D.23}$$

In the other limiting case, when the argument is much smaller than unity, the series expansion simplifies to

$$\text{erf}(x) = \frac{2}{\sqrt{\pi}}\left(x - \frac{1}{3}x^3 + \frac{1}{10}x^5 - \cdots\right), \quad x \ll 1. \tag{D.24}$$

A fairly accurate numerical approximation of the error function is given for arbitrary positive values of x by the following expression:

$$\text{erf}(x) = 1 - (a_1 t + a_2 t^2 + a_3 t^3 + a_4 t^4 + a_5 t^5)\, e^{-x^2} + \varepsilon(x), \tag{D.25}$$

where

$$t = \frac{1}{1 + 0.3275911\, x}. \tag{D.26}$$

The magnitude of the numerical error, $\varepsilon(x)$, is smaller than 1.5×10^{-7}, $|\varepsilon(x)| \le 1.5 \times 10^{-7}$, for any value of x. The values of the numerical constants are the following: $a_1 = 0.254829592$, $a_2 = -0.284496736$, $a_3 = 1.421413741$, $a_4 = -1.453152027$, and $a_5 = 1.061405429$.

Bibliography

Abramovitz, M., and Stegun, I. A., *Handbook of Mathematical Functions*, Dover, New York, 1965.

Banks, P. M., and Kockarts, G., *Aeronomy*, Academic Press, New York, 1973.

Baumjohann, W., and Treumann, R. A., *Basic Space Plasma Physics*, Imperial College Press, London, 1996.

Baumjohann, W., and Treumann, R. A., *Advanced Space Plasma Physics*, Imperial College Press, London, 1997.

Benz, A. O., *Plasma Astrophysics*, Kluwer Academic Publishers, Dordrecht, The Netherlands, 1993.

Bittencourt, J. A., *Fundamentals of Plasma Physics*, Pergamon Press, Oxford, UK, 1986.

Brandt, J. C., *Introduction to the Solar Wind*, Freeman, San Francisco, 1970.

Brasseur, G., and Solomon, S., *Aeronomy of the Middle Atmosphere*, D. Reidel Publishing Co., Dordrecht, The Netherlands, 1984.

Brekke, A., *Physics of the Upper Polar Atmosphere*, John Wiley & Sons, New York, 1997.

Burgers, J. M., *Flow Equations for Composite Gases*, Academic Press, New York, 1969.

Carovilano, R. L., and Forbes, J. M. (eds), *Solar–Terrestrial Physics*, D. Reidel Publishing Co., Dordrecht, The Netherlands, 1983.

Chamberlain, J. W., *Physics of the Aurora and Airglow*, Academic Press, London, UK, 1961.

Chamberlain, J. W., and Hunten, D. M., *Theory of Planetary Atmospheres*, Academic Press, London, UK, 1987.

Chapman, S., and Bartels, J., *Geomagnetism*, Oxford University Press, Oxford, UK, 1940.

Chen, F. F., *Introduction to Plasma Physics and Controlled Fusion*, Plenum Press, New York, 1983.

Cravens, T. E., *Physics of Solar System Plasmas*, Cambridge University Press, Cambridge, UK, 1997.

Foukal, P., *Solar Astrophysics*, John Wiley & Sons, New York, 1989.

Gombosi, T. I., *Gaskinetic Theory*, Cambridge University Press, Cambridge, UK, 1994.

Hargreaves, J. K., *The Solar–Terrestrial Environment*, Cambridge University Press, Cambridge, UK, 1992.

Hasegawa, A., and Sato, T., *Space Plasma Physics*, Springer-Verlag, Berlin, 1989.

Huba, J. D., *NRL Plasma Formulary*, NRL/PU/6790-94-265, Department of the Navy, Washington, DC, 1994.

Hundhausen, A. J., *Coronal Expansion and the Solar Wind*, Springer-Verlag, Berlin, 1972.

Jones, A. V., *Aurora*, D. Reidel, Dordrecht, The Netherlands, 1974.

Kelley, M. C., *The Earth's Ionosphere*, Academic Press, London, 1989.

Kivelson, M. G., and Russell, C. T. (eds), *Introduction to Space Physics*, Cambridge University Press, Cambridge, UK, 1995.

Krall, N. A., and Trivelpiece, A. W., *Principles of Plasma Physics*, San Francisco Press, San Francisco, 1986.

Lyons, L. R., and Williams, D. J., *Quantitative Aspects of Magnetospheric Physics*, D. Reidel, Dordrecht, The Netherlands, 1984.

Nicholson, D. R., *Introduction to Plasma Physics*, John Wiley & Sons, New York, 1983.

Oraevskii, V. N., Konikov, Yu. V., and Khazanov, V. G., *Transport Processes in Anisotropic Near-Earth Plasmas*, Nauka, Moscow, 1985.

Parker, E. N., *Interplanetary Dynamical Processes*, Wiley Interscience, New York, 1963.

Parks, G. K., *Physics of Space Plasmas*, Addison-Wesley, Redwood City, CA, 1991.

Priest, E. R., *Solar Magnetohydrodynamics*, D. Reidel, Dordrecht, The Netherlands, 1982.

Priest, E. R. (ed), *Solar System Magnetic Fields*, D. Reidel, Dordrecht, The Netherlands, 1985.

Rees, M. H., *Physics and Chemistry of the Upper Atmosphere*, Cambridge University Press, Cambridge, UK, 1989.

Rishbeth, H., and Garriott, O. K., *Introduction to Ionospheric Physics*, Academic Press, New York, 1969.

Rossi, B., and Olbert, S., *Introduction to the Physics of Space*, McGraw-Hill, New York, 1970.

Stix, M., *The Sun*, Springer-Verlag, Berlin, 1989.

Størmer, C., *The Polar Aurora*, Oxford University Press, London and New York, 1955.

Sturrock, P. A., *Plasma Physics*, Cambridge University Press, Cambridge, UK, 1994.

Sturrock, P. A., Holzer, T. E., Mihalas, D. M., and Ulrich, R. K. (eds), *Physics of the Sun*, vol. I–III, D. Reidel, Dordrecht, The Netherlands, 1986.

Tascione, T. F., *Introduction to the Space Environment*, Krieger Publishing Company, Malabar, FL, 1994.

Whitten, R. C., and Poppoff, I. G., *Fundamentals of Aeronomy*, John Wiley & Sons, New York, 1971.

Index